RICHA

33474021433634

WITHDRAWN

EVER SMALLER

EVER SMALLER

Nature's Elementary Particles, from the Atom to the Neutrino and Beyond

Antonio Ereditato

Translated from the Italian by Erica Segre and Simon Carnell
Foreword by Nigel Lockyer

The MIT Press
Cambridge, Massachusetts
London, England

© il Saggiatore S.r.l., Milano, 2017
English translation copyright © Erica Segre and Simon Carnell

All rights reserved. No part of this book may be reproduced in any form by any electronic or mechanical means (including photocopying, recording, or information storage and retrieval) without permission in writing from the publisher.

The translation of this work has been funded by SEPS
Segretariato Europeo per le Pubblicazioni Scientifiche

Via Val d'Aposa 7 – 40123 – Bologna – Italy
seps@seps.it – www.seps.it

This book was set in Stone Serif by Westchester Publishing Services. Printed and bound in the United States of America.

Library of Congress Cataloging-in-Publication Data
Names: Ereditato, Antonio, author. | Segre, Erica, translator. | Carnell, Simon, 1962– translator. | Lockyer, Nigel, 1952– writer of foreword.
Title: Ever smaller : nature's elementary particles, from the atom to the neutrino and beyond / Antonio Ereditato ; translated from the Italian by Erica Segre and Simon Carnell ; foreword by Nigel Lockyer.
Other titles: Le particelle elementari. English
Description: Cambridge, Massachusetts : The MIT Press, [2020] | Originally published as: Le particelle elementari by il Saggiatore S.r.l., Milano, 2017.
Identifiers: LCCN 2019029383 | ISBN 9780262043861 (hardcover) | ISBN 9780262358231 (ebook)
Subjects: LCSH: Particles (Nuclear physics)--History. | Particles (Nuclear physics)
Classification: LCC QC793.16 .E7413 2020 | DDC 539.7/23--dc23
LC record available at https://lccn.loc.gov/2019029383

10 9 8 7 6 5 4 3 2 1

To Paola, who supports me, corrects and enlightens

CONTENTS

FOREWORD BY NIGEL LOCKYER   ix
PROLOGUE   xi

1. ATOMS AND BEYOND   1
2. THE INDIVISIBLE AND THE DIVISIBLE   7
3. EVERYTHING IS RELATIVE   19
4. THE CERTAINTY OF CHANCE   37
5. THE FAMILY GROWS   67
6. ENERGIZED PARTICLES   97
7. THREE QUARKS FOR MUSTER MARK!   117
8. MAY THE FORCE BE WITH US!   131
9. THE WORLD THROUGH THE LOOKING-GLASS   151
10. THE MOST BEAUTIFUL THEORY IN THE WORLD   161
11. THE COLOR OF ELEMENTARY PARTICLES   189
12. THE NOVEMBER REVOLUTION   213
13. WEAK, BUT VERY INFLUENTIAL   231
14. THE CHAMELEON   241
15. SEEING THE INVISIBLE   285

16. UNITED WE WIN   307

17. DREAMS, THE UNKNOWN, ADVENTURE   343

18. BE CURIOUS   373

ACKNOWLEDGMENTS   375
INDEX   377

FOREWORD

Enrico Fermi and his associates burst onto the scene during World War II. Physics had just enjoyed half a century of an unprecedented intellectual revolution with the advent of quantum mechanics and relativity, both special and general. Albert Einstein was a household name. World War II ended, and a significant fraction of the intellectual capital of Europe relocated to the United States. Enrico Fermi was surely one of those once-in-a-century geniuses that would change the world forever from his location at the University of Chicago. Important to Antonio Ereditato's story, Fermi and his disciples would be at the vanguard of the "particle revolution," which would span the next half-century and beyond. They would touch the life and career of Ereditato, who weaves his personal interactions with the major protagonists into the storyline of this book. The newly discovered particles would be the entry into particle astrophysics and cosmology, a new field that evolved from the merger of astronomy and particle physics, which combines the big and the small. The twentieth century ended with the emergence of dark matter and dark energy, which have prepared us for another half-century of investigation into nature's closely held wonders.

Ereditato's "Particles" spans this entire period. He intertwines the history of particle physics discoveries with elementary explanations of the physics itself, although he does resort to a few equations, as well as the technology behind the discoveries. Identifying as a Swiss-Italian, he views the period through a European (and even more so, an Italian) lens. Fermi and his disciples are the focal point. "Particles" is the soup-to-nuts menu of the last 100 years in particle physics. It begins with the discovery of the atom components by scientists in the UK, in particular, J. J. Thomson's electron and Ernest Rutherford and his associates' discovery of the nucleus and its proton and neutron components. The atom was complete, or almost—at least that is what many thought. Then a plethora of particles were discovered, and confusion reigned.

Ereditato's story describes a journey where the real fun of discovery begins, and it continues to this day. He describes the history and discovery of the entire spectrum of elementary particles, the pion, the muon, quarks, the W, the Z, gluon, and Higgs boson, and the story behind their discoveries. But we learn that the real protagonist for Ereditato is the neutrino. His story pauses on the way and presents a narrative on the neutrino, a detective story suitable for a Sherlock Holmes novel. His favorite particle, and one he has researched for decades, the neutrino was predicted to exist (albeit with great reluctance) and would prove to be mysterious, ubiquitous, and almost impossible to detect. Ereditato might portray himself as a neutrino hunter, longing to discuss the character of his prey.

The global nature of the field of particle physics became evident after World War II, as well as the crucial role that large collaborations of researchers have played at laboratories throughout the world, devoted to designing, developing, and building impressively complex research infrastructures. Ever-higher energy particle accelerators and the associated sophisticated gigantic particle detectors were a focus. The laboratory cathedrals of particle physics emerged: CERN in Geneva, Switzerland, in the 1950s, followed by the National Laboratory of Frascati in Italy; Deutsches Elektronen-Synchrotron (DESY) in Hamburg, Germany; NIKHEF in Amsterdam, the Netherlands; Rutherford Appleton Laboratory in Chilton, the UK; the Stanford Linear Accelerator Center (SLAC), Brookhaven, and Fermilab in the US; Novosibirsk and DUBNA in Russia; and KEK in Tsukuba, Japan.

This book is aimed at the inquisitive engineer, the prescient high school junior or senior, and science buffs with some background in math and science. There are clear, relatively simple explanations of complex physics topics to keep the reader captivated and intellectually energized. The wonder of nature comes out, but more important, the discovery distance covered in a short period of history has been immense. We are left in awe. The century-long journey brings with it the amenities of modern living: electronics, computers, communications, and advanced medicine. A large body of work, particle physics, has been neatly tied together, including a bow. However, a few gaps exist—maybe many gaps—why anything, why so many particles, why are neutrinos so different from the other particles, and the completely bizarre dark sector. The journey of discovery will continue.

Nigel Lockyer

PROLOGUE

I had the pleasure of meeting Professor Ettore Pancini in November 1974, on the occasion of the first lecture in his course Physics I at the Federico II University in Naples. I will speak of him as a scientist, and of his work as a young researcher alongside two equally young colleagues, as well as of the experiment they conducted in Rome during bombardment by Allied aircraft. It was an undoubtedly fundamental experiment, though one that was not deemed worthy of the Nobel Prize. Pancini was a Partisan commander, and a man of the utmost integrity.

There were 150 of us inside what had been renamed by student collectives the "Salvador Allende Lecture Hall." Pancini welcomed us austerely, in the dry style of a Venetian gentleman, immediately and unequivocally informing us that only a handful of those gathered there would succeed in becoming researchers—and that, of the few survivors of the natural selection process of achieving an undergraduate degree, almost all would end up working in schools or in industry. These were different times. To tell the truth, he was right—and despite the demoralization that came over me at that moment, I was lucky enough in the end to find myself included in that small group of 10. Leaving aside the joy I experienced on successfully surviving the ordeal of his examination, I have many wonderful memories of his course. He was a person from another time, a genius—as we students defined him—in possession of a truly vast array of knowledge that spanned the various discourses of classical and modern physics.

Pancini declared himself to be a *Gestaltist*. He was referring to a philosophical-educational approach, according to which students must only receive minor stimuli from teachers and have to essentially rely on their own attitude and on an autonomous initiative to learn in order to solve problems. This is easily said. But I have to admit that his teaching made a deep impression on me, and that even now, I am grateful to that gruff and even feared professor for his unconventional and extraordinarily beautiful lessons. The department of physics at the Federico II University in Naples eventually was named after him as a tribute to his achievements.

Is there any moral to this story? In fact, there are two. The first is that I was genuinely fortunate because I managed to become engaged in one of the most wonderful occupations there is. To work and play with elementary particles in order to discover the secrets of nature: this is the subject of the book that you hold in your hands. The second concerns my desire to communicate to readers something of what I learned from Pancini. More than providing information, numbers, and formulas, I would like to succeed in stimulating—in providing concepts and some angles of reflection, so as to generate questions and create doubt. In the end, is this not what it really means to be a scientist? Consequently, in this book, you will not find references to original articles and projects: in the era of the Internet, curious and attentive readers can readily access such sources and readily search in greater depth for that which interests them.

This is not a textbook for specialists, and perhaps it is not a popular one either, in the classic sense of the term. As I've already mentioned, it is not my aim to merely provide data to memorize, but to place the main emphasis instead on concepts, methods, and arguments. I imagine the ideal reader of this book to be like my nephew, Francesco, a high school student overwhelmed by information; immersed in social networks and bombarded with news—unfortunately not always very reliable news—and constantly at risk of dilettantism and superficiality. But nonetheless, he is full of a desire to understand, to go beyond appearances or the easy and reassuring explanation. It's an ambitious aim, especially when we are talking about physics, which in the popular imagination is a discipline reserved for a particular caste of individuals with long hair and white lab coats, with a slight German accent and an inability to deal with everyday things. But scientists, through work inspired essentially by their own curiosity, have contributed to leading us out of the caves, and speak to us today about time and space, about extra dimensions and black holes. It's fascinating stuff—but difficult to understand, right?

I should be completely honest with readers on this point. The physics of elementary particles is not a subject that can be learned blithely and with little effort, in weekly installments. You cannot get by without a series of preparatory notions and concepts: mathematics, classical and quantum physics, the physics of the atom and nucleus, special relativity, the technology of particle accelerators and detectors. In a word, when it comes to the science at the frontiers of our understanding, we get to the frontier only by following the arduous road of scientific apprenticeship. This subject, like other scientific matters, requires the completion of a serious trajectory of university studies. Is it "Mission: Impossible," then, to try to make this path accessible to the nonspecialist reader? I think not, and we face a common challenge as writer and reader.

If it wasn't difficult enough already, particle physics also presents us with an extremely vast field. The subjects that we will survey represent a limited (and to a certain extent arbitrary) collection. I hope that my colleagues in physics will forgive the omissions, as well as the inevitable imprecisions and simplifications: it has been my intention to provide an accurate but sufficiently circumscribed picture that makes a panoramic view of elementary particles and their interactions accessible to many, without opening too many parentheses that would weigh down the narrative and the reading experience. I have sought to reduce to a minimum all the necessary collateral ideas, though they will occasionally need to be supplied.

Together, we will try to combine popular science and rigor, conceptual complexity, physical and intuitive understanding, fundamental knowledge, and linguistic lightness of touch. So let's roll up our sleeves and begin!

1 ATOMS AND BEYOND

Everything you can imagine, Nature has already created.

—Albert Einstein

Physics studies the things of our world, in the widest sense of the idea. The origin of the word is Greek—φύσις (*fysis*): "nature." This all-encompassing ambition has always given physics a well-deserved sovereign role among the sciences, perhaps because the topics with which it has been dealing in the past, and is most interested in today, overlap with those of philosophy. Time, space, matter, motion, and energy are concepts treated and studied by humans long before they were addressed in a systematic and operational way by physics. But somehow, over the course of the centuries, the development of physics has also reduced the sphere of influence of myth and religion. We need only think about lightning, the fiery bolts flung by Zeus; about the frequently confused and vague notion of time in classical philosophy, or about the various imaginative theories concerning the origins of the universe (or cosmos) that were devised prior to modern cosmological models.

The realm explored by physics, the queen of the sciences, extends all the way up to the macroscopic world studied by astrophysics and cosmology, and all the way down to the infinitely small ones of atomic, then of nuclear, and for the last few decades, of particle physics. For some years now, always motivated by experimental observations, more unifying theories have led to new disciplines such as astroparticle physics, perhaps indicating that the answers to the big philosophical and scientific questions should not be looked for on a single scale. These days, the connection between microcosm and macrocosm is a solid reality. The physics of elementary particles has naturally evolved from atomic and nuclear physics and is strictly based on the laws of quantum mechanics and on intriguing concepts of symmetry and mathematical beauty. Somehow, on the one hand, physics quantifies our interpretative and predictive limits (theoretical physics), as well as the very strict ones set by experiments, which is our technical ability to probe ever more minuscule structures; while on the other, it contributes to the understanding of phenomena that occur on a cosmological scale.

The first question that we need to ask is about why, in the study of nature, we are immediately compelled to deal with the "ever smaller," to try to divide quantities of matter into progressively more miniscule parts, ideally and concretely, to the point that we are looking for the elementary constituents—the very ones, ultimately, of which our entire universe is composed. It's a bit like disassembling a LEGO toy into its bricks. I believe that behind this reductionist compulsion of science, there lies, in contrast, the desire for a general, holistic and unifying view of nature. To understand the whole and its reasons, however, you need to identify the basic elements. The atoms of Democritus, for example, undoubtedly constituted an intelligent and sensible theoretical idea, even if at the time there was absolutely no possibility of experimentally proving their existence.

An important consideration occurs to us immediately: the definition of *elementarity*, the one that also forms the basis of an "elementary particle," is purely operational and relies, on the one hand, on the need for a theoretical hypothesis, and on the other hand, upon the capacity of physics and technology to verify it. We will be returning to this essential point later, but for now we can affirm that "elementary" is equivalent to "without internal structure," or "not capable of being further broken down or divided." A first classification of nature in terms of elementary and fundamental entities was proposed about five hundred years B.C. by Empedocles, according to whom everything that we are surrounded by is made up of four elementary entities (water, earth, air, and fire), which combine opportunely with each other, interacting through forces of attraction and repulsion. To these four constituents was added a mystical *ether*, an indefinite substance that pervades everything and that, in some way or other, formed the substratum of everything. A good theory, on the whole—one that is remarkably innovative and conceptually not so far from our own present understanding. The situation became more complicated with the discovery of the first elements: iron, gold, lead, and so on. The number of fundamental entities grew until they became frankly too numerous to be elementary. At the end of the nineteenth century, the periodic table of Dmitri Mendeleev and Julius Lothar Meyer provided a reasonable and functional organizational outline of the basic constituents of nature: stated precisely, the *elements*.

But why did the elements seem to be too numerous to be really elementary? The original table even had blank spaces that were to be filled in during the course of the nineteenth century, following new discoveries and to the greater glory of the classification scheme, which perfectly explained chemistry and the combination of the various elements. To be honest, though, there is really no reason why nature should have chosen that its basic constituents should number 215, 1,276, or 12,488 for that matter. According to philosophy, it's just a need to make us ideally prefer a drive for reductionism. It seems natural to think that the set is not complete with 215 apparently independent entities—and that 4 or 5 subentities might have combined to make up the 215. The reflection of this way of thinking, and of reading the world, can also be found in religion. A religion with fewer deities, or better still a monotheistic one, is considered more evolved, or at least more "credible." Likewise, even in modern science, researchers are more satisfied when they can describe a physical, chemical, or biological system in terms of a restricted number of independent variables (parameters). In short, we would be delighted if nature would also reason in this way and choose only a few objects that are truly fundamental—our elementary particles, that is. The atom of Democritus and of Leucippus (from the Greek ἄτομος, "indivisible") contrasts with Aristotle's idea that matter can be subdivided into ever smaller parts by virtue of its observed continuity. In a nutshell, this was the first conflict between classical and quantum physics for many centuries, the lack of a concrete possibility of Popperian falsification of the theory, or more simply of an experimental test, relegated the argument to a mere academic discussion.

At the beginning of the nineteenth century, John Dalton (figure 1.1) developed the first modern atomic theory, based substantially on the improved understanding of chemistry, pioneered by Antonie Lavoisier, on how the various elements combine to form compounds, according to clear numerical rules and through experiments that provide reproducible results—the essence of the method of Galilean science. Dalton's atomic model, like the hypothesis of Democritus, was predicated on a fundamental unit, indivisible and indestructible—the atom—which retained the macroscopic properties of the elements and explained their way of combining according to the laws of chemistry. But why was Dalton's model more credible or advanced than that of Democritus? This is a key point in our story, and in the history of scientific method in general. Dalton's theory had its own experimental phenomenology and was able to provide predictions that could in turn be verified.

FIGURE 1.1
John Dalton, the founder of modern atomic theory.

It's as if nature had drawn a beautiful and complex painting in the form of an immense mosaic. For some hundreds of years, every so often, scientists have found tesserae from it, and with great effort they have tried to recompose them into a coherent picture. When we think that a part of the picture has been properly reconstructed, we develop an interpretive theory based on what little we have of it; there are also those who dare more and try to imagine the great overall design. You can easily understand from this how many times a perverse combination of tessarae can cause us to imagine something plausible but erroneous; only the collection of more of them makes it possible to progress and to acquire a better understanding, though alas the general design remains unknown. I will return to this topic later, but let me say now that in science, we must try to have as few certainties as common sense will allow. A particular tessera can make us see, at any time, that the flower that we thought we recognized in the mosaic is really a mushroom. Scientists must be full of doubt—not of certainty, or worse still, of preconceived ideas.

Dalton's atom was, for his time, the best possible interpretation of the manifestation of nature; it pointed to a more orderly and simpler understanding of the structure of matter. It's a pity that this illustrious English scientist did not go further. He assumed that atoms of iron were different from those of gold, and that one could not be turned into another, in keeping with the drastic laws of chemistry and the famous principle that "nothing is created, nothing is destroyed." The aforementioned table of Mendeleev and Meyer, which could take into account the regularity of the various elements and their chemistry, represents a clear success for Dalton's atomic model. The only problem was that no one had a clue as to how an atom was made and, in a more profound way, how phenomenology in physical chemistry could prefer one atomic structure over another. The strength of chemical bonds between different atoms, their capacity to form what we call molecules, was already intuited by Democritus and Epicurus, who believed that the properties of the various compounds—their solid or liquid nature and even their taste or appearance—depended on the shape of elementary atoms and their specific types of bond (forces). A more advanced understanding of molecules was developed toward the end of the seventeenth century by Robert Boyle, who spoke of corpuscles (atoms) that were capable of aggregating into groups through chemical bonds. It was Daniel Bernoulli, at the beginning of the next century, who introduced a modern definition of molecule through the systematic study of the properties of gases. He affirmed that a gas consists of numerous minuscule molecules given to chaotic and random movements, which cause them to collide elastically with each other and with the surface of their container, thus exerting a pressure. The "amount of heat of the gas" is linked to the kinetic energy due to the motion of the molecules. Today, we know that it is not correct to describe the heat of a body as one of its properties; instead, heat is just a form of energy that is transferred from one system to another thanks to the temperature difference between them.

The study of the gases and of the chemical properties of the various compounds developed rapidly during the course of the nineteenth century, stimulated in part by their burgeoning number of industrial uses. Various scientists contributed to the atomic/molecular theory with brilliant intuitions and ingenious experiments. Famous for formulating the equations that describe all of the electric and magnetic processes in a unifying framework, James Clerk Maxwell published a celebrated article in the journal *Nature* in 1873, in which he orders the concepts of and knowledge about atoms and molecules in a comprehensive and very convincing way. He points out, however, that no one as yet had been able to observe a molecule, to experimentally prove its existence, nor had this been achieved even for the atom! It was Albert Einstein who would provide the final proof of the existence of molecules, and ultimately of atoms themselves, in one of the most monumental and rightly famous scientific articles, published in 1905 while working in the patent office in Berne.

Brownian motion takes its name from the Scottish botanist Robert Brown. In 1897, he had observed through a microscope that minuscule particles, such as grains of pollen suspended in water, exhibit chaotic movement, as if they were continuously being bumped into by invisible (and therefore even more microscopic) entities. In his article, Einstein explained simply that such abrupt movements of the pollen are caused by the collisions with molecules of the medium (water), in rapid motion due to their own kinetic energy, and he proposed a physical and mathematical model that could quantitatively interpret these features. A solid experimental confirmation of Einstein's theoretical model was provided in 1908 by the French physicist Jean Baptiste Perrin. In 1926, he received the Nobel Prize in Physics for his work on the discontinuous structure of matter. This was also an extremely important discovery for the story that we are telling: matter is not continuous, but rather is composed of microscopic and discrete entities, atoms, which aggregate into molecules to form the innumerable chemical compounds that we observe in nature. The direction and the intensity of the collisions between molecules changes randomly to produce hard collisions in one direction or another. This fact makes it easy to understand a further fundamental and mysterious characteristic of the microcosmic world: the randomness of processes, contrary to the deterministic assumption of classical physics, must be advocated to interpret the motion of bodies and of their interactions. An enormous revolution was around the corner. At this point, the Great Question becomes legitimate: are our atoms truly elementary, or do they have an internal structure? We are therefore ready to begin our journey among elementary particles and their forces. It is a journey that will take us far.

2 THE INDIVISIBLE AND THE DIVISIBLE

There are only two possible outcomes:
—if the result confirms the hypothesis, then you have made a measurement;
—if the result is contrary to the hypothesis, then you have made a discovery.

—Enrico Fermi

The year 1897 witnessed another great discovery as well. Joseph John (J. J.) Thomson (figure 2.1), a professor of physics at Cambridge, was convinced that atoms had an internal structure—something of a contradiction in terms, given the meaning of the word. In those years, at the end of the century, he was conducting experiments with so-called cathode tubes, the precursors of the now-obsolete cathode-ray tube (CRT) televisions. Thomson used a long glass cylinder, in which he made a vacuum and placed a cathode set to positive voltage (figure 2.2). At this time it was already known that between the two electrodes, a current flow is created. The latter could be either induced by a hypothetical electromagnetic radiation—like the waves of Heinrich Hertz—or even, as Thomson hoped to show, consisting of some kind of material rays made of unknown matter corpuscles. Thomson was convinced of this "particle" interpretation of the rays, in contrast to most of his colleagues, who were inclined toward what we call a "luminal" explanation. It was thus necessary to devise an experiment that would definitively clarify the nature of cathode rays, and to see if negatively charged particles were actually emitted from the cathode and directed toward the anode.

Thomson eventually demonstrated that it was possible to deflect the cathode rays with an electric field transverse to the direction of the beam. This was the proof that the rays were electrically charged, and more specifically that they were of negative charge. Any kind of electromagnetic waves would not be deflected by the electric field, as the wave itself—or, as we

FIGURE 2.1
Joseph John Thomson, discoverer of the electron.

would now say, the photons that constitute it—was electrically neutral. For the experiment to succeed, it was crucial that Thomson created a strong vacuum in the tube. In a prior attempt, because of an imperfect vacuum, Hertz had not been able to reveal any deflections of the beam. In this case, however, we can surmise that the positive ions produced by the interaction of the beam with the residual air atoms shielded the electric field, preventing it from deflecting the beam. We will see later on that ions are atoms for which the number of electrons and protons is not the same.

This Cambridge physicist then came up with another very astute idea. By means of a magnetic field B, which as we know bends the trajectories of moving electrically charged particles, he succeeded in neutralizing (i.e., counterbalancing) the effect of the electric field E. In this way, the electric force $F_{ele} = Eq$ (where q is the charge of the deflected particles) is matched by the magnetic one, $F_{mag} = Bqv$ (where v is the speed of the particles), hence obtaining $Eq = Bqv$. From knowing the values of E and B, Thomson determined the velocity of his corpuscles. Finally—by knowing the value

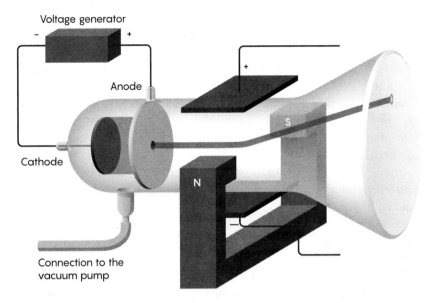

FIGURE 2.2
Thomson's cathode tube. The beam of electrons coming from the cathode is accelerated toward the anode. The latter has a slit to allow the beam to continue its path toward the fluorescent screen. The trajectory of the electrons can be bent by both an electric and a magnetic field.

of v—by turning off the magnetic field and therefore measuring only the deflection produced by the electric field—Thomson also succeeded in determining the q/m ratio, where m is the mass of the hypothetical particles that make up the cathode rays. All of this occurred thanks to simple, classical kinematic equations that will not be given here.

The outcome of the experiment was truly spectacular. It proved the existence of particles with a negative electric charge—namely, electrons, extracted from the atoms of the cathode and characterized by a q/m ratio 2,000 times larger than that of the hydrogen atom, the lightest known particle at the time. Thomson also proved that electrons could pass through thick layers of metal and that their properties are not altered by changing the material of which the cathode is made. The electrons emitted from various elements were identical, regardless of the element from which they came. The English physicist received the 1906 Nobel Prize in Physics for this revolutionary discovery—the first particle—and one that is still elementary today …

In 1909, the absolute value of the electron charge was measured by the American physicist Robert Millikan. Today, we know that a hydrogen atom consists of a proton with an electron orbiting around it, though not in a manner described by classical physics, as we shall see; the proton is about 2,000 times heavier than the electron and hence comprises almost the entire mass of the atom. More precisely, the electric charge of the electron is 1.6×10^{-19} coulombs, and its mass is truly minuscule: 9.1×10^{-28} grams (0 followed by 27 zeroes after the decimal point and before the 91!). The absolute value of the electric charge of the electron is that of *almost* all charged elementary particles ... we will have more to say about what *almost* means here later.

On the one hand, the discovery of the electron opened a Pandora's box from which future elementary particles would emerge; and on the other, it required physicists at the end of the twentieth century to come up with a reasonable model of the atom that could take into account the existence of the new particle inside it. It was Thomson himself who took the first step in this direction. His model, called the "Plum Pudding" after an English dessert, supposed that the electrically neutral atom is made up of a rigid sphere that is positively charged, in which there are distributed here and there negatively charged electrons, which neutralize the positive charge of the sphere. This can be thought of as well, as being a little bit like a panettone, the Italian Christmas cake, in which the raisins are the electrons (figure 2.3).

In parallel with the research undertaken by Thomson and his colleagues, something equally important was happening in other physics laboratories,

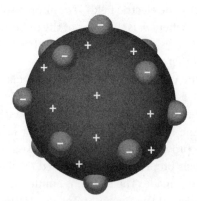

FIGURE 2.3
Thomson's model of the atom. The negatively charged electrons are dispersed in a positively charged sphere. The total electric charge is null, guaranteeing the atom's neutrality.

where the study of elements with completely unexpected properties had begun, opening the door to the revolutionary new physics of the twentieth century. In 1895, Wilhelm Röntgen discovered the X-rays, nowadays considered an unsurpassed diagnostic tool in medicine. (The *X* in the name was appropriate, in that it underlined the unknown nature of this penetrating radiation.) In 1896, Henri Becquerel discovered natural radioactivity, which is the capacity of certain elements to emit rays or corpuscles of matter that can interact with detectors consisting of photographic films placed a certain distance away. Soon after this, the husband-and-wife team of Maria Skłodowska Curie and Pierre Curie (figure 2.4) made fundamental and systematic studies of radioactivity, discovering the relevant role of elements such as radium—from which *radioactivity* gets its name—and identifying a new radioactive element: polonium.

FIGURE 2.4
Pierre Curie and Maria Skłodowska Curie.

FIGURE 2.5
Ernest Rutherford, discoverer of the atomic nucleus and of the proton.

The great New Zealander Ernest Rutherford (figure 2.5), one of the fathers of atomic and nuclear physics as well as a doctoral student of Thomson at Cambridge, showed that there are two types of radioactivity, which he baptized with the first two letters of the Greek alphabet, alpha and beta (α and β). When a third kind of radiation was discovered, with little imagination but with taxonomic consistency, it was called gamma (γ). The Curies soon discovered that β-rays were nothing more than Thomson's electrons—the only particles that were known at the time, together with the atom. Rutherford noticed that α-radiation emitted by radioactive elements very much resembled high-energy helium atoms. He was right. Today, we know that α-particles are just helium nuclei—that is, two neutrons and two protons held together by particular forces, which we call *nuclear* and which will be discussed later. The third kind of radiation, γ, is constituted of high-energy electromagnetic waves. This work brought Rutherford a Nobel Prize (albeit for chemistry) in 1908.

The reader may already know that electromagnetic waves, such as visible light, radio-waves, or X-rays, can be interpreted both as oscillating electric and magnetic fields that propagate even in vacuum, and as corpuscles of energy called *photons*. As we will see in more detail in the following, seen in corpuscular terms, X-rays and γ-rays consist of photons that are much more energetic than those of visible light. Particular and well-deserved attention was immediately given to β-radiation. The reason for this interest was simple. When an atom of a particular radioactive element emits a β-electron, it is transformed into an atom of the next chemical element in the periodic table (e.g., from cobalt to nickel). In this vein, remember that John Dalton said that it was not possible for one element to be transformed into another. Yet what if the medieval alchemists had been right after all? The radioactive β-process was therefore interpreted as if the atom of cobalt was transformed into two distinct objects: an atom of nickel and an electron. It's easy to understand: the "cake" comprised of the total energy of an atom of cobalt had to be split between the two participants, the atom of nickel and the electron, and this always produced the same electron energy for each radioactive event, which was nearly equal to the difference between the mass of the initial and the final atom. This is, as we shall see, in full accord with Einstein's theory of special relativity. As is said in scientific lingo, the energy spectrum of the electron, accurately measurable by skilled scientists at the beginning of the twentieth century, has to exhibit a "line"; that is, to always show the same numerical value, apart from the dispersion due to the limited resolution of the detectors. Recall that accurate measurement means that the measured value is close to the real one, whereas precise measurement implies a small margin of uncertainty: the experimental error. If a table is 1 meter long—a value unknown to the experimentalists—a measurement that gives the result of 1.01 ± 0.70 meters is accurate, but not precise. And the reverse is true: a measurement yielding 0.78 ± 0.01 is hardly accurate, but very precise. This simple consideration of elementary physics will turn out to be useful later.

Lise Meitner (figure 2.6)—a physicist of outstanding ability but one who received insufficient recognition in her time—James Chadwick, and others managed to indisputably demonstrate that the energy spectrum of β-electrons was really continuous, and not a "line" as expected. The electron, that is, could assume all of the energy values between zero and precisely the value that it should always take according to the theory accepted

FIGURE 2.6
Lise Meitner, one of the protagonists of physics at the beginning of the twentieth century.

at the time. The electron never receives the same energy and often has to content itself with crumbs—or even less than crumbs—on account of a third player, which we know today as the neutrino. We will be dealing fully with this fascinating particle a bit later in this book.

As for Rutherford, his contribution to physics is certainly not limited to identifying the three types of radiation. This brilliant physicist discovered that in addition to the tiny electrons, the atom contains a nucleus, which is almost as heavy as the whole atom but much smaller in size. To arrive at this discovery, he conducted and analyzed a series of experiments that are among the most famous in the history of physics. But let's take things in their proper order. Around 1910, the young researcher Hans Geiger (who invented the well-known Geiger counter) and the student Ernest Marsden were working in Rutherford's group in Manchester, where the professor had moved a few years previously. Under Rutherford's guidance, they devised and conducted an experiment, the primary purpose of which was to verify Thomson's atomic model. The experimental apparatus is shown schematically in figure 2.6: a radioactive substance emits heavy α-particles, which had relatively high energy for the time, although we don't yet have the instruments to quantify it. For now, you just need to know that the speed of these particles was around 10,000 km/s! The α-rays hit an extremely thin sheet of gold. Gold is a very dense material and contains heavy atoms; the extreme thinness of the target guarantees that each particle that passes through it collides with a small number of atoms (ideally only one). Behind the target, a detector consisting of a semicircle of phosphorus displays the impact of the particles, allowing the measurement of the deflection angle of the projectile—the positively charged α-particle—after its interaction with the atom. Rutherford, Geiger, and Marsden expected the two positive charges (of the projectile and the target) to repel each other and to produce only a small deflection of the incoming particles on account of the "dilution" of the positive charge of the target, as supposed to be made of Thomson's atoms.

But the result of the experiment was entirely unexpected. There were many collisions, in which the α-particle was deflected just a little, or not at all; but from time to time, in the darkness of the laboratory, the diligent researchers began to observe sparks in the phosphorus at very large angles—of almost up to 180 degrees—with respect to their incoming direction. Rutherford was so surprised that he wrote: "It was almost as incredible as if you fired a 15-inch shell at a piece of tissue paper and it came back and hit you!" In effect, these very large deflection angles could be only explained if the positive electric charge that interacts with that of the α-particle through the Coulomb force had all been concentrated within a minimal space inside the atom and that, consequently, this was in effect empty—the approximate dimensions of the atoms of various elements were known from their chemical properties. Thomson's scheme, therefore, was inadequate. Rutherford then proposed his own new atomic model (figure 2.7), which portrayed the atom as having a very dense center (the nucleus) containing all of its positive charge and at least 99.9% of its mass, surrounded by very light, negatively charged electrons. These were imagined to be bound to the nucleus by the Coulomb force in a sort of planetary system, just as the Earth is bound to the Sun by the gravitational force. Rutherford subsequently developed further calculations by assuming a Coulomb interaction between α-particles and the nucleus: his predictions were in excellent agreement with the measured angular distribution of the scattered particles. It was a truly tremendous result: nuclear physics was born.

Rutherford's model does not differ much from our current understanding, especially in its fairly accurate predictions of the dimensions of the nucleus. To this end, it is instructive to scale atomic dimensions so that they have more consonance with the software of our brain ... So, if the circumference of the hydrogen atom, defined by the average orbit of its electron, was equal to that of the San Paolo soccer stadium in Naples, then

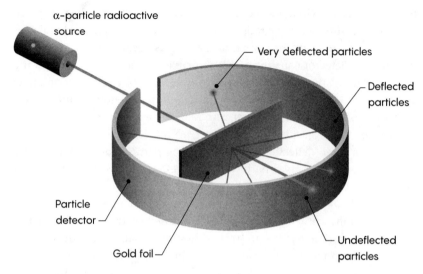

FIGURE 2.7
Diagram of the Rutherford-Geiger-Marsden experiment, which enabled the discovery of the atomic nucleus.

the electron would be the size of a speck of dust, and the nucleus of the atom the size of a peppercorn placed in the center of the pitch! In simple terms, even though it often appears to be very "full" and rigid, matter at a microscopic scale is essentially empty. Numerically, the diameter of an atom is on average around ten-billionths of a centimeter (10^{-8} cm), and that of a nucleus is a thousandth of a billionth of a centimeter (10^{-12} cm). After a series of progressively refined experiments, Rutherford also managed to discover the proton (i.e., the simplest atomic nucleus, that of hydrogen). By bombarding atoms of nitrogen gas with α-particles, he succeeded in extracting hydrogen nuclei (the protons), which, being about four times lighter than α-particles, were pushed forward and revealed at relatively small angles.

Following the fundamental experiments of the Manchester group, and then of the one at Cambridge where Rutherford had relocated in 1919 as director of the prestigious Cavendish Laboratory, the atomic nature was increasingly clarified. Around 1920, it was known that the atom has an internal structure, so therefore, it is not an elementary particle. Other entities appeared to be elementary instead: the proton, so named by Rutherford from the Greek πρωτον ("first"), and then the electron. Protons join together to form the atomic nucleus, and the electrons orbit around it just

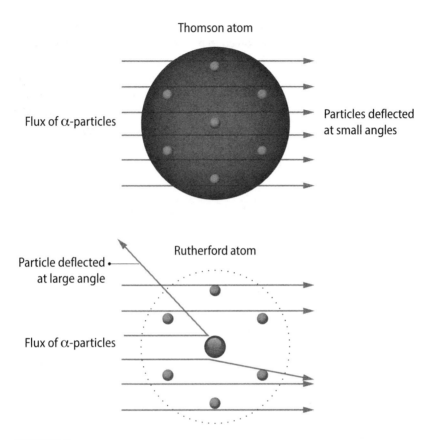

FIGURE 2.8
Comparison of the atomic models of Thomson and Rutherford. The atom is exposed to a flux of α particles.

like in a solar system. This appears to be a fine idea, but frankly, it was soon considered not quite satisfactory. Leaving aside the question of how it was possible that the positive charge of the protons would remain united to form the nucleus despite the repulsive electric forces, it soon became clear that whereas a planet subject to gravitational interaction continues its rotation undisturbed, and with the precision of a Swiss watch for billions of years, the same cannot be true of electrons. It was known, in fact, that an electron, when forced into its orbit by the centripetal electromagnetic attraction with the charged nucleus, is an accelerated electric charge. But for the laws of electromagnetism, accelerated charges must radiate—that is, they emit photons (or equivalently electromagnetic waves). This causes the electrons to lose energy, which, after an amount of time of the order of a nanosecond (10^{-9} s), would be forced to plunge into the nucleus, destroying the atom. But our world has been here for billions of years, and we are here to tell the tale of it. So at this point, was there something that was escaping us? There certainly was. The time was ripe for the quantum mechanics revolution. We will not talk about it yet, however. First, we will introduce the other tool needed for our purposes: Einstein's special relativity.

3 EVERYTHING IS RELATIVE

Even God can't change the past.

—Agathon

"There is nothing new to be discovered in physics now. All that remains is more and more precise measurements." With these words, William Thomson, better known as Lord Kelvin, addressed his colleagues in physics at the British Association for Scientific Progress in 1900, just a few months before Max Planck proposed a revolutionary solution to the problem of black-body radiation, in effect opening the way to quantum physics. To be fair, it seems that Lord Kelvin never actually uttered the phrase that turned out to be such a poor attempt at prophecy, but it was true enough that at the end of the nineteenth century, the successes of classical physics, apart from a few details that still needed clarification, had been astounding. Mechanics, thermodynamics, and the theoretical description of electromagnetism all provided scientific support to positivism in Europe, thanks to the enormous and growing confidence in the contemporary technological and industrial revolution.

Among the things not yet clarified was the unexpected result of an experiment by Albert Michelson and Edward Morley. In 1887, the pair had demonstrated the nonexistence of ether, the hypothesized impalpable medium (with nevertheless well-defined physical properties) that was considered necessary for the transmission of electromagnetic waves. But mistaken ideas do not die easily, as is demonstrated by the fact that today, more than a century after the fact, some still speak of "transmissions via the ether" when referring to the airing of a radio or television program. Let's start here, and return to Planck later.

Maxwell's equations describe all electric and magnetic processes and, among their solutions, predict the existence of varying (oscillating) electric and magnetic fields that propagate in time and space; these are electromagnetic waves, which include visible light. The four famous equations contain a constant c, which has the dimensions of a speed, obtained by combining two other constants that are typical of electric and magnetic phenomena, respectively; that are, ε_0 (the dielectric constant of the vacuum) and μ_0 (the magnetic permeability): $c = 1/\sqrt{\varepsilon_0 \mu_0}$. Our c is a huge speed of nearly 300,000 km/s, enough to reach the Moon in one second or the Sun in 8 minutes. It's the velocity of electromagnetic waves in vacuum: *the speed of light*. It's important to emphasize the phrase *in vacuum* because electromagnetic waves, including light, propagate at a speed lower than c in a medium such as water. However, the mechanics of Galileo and Isaac Newton teaches us that every speed is *relative* to the coordinate reference frame in which the object is in motion. Everyone knows that if the speed of my car on the motorway is 100 km/h compared to the gas station at the side of the road, it's much less—possibly as little as nothing—compared to the car that comes up behind it before overtaking it. Galileo also tells us that the laws of mechanics are the same (i.e., invariant) in all inertial reference frames; an *inertial frame* is that for which a body completely isolated from any other body either stays at rest or is traveling at a constant speed. All other frames that move with respect to it at a constant speed are also inertial.

To pass from one inertial frame to another, one has to apply coordinate transformation rules, called *Galilean transformations*, to take into account the relative speeds of the reference frames. Essentially, taking up the arguments of the previous example, if I find myself in the carriage of a train traveling at 100 km/h (constant) relative to the station platform, and begin to walk at 5 km/h toward the front of the train, my speed compared to the station will be 105 km/h: something that is completely natural and obvious, even common sense. The demand for invariance leads us to add that any mechanical experiment (motion of bodies) that I can make in the carriage produces the same results as an analogous experiment conducted at the station. The only condition is that the reference frame should be inertial (i.e., not subject to accelerations). For those non-inertial frames, even our everyday experience and simple common sense tell us that the results *must* be different. Try dropping an object while driving a Ferrari that is accelerating rapidly. The object will move toward you as it falls. However, you would get a very different result if the experiment were conducted with the car parked in a garage or if you were driving the car at a constant speed.

At this point, it is natural to request the laws of electromagnetism—and why not, all of the laws of physics—to hold Galilean invariance. But this does not happen for Maxwell equations, which are patently not invariant following the application of Galilean transformations. For James Clerk Maxwell, in fact, the speed c of the electromagnetic waves is such only in a given reference frame: the one, that is, in which the hypothetical ether is stationary. Hence, it is possible to understand the key role that the demand for invariance of the laws of electromagnetism played in the elaboration of the theory of relativity. Einstein's famous article on special relativity was entitled "On the Electrodynamics of Moving Bodies." To achieve its purpose, in the first years of the twentieth century (and in the quiet of Berne, which I know so well), Einstein began to demolish principles and concepts that were apparently natural and indestructible, as they still appear to be for many. Newton's concept of time was the first to fall due to Einstein's action.

The concepts of space and time have been widely discussed in various fields for thousands of years. The physicist John Wheeler has argued: "Time is Nature's way to keep everything from happening all at once." It would be really terrible for us physicists (among others) to find ourselves in a situation in which everything happened at the same moment ... Aristotle, a man of genius despite working without the support of experimental instruments, gave a definition of time that is still considered to be very modern. The great philosopher linked the notion of time to movement as follows: "Time is movement, time is the justification of movement and movement happens through time." What this provides is an operational definition. Imagine a situation in which there is nothing that moves: we would have no possibility of defining time operatively.

With Newton (figure 3.1), the first real revolution occurred. Unlike Aristotle, Newton had the advantage of scientific tools such as his infinitesimal calculus. For him, time had become part of the exact sciences. The English scientist believed that the motion of the planets and of all objects on Earth was connected in some way to time, but that time was not tied to it. This was an independent time that flowed in a uniform way—imperturbable and independent of every mechanical event, indeed of all events. This interpretation of Newton's was truly powerful, not least because the formalism of his mechanics worked so well. It explained the motion of the planets, of the apple that fell, and ultimately of everything that surrounded him. Time for Newton was not static, but it was still completely unconnected with mechanical events. Newton's vision contributed powerfully to the development of the determinism and the mechanicism of the next century, in philosophy and, above all, in physics. Perhaps you have heard the following anecdote. Napoleon was talking with Pierre-Simon Laplace, the distinguished French scientist who contributed greatly to the mathematical

FIGURE 3.1
Sir Isaac Newton.

formulation of the laws of mechanics. The emperor said, "According to your equations, I can determine the motion of any object in the remotest future, provided I know the conditions of it at a given moment. I can know how it behaved in the remote past and how it will behave in the future." And Laplace was not wrong; his determinism explained the eclipses, the orbits of the planets, and the motion of every terrestrial object subject to mechanical forces. Finally Napoleon asked, "So what is the role of God in all this, if everything is determined?" To which the scientist bravely replied: "God is an unnecessary hypothesis." In reality, beneath the trenchant surface of Laplace's remark, he was implicitly saying that time itself would no longer have a primary role because motion is completely determined by the equations that describe it—equations in which time is just a variable like the others.

Albert Einstein (figure 3.2) achieved a second great revolution, questioning in an operational way concepts that appeared to be altogether outside of scientific discourse. He began to critically address the meaning of the simultaneity of events and their observation, and then continued demanding the greatest degree of generality possible for the validity of physics laws. According to Einstein, time loses its absolute nature, becoming a quantity that is totally relative to the state of motion of bodies—either if they have a speed or if they are stationary. There is no longer an absolute Newtonian

FIGURE 3.2
Albert Einstein.

time, but rather a relative time that is different for each of us. But be careful—this Einsteinian time, which is different for everyone is *not* tied to differing subjective perceptions of the passing of the hours. No, Einstein speaks specifically of the ticking of clocks, of the oscillation of atoms. Einstein's time, which is the time of our world in extreme conditions and at high relative speeds, is a time that is also completely relative. Einstein's arguments, supported by the success of his theory of relativity, have become indispensable to all subsequent discussions of time and space. We cannot do without the hypothesis behind relativity—either from a scientific point of view or, for that matter, from a philosophical one. The evolution of the concept of time has gone from a terrible and somewhat chaotic pre-Newtonian subjectivism, in which in essence everyone had their own opinion, to an absolute Newtonian time, and back again—to a relativism that is absolute in kind (if I may be allowed that contradiction). And consequently, therefore, it takes a new form of subjectivism based on scientific and verifiable factors. Absolutism, or rather the absolutism of time, has shifted with Einstein to spacetime. Instead of three space coordinates x, y, and z, and one time variable t, we need to have four interconnected spacetime coordinates x, y, z, and t. This was the true nature of his revolution.

At this point, however, before proceeding with a discussion of relativity, it is timely to reflect on a matter that will remain useful throughout the rest of this book. Einstein's theory of relativity was certainly not conjured from nothing. Its grounding relies on important theoretical works developed previously by other scientists, the most important of these being the Dutchman Hendrik Lorentz and the Frenchman Jules Henri Poincaré (figure 3.3). The conclusions that Einstein would formulate were already gestating, at least in part, in the works of his predecessors. Einstein's achievement is truly outstanding, of course, and rightly celebrated. Nevertheless, our reading of history—and the story of physics in particular—is sometimes presented in a simplified and perhaps excessively personalized way—including by myself! Readers need to be constantly aware of this. Historical and scientific events are much more complex than they appear to be in an official, necessarily synoptic narrative. Today, the contribution

FIGURE 3.3
Hendrik Lorentz (left) and Jules Henri Poincaré, the authors of important contributions to the foundations of special relativity.

of the unassuming Poincaré is well known only to specialists, whereas Einstein's fame is universal. The message is simple: we must always try to go beyond a cursory understanding of the facts, and to engage in more depth with less obvious aspects. Just as my colleague Sandro Bettini recommends, it is always a good time to go to the sources, to the original works, and never to be satisfied by hearsay or commonplaces.

Returning to the matter at hand, the theory of relativity has two main parts. The first, published in 1905, is that of special relativity, which I will give an account of here because it is very relevant to our discussion of elementary particles. The other, from 1915, is that of general relativity, an extension of the first theory to non-inertial frames and to gravity, and therefore specifically pertaining to the macrocosm and to the concepts of space and time extended onto a very large scale. General relativity is one of the most beautiful and powerful theories in all physics—beautiful for its simplicity and mathematical elegance, as well as for its profound physical meaning; and powerful because today it is the reference point for the study of astrophysical and cosmological bodies and large scale structures. At this point, it would be completely appropriate to mention the observation of gravitational waves— the existence of which was predicted a century ago by Einstein— for the first time in 2016 by the LIGO and VIRGO experiments, respectively, in the United States and in Italy, and presently together in a single scientific collaboration. This was a historic result that should make us feel fortunate to be living at such a period in history. Unfortunately, however, given the particular focus of this book, we will not be able to discuss it here. It is nevertheless worth emphasizing that general relativity, as opposed to special relativity, is incompatible with quantum mechanics in the extreme conditions of matter at the very beginning of time, shortly after the Big Bang or, equivalently, when we consider infinitesimal distances between interacting particles. This is one of the great unsolved problems of particle physics; though it is perhaps too complex for our present purposes, we will touch briefly on this topic at the end of the book.

Philosophy aside, Einstein replied to the following question: the value of the speed c that appears in Maxwell's equations, the speed of light, is so big in relation to which reference frame? If I go toward a light ray on the back of another ray, will the relative speed of approach be $2c$? If this were the case, as we have seen, there would be only one reference frame, among the infinite ones available, in which Maxwell's equations would be valid. This is the privileged frame where the ether is at rest. The ether is the invisible medium without mass through which electromagnetic waves propagate, even if apparently in a physical vacuum, such as that of cosmic space. The latter may be empty of matter, yet full of ether, with respect to which stars, galaxies, planets, and astronauts move at different speeds. Hence, light travels with speed c only and exclusively in the reference frame in which ether is stationary, in stillness. Farewell to the invariance of the laws of electromagnetism for Galilean transformations!

Michelson and Morley, however, devised and carried out their fundamental experiment, in which an interferometer combined light rays on an optical bench, the orthogonal arms traveling together with the Earth with a given speed with respect to the hypothetical ether. The interference pattern observed at the end of the path of the light rays, therefore, should be different if we varied the direction of motion of the interferometer (or of the Earth) with respect to the stationary ether (day/night, summer/winter). But no variation, no different value of the speed of the light, was observed. The result of the experiment was altogether compatible with the constancy of the speed of light—equal to c, independent of the reference frame—and thus proved the nonexistence of ether. It was definitively discovered that something for which there was no experimental proof did not exist. If I may be allowed the simplification, it was as if it had been proven that pigs do not fly. If truth be told, even today physicists spend time, and work hard, to dismantle theories that are fanciful but nonetheless have a certain degree of plausibility. This is part of the "game" of scientific research. The Michelson and Morley experiment provided solid grounds for the subsequent development of Einstein's theory of special relativity, although the ingenious theoretical physicist made it understood that the results of the experiment had not had any decisive influence on his work. Einstein started to develop his theory on the basis of two assumptions: (1) all the laws of physics must be invariant, whatever the choice of the inertial reference frame; and (2) the speed of light has the same value c in each of these frames. The attentive reader has understood this correctly. If, on the back of a photon, I go toward another photon, then the relative speed will still be c, not $2c$! In reality, such a ride is impossible in principle, but the result would not change if I were on the shoulders of an object traveling at 99.9999 percent of c. The consequence of these two assumptions is catastrophic: Newtonian mechanics is incorrect, and the kinematical transformations that make it possible to pass from one reference frame to another, which moves at speed v in relation to it, are no longer those of Galileo. Considerations of absolute or relative time, of simultaneity or lack thereof of events, and so on led Einstein to replace the transformations of classical mechanics for those developed by Lorentz—mathematical formulas from which the relativity of lengths and of intervals of time for different observers emerge—as well as all the incredible associated effects which, from this moment onward, have appeared in works of science fiction and tantalized the popular imagination. One needs only think in this regard about the frequency in such books, movies, and so forth, of traveling back and forth in time. What must be underlined, however, is that special relativity does not really cancel out Newton's mechanics; rather, it includes it in a more extensive scheme. It is not only the laws of mechanics that are invariant for transformations from one inertial frame to another, but electromagnetism as well. The price to pay is the need to use Lorentz's transformations rather than Galileo's. The former, however, are reduced to the latter for speeds that are relatively slow compared to that of light, a condition that applies to the majority of events that occur in everyday life, as well as in many experiments that we conduct in physics laboratories. Relativity becomes necessary only when we enter into the magical world of the microcosm—the world of our elementary particles. Then the rules of the game change, and we are forced to replace Newton's vision with the more inclusive and complete version advanced by Einstein. Let's try to clarify this point, which is of utmost importance to what will be discussed later.

According to the Lorentz transformations, if I measure the length L_0 of a table in my laboratory, a colleague of mine who moves in relation to me at a constant speed v will observe for the same table a contraction along the direction of his motion; his measured length L will be a smaller quantity, which is a function of the relative speed between us:

$$L_0 \times \sqrt{1 - \frac{v^2}{c^2}}.$$

From this formula, it appears to be clear that the difference between L and L_0 increases as v approaches the speed of light, c. On the contrary, for a small relative speed, L and L_0 tend to become equal, as predicted by the classical mechanics of Newton (and by common sense). In a similar manner, if in my laboratory I measure the duration of an event—say, the length of a day equal to T_0 (24 hours)—, my colleague moving at a speed v will obtain for the arc of time of my day T a result that is difficult to understand: $T = T_0 / \sqrt{1 - (v^2/c^2)}$. For him, time "dilates." What for me is a day appears longer for him, and this effect increases the closer to c our relative speed v is. Let's try to clarify this scenario with a few concrete numbers. If $v = 0.3\ c$ (around 90,000 km/s) for my colleague, the table will be 5 percent shorter, while my day for him will last not 24 hours, but 24 hours and 72 minutes! Consider this an excellent way of keeping young—or, if you prefer, of taking journeys into the future. Think of a spacecraft with which we could journey for a year at 99 percent of the speed of c. Well, due to the fault—or merit—of the relativistic dilation of time, when we get home, the friends waiting for us on Earth will be seven years older than when we left. Obviously, I have bracketed the small matter of being clueless as to how we would travel at approximately the speed of light, but we can legitimately think of that as a technological limitation rather than a theoretical one. Someone, however, may wonder why—for pure symmetry arguments—when the distance between us and the astronaut increases, at the end of the journey he looks younger than us! For symmetry, the opposite could be true: he looks old and we on Earth look young. This is the so-called twin paradox. The answer is that our two situations are not equivalent. To increase the relative speed between the traveler and the Earth base, it is necessary to first accelerate and then decelerate the astronaut when he returns. This is achieved by applying forces on him, for example, by means of the propellant of the spaceship. Only the astronaut will perceive an acceleration (and also a strong one!) to reach the cruising speed and not us, quietly at rest on our planet.

Einstein's space and time are merged together in an indivisible way. The Lorentz transformations and the bizarre effects that they produce are the price that must be paid to have *all* the laws of physics invariant in *all* inertial reference frames. We could continue to develop this fascinating analysis, but by doing so, we would be going beyond the scope of our objectives here. The message, however, is clear: special relativity produces effects that diverge from the predictions of Newtonian mechanics only for relative speeds that are close to that of light. In all other cases, Newton's theory works perfectly. For example, think about the possibility of accurately predicting the arrival of a comet and of defining with precision the trajectory of spacecrafts and satellites. The speeds in question in such cases are indeed high for us, but they still are small compared to c. The case is very different, however, for the particles of the microcosm, for which traveling at 99.99999 percent of the speed of light can be "normal." In this regime, all of the other quantities—kinematical and dynamical, functions of space and time—are modified. This assertion deserves some additional attention, so we will examine it next.

Let's consider the case of the energy and momentum of a body, two physics quantities for which a conservation principle holds. In any process or reaction that occurs in a closed system, the total energy and momentum of the system before and after need to be identical. Classically, these quantities are a function of the speed and mass of the physical macroscopic object or of our elementary particle. One component of the total energy of a body is the kinetic energy or movement. The two definitions of kinetic energy and momentum are $E = 1/2 mv^2$ and $\boldsymbol{p} = m\boldsymbol{v}$, respectively. The variables in bold denote physical vector quantities, and hence are defined by numerical values or magnitude and direction—such as velocity, and consequently momentum. The kinetic energy, on the other hand, is a scalar quantity—that is, it is wholly determined by a number given by the product of the mass of the particle and the square of its velocity divided by 2 (v^2 is also a scalar quantity). The potential energy (e.g., that due to the position of the particle in the Earth's gravitational field) is eventually added to the kinetic energy. In the theory of special relativity, something incredible happens. Not considering the potential energy, we have an expression for the total energy (squared) of the particle:

$$E^2 = p^2c^2 + m^2c^4. \tag{1}$$

The first term depends on the momentum p, and hence on the particle's motion. The second term is wholly unexpected and derived from the relativistic correlation between space and time. A form of energy appears that is due only to the mass of my particle. What happens? Even if the particle is at rest, without speed or momentum, in contrast to the classical case, it nevertheless possesses an amount of energy: $E^2 = m^2c^4 \rightarrow E = mc^2$ (c^2 being a very large number). This is one of the most famous formulas in physics: the equivalence between mass and energy or, analogously, the possibility of transforming the former into the latter, multiplying it by the extremely big value of the speed of light squared. Mass, a kind of concentration of energy, becomes an important element in the energy portfolio of a material object or of a particle. The contrary is also the case. Small values of mass can give rise to enormous quantities of energy. Think of the atomic bomb or of nuclear reactors, where minuscule fractions of uranium disappear (in the literal sense of the term) to reappear as pure energy and other particles. Imagine that in the complete transformation of a single gram of matter into energy, we obtain an amount equal to approximately 10^{14} joule (100,000,000,000,000 joules) or, if you prefer, equivalent to the energy unleashed by the detonation of more than 20,000 tons of trinitrotoluene (TNT), which in turn is sadly close to the power released in the atomic bomb dropped on Hiroshima, corresponding to 15,000 tons of TNT. For our present purposes, it is quantitatively important instead to run Einstein's famous equation in the opposite direction. This is hardly an insignificant detail if we talk about the peaceful use of energy. As physicists, this is what we do in our labs on a daily basis (e.g., at CERN, the particle physics laboratory near Geneva, Switzerland): we accelerate particles, supplying them with high kinetic energy (speed), and we make them collide. In the collision, a part of this energy is transformed into mass, and other particles are created from nothing, once again in the literal sense of the term. Of course, some of the mass of these particles sometimes retransforms back into energy, but as their masses are no more than microscopic, the production of so-called secondary energy is virtually negligible. Hence, there is no risk of explosions.

Our discussion of elementary particles will frequently lead us to encounter certain (relativistic) quantities, and it is worth pointing three of them out here:

$$\beta = \frac{v}{c}; \gamma = \frac{1}{\sqrt{1-\beta^2}}; p = \gamma mv.$$

The first formula defines the β ratio between the speed of the particle and that of light; β, therefore, can take any values between 0 and 1. Consequently, in the second relation, the quantity γ can range from 1, for a particle at rest, to infinity, for a particle with arbitrarily high energy. Finally, the third formula expresses the value of the relativistic momentum. From these relations, we derive the following four formulas, and they too will be very useful for our next, simple calculations:

$$\beta = \frac{pc}{E}; \gamma = \frac{E}{mc^2}; L = \frac{L_0}{\gamma}; T = \gamma T_0. \tag{2}$$

Let's give a few examples. From equation (1), it follows that if a particle has vanishing mass, then $E = pc$. This, for the first equation of (2), implies that $\beta = 1$; that is to say that its speed is equal to c—in other words, the speed of light in a vacuum. This is the case for the photons that constitute the corpuscles of the energy carried out by electromagnetic waves, which are energetic but without mass. On the contrary, a particle endowed with mass will never be able to travel at the speed of light, much less overtake it.

We shall see in what follows that many of our elementary particles have a very brief existence, the duration of which is distributed around an average value. They die by decaying; that is, by transforming into other particles of lesser mass. This is not the case for the proton or electron that ordinary matter is made of—and which, to the best of our knowledge, appear to be wholly stable. Consequently, the matter that surrounds us is stable as well. The reason for this stability is that the decay into lighter particles would violate solid principles that appear to be rigorously respected by the physics laws. For unstable particles, on the other hand, there is a time scale involved that is typical of our microcosm. We speak of microseconds, nanoseconds, and even smaller intervals of time. This makes the identification of the particles difficult and problematic from an experimental point of view. In effect, notwithstanding that they propagate themselves at almost the speed of light, such particles can travel only for extremely short distances. Let's take a hypothetical particle with mass m that lives on average for 1 picosecond (10^{-12} seconds). Classically, even if its speed was equal to c, it would travel in the imaginary detector in our laboratory for a distance of only 0.3 millimeters. But if we exploit the effects of special relativity, we observe that the experimental situation improves significantly. Accelerating the particle until it is supplied with a momentum p, such that $pc = 10\ mc^2$, for equation (1), we obtain

$$E^2 = (10\,mc^2)^2 + m^2c^4 \rightarrow E = \sqrt{101\,m^2c^4} \approx 10\ mc^2.$$

Inserting this value in the second of the equations (2), we obtain $\gamma = 10$. This implies that the particle, for the fourth equation in (2), now will not live for 1 picosecond, but 10, therefore traveling 3 mm in the detector, making it easier to identify. Now someone is sure to ask what the particle will make of living so long as to travel a distance that is 10 times greater. It's not a problem—the situation is altogether symmetrical for the principle of relativity. The particle will not live for any longer in its own reference frame. Instead, it will measure—if it were ever capable of measuring—a travel distance shorter by a factor of 10 with respect to the distance measured by us in the laboratory—for the third equation in (2)—and will cover that length calmly in the only picosecond of life that it has at its disposal.

The preceding considerations are applied analogously, with staggering and intriguing results, to the interstellar voyages of the astronauts. Think of the previous example. For astronauts, as well as for relativistic particles, the seven light years that the spacecraft took to reach the hypothetical distant planet would seem contracted and would correspond to just 1 year, for a speed 99 percent that of light. One should remember that a light year, the distance traveled by light in 365 days, amounts to the astronomical distance of nearly 10,000 billion kilometers. In this way, the journey to the planet would take much less time for the members of the crew than for those in the control room in Houston (and for all Earthlings). The reduction in journey time is much greater as the speed of the spacecraft approaches c: 99.9 percent, 99.99 percent, and so on. In principle, the seven light years could last a single day! Even crossing the Milky Way in its entirety, a distance of more than 100,000 light years, could be done in an arbitrarily brief time if the technological limitations on supplying sufficient energy to the spacecraft in order to propel it to close to the speed of light could be overcome. What a pity, though, that the astronauts would not be able to report back the marvels they had encountered on their journey, because upon returning home they would find that everyone there had aged ... by 200,000 years! These are phenomena at the limit where science and science fiction meet. Nevertheless, we may be pleased to discover that at least in principle, we can dream of covering the abyss-like distances of the cosmos, without hope of returning to our world but with the same spirit of adventure shown by Ulysses when he passed through the pillars of Hercules.

To complement what we have affirmed thus far, the reader should observe that, in practice, the speed does not vary a great deal in our ideal experiments. Passing from 99 percent of the speed of light to 99.9 percent is no great matter; what changes things instead (and by a good deal) is the increase in energy, and with it the lifetime of the particle (or astronauts ...) with respect to the laboratory frame. In the CERN accelerators, particles soon reach, in essence, the speed of light, but their energy can still continuously and appreciably increase with further energization. This fact is illustrated in figure 3.4. The role of the mass—actually its value when the body is at rest—is very important in particle physics. It is an invariant quantity; that is, it assumes the same value in all inertial reference frames and is decisive in characterizing and distinguishing one particle from another. Invariance is particularly useful when studying processes (and reactions) that take place in the microcosm. We have already spoken of the decays of unstable particles. The other, highly important process is that of the collisions among high-energy particles. In this case, with the principles of conservation of momentum and total energy in the reaction before and after the collision remaining fixed, nature and its laws have the freedom to transform the total available energy (E) into momentum (p) or into mass (m) of the particles, some of which are created ex novo.

Speaking of experiments, of energy, and of the mass of elementary particles, it is advisable to use more practical units to avoid having very big or very small numbers in our calculations and formulas. Therefore, we define an *electronvolt* as the energy acquired by an electron accelerated by a voltage of 1 volt, a concise way of indicating 1.6×10^{-19} joules. A million (MeV) or even a billion (GeV) electronvolts constitutes a negligible amount of energy at macroscopic scale, insufficient to flap the wing of a mosquito but enormous if concentrated on a single, smaller than minuscule particle. In this case, the density of energy may be huge, crammed into volumes of space that are truly microscopic. It is enough to think that in our most powerful particle accelerator, the Large Hadron Collider (LHC) at CERN, we can reach an energy of up to 14,000 GeV, still negligible at macroscopic level but indicative of the density of the hyperheated universe when it only had scarcely a thousandth of a billionth of a second of existence. At that time, all of the energy-mass existing today in the billions upon billions of galaxies was concentrated into an infinitely smaller universe. It can easily be seen,

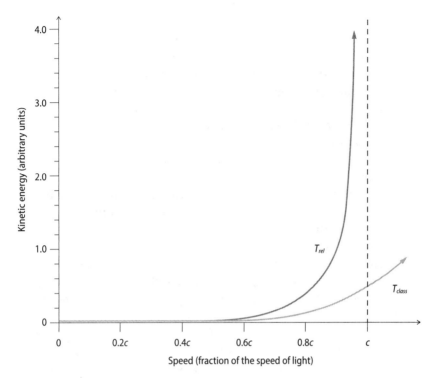

FIGURE 3.4
The curve T_{rel} shows the progression of the kinetic energy of a particle as a function of its relativistic speed. When the latter approaches c, the energy of the particle increases dramatically. On the contrary, classical Newtonian energy T_{class} increases much less, and moreover, the speed is not limited to c. Another way of interpreting this trend is that however much energy I supply to my particles, I will never reach the speed of light, or of any particle with null mass.

perhaps without too much simplification, why we say that at CERN we are re-creating in the laboratory miniature Big Bangs. This is a very important aspect because it illustrates the paradigm that provides the basis of our study of elementary particles. It is true (as we shall see) that increasing the energy of the collisions makes it possible to probe ever-smaller dimensions and to produce hypothetical particles of ever-larger mass. But this is not all: by doing this, we recreate the physical conditions of the universe ever closer to the beginning of everything—to the Big Bang itself. The physics that we know today, valid for billions of years, originates from the physics that predominated for negligible intervals of time—and that is still largely unknown today—in the first instants in the life of the universe, when its energy density decreased incredibly quickly with time. Every tiny step back in time, such as from 10 billionths of a billionth of a second after the beginning of time to 1 billionth of a billionth of a second, opens up unknown scenarios with a wealth of information of incalculable value, precisely due to the very different energy density, or equivalently, temperature.

The really important facts in the development of the universe happened after its first birth cry. From that instant until now, 13.8 billion years have elapsed, but little really new has happened since the first moments of life, given that the extremely rapid expansion of space had already drastically reduced the temperature of the universe. Of course, the laws of physics came into operation, creating atoms, stars, galaxies, planets, living beings, and the self-consciousness of *Homo sapiens*, at least on Earth. But to understand how all of this happened, how it developed and how it will end, we need to negotiate the extremely steep slope back, which at the expense of extremely arduous effort carries us into the vicinity of the summit at $t=0$, into the extremely young and hyperenergetic universe. This is a journey into the past that is fantastic and demanding. After each fraction of an instant gained, a new scenario opens up, with different experimental conditions and perhaps with new particles, the parents of those we know today. This scenario is similar to a videogame in which the most advanced players go further into new rooms, finding themselves confronted with unexpected situations. It is of little consequence that the geological eras of the universe last for only fractions of a nanosecond. We have an exceptionally good method for exploring them and searching for unexpected fossils, and we do not neglect the opportunity to do so, even if, just as in climbing a mountain, it becomes harder and harder as we approach the top.

But let's return to Earth, and to our attempt at basic technical insight. Given that mass is a form of energy, we decide logically to express the very small mass of the elementary particles in units derived from those describing the energy that have just been defined, paying attention to the factor c^2 (eV/c^2, MeV/c^2, GeV/c^2, etc.). When we talk of momentum, then, the units are eV/c, MeV/c, GeV/c, etc. Thus, an electron has a mass of 0.5 MeV/c^2 (500,000 eV/c^2), a proton around 1 GeV/c^2 (1 billion eV/c^2), and so on. If we were to operate the LHC accelerator at its highest energy of 14 TeV (14,000 billion eV), then in principle we could produce something like 14,000 particles of 1 GeV/c^2 from pure energy/matter conversion. This is why the table in figure 3.5 is very instructive, showing the relations between the mass of particles and their energy and speed. Mass and momentum combine to define the relativistic energy of the particle in keeping with equation (1). Let's imagine, for example, that we are supplying a proton, the exact mass

Energy (electronvolt)	Energy (joule)	Electron speed (electron mass: 511 keV/c^2)	Proton speed (proton mass: 938 MeV/c^2)
1 eV	1.6×10^{-19} J	593 km/s 0.002 c	14 km/s 0.00005 c
1 keV	1.6×10^{-16} J	18730 km/s 0.062 c	438 km/s 0.0015 c
1 MeV	1.6×10^{-13} J	282128 km/s 0.94 c	13832 km/s 0.046 c
1 GeV	1.6×10^{-10} J	299792 km/s 0.9999998 c	262338 km/s 0.88 c
1 TeV	1.6×10^{-7} J	299792 km/s 0.9999999999998 c	299792 km/s 0.9999996 c
7 TeV	1.1×10^{-6} J	299792 km/s 0.999999999999997 c	299792 km/s 0.999999991 c

FIGURE 3.5
Energy and speed of an electron and a proton as a function of their kinetic energy, which add up to their mass values. Energy is given in multiples of electronvolts, while the speeds are given in km/s and as a fraction of the speed of light c, as well. It is evident that when increasing the kinetic energy, both the electron and the proton soon become relativistic particles, namely with speeds close to c.

of which is 0.938 GeV/c^2, with a momentum of 2 GeV/c. The total energy of the proton will then be

$$\sqrt{(0.938^2 + 2^2)} \text{ GeV} = 2.21 \text{ GeV}.$$

Finally, I would like to end this chapter by returning briefly to the time intervals generally associated with the life of elementary particles. We have talked about more than microscopic dimensions, about fantastic speeds and about time intervals that are in effect often negligible as far as our human perception is concerned. It is almost as if the world of particles had its own spatiotemporal scale, very different from our own. This is not such a mistaken idea. With an effort at abstraction, we must consider that a particle that lives for a nanosecond has enough time to "experience" its often complex life, made up of interactions and correlations with other particles, of events that happen on an even more reduced temporal scale, and at frenetic speeds. Here I'll anticipate information that will seem more intelligible and justifiable subsequently. There is a particle, the positron, which in every possible respect is the same as an electron—except that it has a positive electric charge. Got that? Good. When an electron and a positron meet sufficiently close to each other, they self-destruct (annihilate) emitting energy in their collision, precisely according to the laws of matter to energy conversion previously described. But it can also happen that their meeting does not have an immediately negative outcome, and instead, they can even start to orbit around each other, in a kind of cosmic dance. This ballet can last an extremely long time on the temporal scale of particles—for hundreds of nanoseconds, even! The surprising thing is that during this time, the two dancers manage to exchange reciprocal information, decide to execute particular steps, and coordinate in assuming specific speeds, even to the extent of sending us other particles as witnesses to their energy levels or, to continue with the analogy, of their degree of involvement in the dance. Here is a life, then, that's rich in events and situations! During the course of the fantastic voyage in the world of the microcosm, such considerations will have to be our constant companions to facilitate a complete understanding of the facts. And furthermore, if a nanosecond is something of little consequence to us, we should not forget that a picosecond is a thousand times shorter (just as a year is that much shorter than a millennium). Everything is relative.

4 THE CERTAINTY OF CHANCE

I think I can safely say that nobody understands quantum mechanics.
—Richard Feynman

To physicists at the beginning of the twentieth century, the need to use special relativity in the study of the microcosm of elementary particles did not seem compelling, given that the speeds of the particles with which they were dealing were not relativistic. But the situation subsequently underwent a drastic change, as we shall see, with the discovery of particles originating in space and the first particle accelerators. However, the urgent need for a new interpretative vision of natural phenomena was not just limited to relativity. As we have already discussed, at the beginning of the twentieth century, certain problems of physics that could not be resolved suggested that for the description of the motion and of the behavior of atomic and subatomic systems, it was no longer possible to employ the laws of classical physics. As is often the case, a strong indication that something was not going in the right direction was provided by an apparently marginal problem not pertaining to our specific field of study of the microcosm. This was the problem of the black-body energy spectrum, a subject of classical thermodynamics. A scorching hot oven, the Sun, and a closed box with a small aperture are three examples of a black body. Simplifying still further, a black body is an object that absorbs all the electromagnetic radiation which is hitting it, without reflecting any of it, and therefore should appear black on the chromatic scale. A blue body only reflects light of the frequency that corresponds to the color blue, while a white object reflects all wavelengths.

Well, you might argue the Sun does not look conspicuously dark ... The reason for this is that a black body can also have its specific emission of radiation (not coming from reflection). Further, all bodies emit electromagnetic

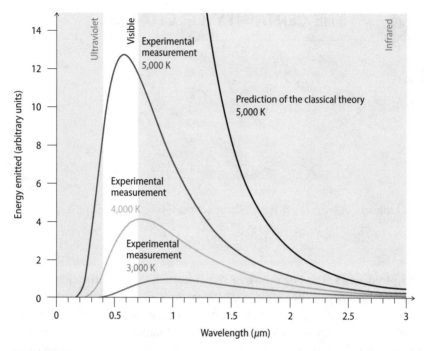

FIGURE 4.1
Energy spectrum of the black body radiation at different temperatures measured in kelvin: this is a temperature unit similar to degrees centigrade, but assumes that its 0 is at the so-called thermodynamic absolute zero, the lowest imaginable temperature, corresponding to nearly −273 degrees centigrade. In the spectrum in the figure we have delineated the short wavelength zone (of high, ultraviolet frequency), the one corresponding to visible light (where the intensity is near maximum), and the zone of long wavelength (low frequency) relative to infrared radiation. Also shown is the "classical" prediction. The lack of agreement of the latter with the corresponding experimental data is striking.

waves at a given temperature. Let's think about the filament of an electric stove, which transmits the greater part of its energy in the form of light and infrared radiation (i.e., heat). It can be demonstrated that the radiation intensity emitted by the black body (its own), expressed as a function of the radiation frequency, depends only on the temperature of the body, not upon the material of which it is composed or on its particular geometric structure. This is a good reason for wanting to derive a formula that can take account of such universality. Measurements of the energy radiation spectrum of the black body, however, did not yield the results expected by the physicists at the end of the nineteenth century. As can be seen in figure 4.1, the data collected for various temperatures showed a distinct discrepancy in relation to the theoretical prediction made with the notions of so-called classical physics. Something, clearly, was amiss.

The problem of black-body radiation undoubtedly marks the borderline between classical physics and what would soon become quantum physics. The lack of convergence between the experimental data and the theories that followed between the end of the nineteenth century and the first years of the twentieth was embarrassing, especially given the apparent simplicity of the physical system being examined. On December 14, 1900, after several unsuccessful attempts, the solution to a problem that was anything but academic was finally proposed by Max Planck (figure 4.2). This was a historic landmark, marking the birth of nothing less than the new quantum physics. The German physicist's hypothesis was drastic and revolutionary. He found himself obliged to admit, albeit with enormous reluctance, that the energy of an electromagnetic wave, specifically inside a black body, is not manifested in nature as a continuous flow, but rather is quantized in a discrete way in microunits, each one corresponding to a quantity of

FIGURE 4.2
Max Planck.

energy—and here's the beauty of it—that is proportional to the frequency of the electromagnetic wave itself: $E = h\nu$. The proportionality constant h is an extremely small number, equal to approximately 4.1×10^{-15} electron-volts × second, which we have known ever since as the *Planck constant*. It plays a key role in nature, and especially in the microcosm of particles, every bit as much as the speed of light and the elementary charge of the electron do. As we will see in the following, the elementary and discrete energy "packets" are the photons, each of them carrying a quantized and small amount of energy defined by the Planck relation. Obviously, this discrete nature of radiation does not appear in the macroscopic world—given the much higher energy associate to its events—just as relativistic effects are wholly insignificant in everyday life.

As if by a miracle, Planck's theory explained the energy spectrum of black-body radiation. The reasons for this are complex, but in a nutshell, we can say that the specific mathematical form of the energy-frequency dependence proposed by Planck considerably reduces the large quantity of energy that was calculated would be emitted at short wavelengths—that is, high frequencies—in the classical interpretation of a black body, to a degree that had been experimentally observed (figure 4.1). In practice, the emission of photons with high energy—high frequency or short wavelength—turns out to be very improbable. The price to be paid for such success would be very high indeed. The very same Max Planck who in 1918 had deservedly received the Nobel Prize for this discovery was well aware of his theory's enormous implications—including the philosophical ones. The quantization of energy demanded a new vision of the world, and he himself had said from the start that his was only a formal hypothesis that would soon become reconciled with classical physics. But things did not turn out that way. The dice had been thrown, and it was Albert Einstein again who now went into action.

The year 1905 in Berne has rightly been called by many the *annus mirabilis* of physics. That year saw the publication of the theories of special relativity and Brownian motion, as well as the third fundamental article by Einstein related to the explanation of the photoelectric effect in quantum terms. This latter effect is produced when an electromagnetic radiation that hits the surface of a metal succeeds in extracting some electrons from the metal itself. These are called *photoelectrons*. Thanks to the pioneering work of Heinrich Hertz, it had become well known that photoelectrons are emitted only if the frequency of the radiation is higher than a certain value. For every metal, there is a sort of threshold frequency, below which no electrons are produced, regardless of how intense the incident radiation might be. Einstein's explanation, which was barely subsequent to Planck's revolutionary hypothesis and extended it significantly, affirmed that light, and any other electromagnetic wave, should not only be considered as a wave, but also as a set of quanta (photons), each with an energy proportional to the frequency of the wave, in keeping with Planck's formula $E = h\nu$. It consequently became clear why, if the frequency of an electromagnetic wave is not sufficiently high, the photons associated with it do not have the necessary energy to overcome the potential barrier that keeps the electrons tied to the atoms of the metal. Einstein's model explained, furthermore, how the increase in frequency increments the energy of the photoelectrons: the cause lies in the higher (relativistic) energy that becomes available in the collision between photon and electron. The first scientist to verify this audacious but plausible theory of Einstein's was Robert Millikan, in 1914. It's worth reflecting on the fact that, during the same dramatic years that the world was careening into the conflagration and lunacy of World War I, a number of great scientists were prompting humanity to take giant leaps forward with their discoveries. Einstein was awarded the Nobel Prize in 1921, in recognition of his contributions to theoretical physics, especially for his discovery of the law governing the photoelectric effect—and oddly not for the theory of Brownian motion, still less for that of relativity. Millikan earned the Nobel in 1923 for the first measurement of the electric charge of an electron.

It was clear to everyone by then that the hypothesis of energy quantization was much more than just a formal mathematical expedient: it was the way in which nature manifested itself at the microscopic level. In the meantime, the standard model of the atom was still that proposed by Ernest Rutherford. Hanging over it, however, like the sword of Damocles, was the awareness of the weakness that its instability represented from the outset. As we said in chapter 2, this is due to the almost instantaneous loss of energy by the electrons classically assumed to be orbiting around the nucleus. The need to come up with a solution to this problem, as well as the success of modern quantum ideas, led the Danish physicist Niels Bohr (figure 4.3) to propose an atomic model of his own, an intermediate step between the pure classical vision and what would go on to become the interpretation based on quantum mechanics. Bohr's model was introduced in 1913. It envisaged that the electrons would circle around the nucleus, but only in orbits that were quantized and stable, given that the transitions toward corresponding orbits at higher or lower energies could not occur in a continuous manner, but only in a way that was discrete and selective, by means of the emission or absorption of photons quanta of energy equal to $h\nu$

FIGURE 4.3
Niels Bohr as a young man.

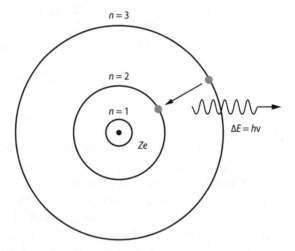

FIGURE 4.4
Schematics of the Bohr atom. The electrons revolve in a stable fashion on orbits fixed around a nucleus of positive charge Ze, and on which they do not lose energy. They can jump from one atomic level to another, emitting or absorbing photons (quanta).

(figure 4.4). The model immediately enjoyed remarkable success, not least because it explained the atomic spectroscopy that was being rapidly developed at this time. Atoms, in effect, can be excited if subjected to a flux of external electromagnetic radiation. Their electrons jump at energy levels (Bohr orbitals) that are higher, and in the subsequent deexcitation, they emit well-defined energy photons that precisely correspond to the energy differences between the levels. The model was capable of interpreting the observed, discrete, and noncontinuous (!) energy spectrum of the emitted photons, for the glory of the great Danish scientist. It should be noted that dealing with bound states, the energy of the electrons corresponding to the various quantum levels is assumed to be negative. Then, the ground level of the hydrogen atom corresponds to −13.6 eV. The excited levels have higher energy (i.e., they are less negative) but with decreasing absolute values, as follows: −3.4 eV, −1.5 eV, −0.85 eV, and so on. Levels with positive energy belong to free electrons, not bound to the nucleus.

Although not wholly satisfactory, as it lacked a true fundamental physical theory of the kind that would arrive with quantum mechanics, Bohr's proposal nevertheless represented the official introduction of quantum concepts into atomic physics. In the wake of this, the nuclear-atomic model was further refined when it was understood that the number of protons in the nucleus was equal to the number of electrons. In parallel with this development, the nature of the chemical bonds among the various elements was beginning to be understood, essentially due to the way in which different atoms attach to each other, sharing some of their electrons or transferring them from one atom to the other to form molecules, and favoring the creation of systems that are largely stable as far as energy is concerned. In short, the situation was clarifying.

Another milestone in the brilliant advance of the physics of that era was the hypothesis of "matter waves" proposed in 1924 by the French aristocrat and physicist Louis de Broglie. Electromagnetic waves allow signals to be transmitted and energy to be transferred from one point to another by means of the undulatory propagation of electric and magnetic fields oscillating at a given frequency (radio, infrared, visible, ultraviolet, X-rays, etc.). In a wholly equivalent way, we can interpret an electromagnetic radiation of frequency v as being due to the propagation of corpuscles (photons) that transport energy in a quantized way according to the Planck-Einstein formula $E = h v$. Posing the symmetrical question was completely legitimate for de Broglie: is it possible to conceive that, associated with the motion of a material object of the microcosm (an electron, a proton, or a nucleus), there can be a matter wave? Is it possible that particles exhibit an undulatory behavior, just as the photon can equivalently be considered as a wave? The French physicist proposed that this wave-corpuscle dualism can be substantially verified for microscopic particles purely as a matter of principle, and therefore it was irrelevant for macroscopic objects.

The point of departure for de Broglie, who developed his model starting from the dualism between geometrical and wave optics, is the Planck relation for the photon. By recalling the formula that links the wavelength λ, frequency ν, and the speed of the wave, and knowing that the photon always travels at the speed of light c, we have $\lambda = c/\nu$; $E = h\nu$; $E = pc$. From these relations, it follows that $\lambda = h/p$. The hypothesis of the French physicist consisted of assuming that the same double undulatory and corpuscular behavior also applied for the particles endowed with mass, such as the electron. In principle, this was a reasonable and unifying idea. For the electron too, therefore, we define a wavelength associated with it—or better still, with its motion—as $\lambda_{elect} = h/p_{elect}$. It is important to remark that such a wavelength is inversely proportional to the momentum of the particle. From the preceding equation, it follows that the de Broglie wavelength for an electron traveling at a speed 100 times slower than the speed of light is approximately equal to 2.5×10^{-8} cm, which is compatible with the typical atomic dimensions. But if we make the calculation for a car running along a motorway at 100 km/h, then the wavelength will be approximately 10^{-36} cm (i.e., 10^{23} times smaller than the dimensions of a proton). Beyond the appeal of the generalization, a problem persisted: a reasonable interpretation of this matter wave was still lacking. However, de Broglie's hypothesis could be falsified by carrying out an experiment capable of revealing (or not) some aspects associated with the hypothetical undulatory nature of the particles. For instance, for photons or for electromagnetic waves, we can carry out optical experiments such as the one shown schematically in figure 4.5. The appearance on the screen of an interference signal is evidence that we are dealing with waves, where the photoelectric effect indicates, correspondingly, the existence of photons as corpuscles.

Two independent and conclusive experiments were conducted by the American researchers Clinton Davisson and Lester Germer, and by the Englishman George Thomson (the son of Joseph John Thomson, the discoverer of the electron). Carried out between 1926 and 1927, these proved the wave-particle dualism for electrons. The experiments involved sending a beam of particles onto a crystal and measuring the angles at which they were scattered after the collision. Instead of a distribution with little dependence from the emission angle, the physicists observed one that was wholly compatible with the shape of interference—an indication that the electron

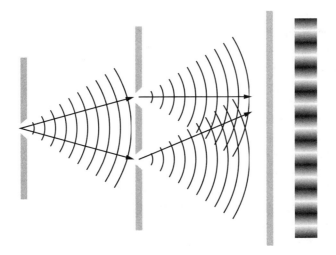

FIGURE 4.5
A radiation consisting of electromagnetic waves passes through two slits of width comparable to its wavelength. The outgoing waves interfere, producing on the screen of the detector the characteristic fringes of interference, proving the waveform nature of the radiation.

behaves like a de Broglie matter wave. This was an outstandingly important result, in recognition of which Davisson and Thomson were awarded the Nobel Prize in 1937. Naturally, de Broglie himself had also achieved that well-deserved recognition, in 1929, for the discovery of the undulatory nature of electrons. It is nice to observe that father and son, J. J. and George Thomson, were both awarded the Nobel Prize, the first for discovering that the electron is a particle, and the second for demonstrating that the electron behaves as a wave!

Louis de Broglie would go on to enjoy a long career as a scientist. After World War II, he was among the most committed proponents of the creation of a European research center for nascent particle physics, which became CERN, in Geneva. He made the first official proposal for the constitution of such a laboratory, at the European Cultural Conference of Lausanne, on December 9, 1949. Such vision and foresight were also manifested by other great physicists who were active in the reconstruction of Europe after the war: Edoardo Amaldi, Pierre Auger, Raoul Dautry, Lew Kowarski, and Niels Bohr. All of these men were truly visionary in wanting to rebuild through peaceful means international scientific collaboration in a war-racked continent reduced to ruins—and to do so through nuclear physics that was still fresh in world history as being directly responsible for the development of nuclear weapons used in the tragic destruction of the two Japanese cities of Hiroshima and Nagasaki.

In any case, back to 1925: the official birth of quantum mechanics was imminent, though as sometimes happens in science, it is difficult to attribute ownership and assign a precise date of birth to an idea, a theory, or even a discovery. In this particular case, we have seen that it was more like a mosaic constructed by many scientists, each one capable of providing the required and frequently fundamental pieces. It was a propitious time for a clear solution that was radical and general, and that would take into account the existing phenomenology and would be able to frame the experimental scenario within a genuine organic theory. The subject is vast, and its scientific aspects are many and complex. Therefore, I will attempt to focus on a few key points pertaining to the subject of this book—the elementary particles.

The term *quantum mechanics* was coined in 1924 by Max Born, a German physicist, mathematician, and author of important theoretical contributions. Between 1925 and 1926, Niels Bohr in Denmark, Werner Heisenberg in Germany, and Erwin Schrödinger in Austria (figure 4.6) laid the physical-mathematical foundations of the theory with two separate formulations that subsequently turned out to be equivalent. Heisenberg developed a formal and mathematical model called *matrix mechanics*, and Schrödinger proposed *wave mechanics*, extending de Broglie's hypothesis. A founding principle of quantum mechanics is the one that we have just discussed, relating to the double undulatory-corpuscular nature of the objects of the microcosm. In some experiments, they behave like material particles, in others they are like waves. It was Schrödinger who came up with a key equation: one that today bears his name. It applies to some mathematical functions (wave functions) that describe the state of the particle, and that determine its successive spacetime evolution, in contrast to classical mechanics wherein the state of a particle is entirely defined by its coordinates x, y, z, and t, and by the corresponding values of the associated physical quantities. Just as in the formulas of classical physics, Schrödinger's equation makes it possible to predict the way in which the wave function will appear at some time in the future in a deterministic manner. Less predictable is the way in which the wave function will change after a measurement performed by an observer. The importance of this point is crucial. While in classical physics, the action of the experimenter on the object under measurement is generally minimal and does not disturb the future evolution of the system, the same cannot be said of the microcosm. The

FIGURE 4.6
Werner Heisenberg (left) and Erwin Schrödinger, among the fathers of quantum mechanics.

quantum observer-object relationship is essential. Einstein forced himself to ask, provocatively: "Is the Moon still there, if I don't look at it?"

The scientific debate on the role and significance of wave functions was immediately intense and fomented many outstanding scientific advances. Many physicists made their contribution in those truly creative and exciting years. The so-called Copenhagen interpretation, from the name of the Danish school that developed around Heisenberg and Bohr, is the one that is mostly followed nowadays. Before a measurement, physical systems have only defined probabilities for each of the possible values of the associated physics quantities, such as energy, angular momentum (I will define this quantity shortly), momentum, etc. It might look academic, but the particle before the measurement *is* to be considered simultaneously in all the possible states, though with different probabilities. In the act of measurement that influences the system—certainly in the microscopic world—we have the so-called collapse of the wave function in one of the possible values for each physical variable. Even today, the discussion is open on these subtle issues, but the complexity of the subject is beyond the scope of our present study.

As an example, we could think of the concept of *entanglement*, which hides mysterious and yet to be resolved aspects of wave mechanics. The phenomenon, exquisitely quantum in kind, implies that in certain cases, the state of a given system (e.g., of a particle) cannot be described individually; rather, it is intimately tied to the conditions of other systems (other particles). As a result, in a counterintuitive sense, the measurement of a certain variable of a particle entangled with another can have an *instantaneous* impact on the value that the same variable assumes for the other particle, even if arbitrarily distant from the first, apparently violating the condition of the necessity for a finite time (not vanishing) for the transmission of information! One should add that the remarkable success of the theory is beyond doubt. It not only perfectly explains what happens in the atomic and subatomic world, but also provides quantitative predictions that can be verified experimentally. In short, quantum mechanics was in reality the culmination of many years of attempts at interpretation made by nascent atomic and nuclear physics, and also an extremely powerful instrument to proceed still further, to tackle that which nature subsequently would have in store for us. Today, quantum mechanics not only describes our elementary particles, it also forms the basis of atomic physics, of solid-state physics, of nuclear physics, and also of chemistry. Frequently, the calculations for illustrating and predicting the corresponding physical processes are so complex that is necessary to employ approximations or simplified models. But the first principles are always the founding ones of quantum mechanics—without doubt, the most powerful tool that we have at our disposal for the study and understanding of nature at its deepest level.

Let us now develop a little more formally (but not too much!) the concepts that have just been outlined. Our wave function $\Psi(x, y, z, t)$ completely describes a given quantum system (e.g. an electron, an atomic nucleus, or an entire atom) and makes it possible to determine the values of all the various physical quantities that characterize it. The $\Psi(x, y, z, t)$ is a mathematical function in an abstract space that is not "physical." That said, in the current interpretation of quantum mechanics, its square $\Psi^2(x, y, z, t)$—in reality, its modulus squared, as it involves a function of complex numbers—is proportional to the normalized probability density of finding the particle in a small volume of the space (identified precisely by x, y, z) at the time t (figure 4.7). Considering the spatial variables, this probability extends in principle across the entire universe. This property, negligible for macroscopic objects that we always know the position of "with certainty" at any given moment, produces appreciable effects for atoms, nuclei and elementary particles—you will remember that de Broglie's matter wave in association with a macroscopic entity is virtually nonexistent, which is what one would expect.

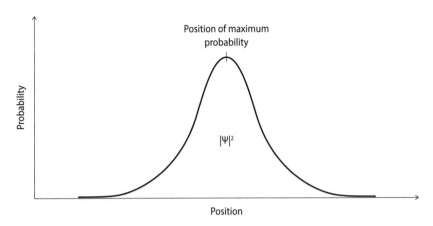

FIGURE 4.7
Example of a wave function associated with a particle. For the sake of simplicity, let's assume that Ψ is only a function of the space coordinate x. The value of the square of the function is proportional to the probability of finding the particle at that point in space. Nevertheless, given that it is a probability and not a certainty, other positions in space are possible at the same time. In principle, the probability function extends throughout the whole universe.

It is worth underlining that an important consequence of the formalism of quantum mechanics is that before the measurement, we cannot define with exactness the position of an electron around an atomic nucleus—but only regions where the probability of finding it has a certain value (let's say 10 percent, 1 percent, or even one in a million) normalized to the 100 percent probability (obviously) of its existing in the entire universe. In a corresponding manner, a probability exists—frankly derisory, but not completely null—that the electron that we suppose is orbiting the nucleus of an atom of the laptop on which I am writing this book, at a certain moment in its incessant quantum journey, may find itself wandering on the surface of Proxima Centauri! Following the same principles, a quantum particle (described by a wave of probability) can go through a huge barrier, like a wall, and actually be found without any problem on the other side thanks to the well-known *tunnel effect*—an absolutely real process, although apparently nonsensical.

This effect, albeit not the most interesting topic of all quantum physics, is certainly one of those that most affects our imagination as deterministic animals. If an elementary particle is precisely described by a diffuse cloud of probability, we said that even with a very low relative probability, it may be that it can also be found (simultaneously!) where it should not be. For example, beyond a wall or a potential barrier so high that it is well represented by a very thick wall. Think of a ping pong ball in a top hat that even if we shake it from side to side will always remain confined inside the cylinder. Here, if the ball is an elementary particle, there is a low probability, though not vanishing, that it will be found outside the cylinder without shaking it. This is the reason for the name *tunnel effect*. It is as if the microscopic object crossed a mountain without climbing it but passing through an imaginary and therefore nonexistent tunnel, which is created instantly and just as instantly closes, as in the biblical crossing of the Red Sea. All this can only happen by virtue of the double corpuscular and undulatory nature of matter at its most intimate level, that of elementary particles: while a corpuscle of matter is reasonably identified by its position in space or time, its undulatory counterpart may well extend into an arbitrarily vast area, and therefore even beyond a potential or physical barrier such as a wall or the sides of a hat. Obviously nothing similar happens in our macroscopic world; however insane a scientist may be in taking a run and trying to pass through a concrete wall, the probability that he will find himself on the other side, without breaking the wall ... and his nose, is unfortunately slim in our macroscopic world. It is emblematic, however, that some of the electronic devices that we use on a daily basis, such as electronic memories and many of the processes related to the natural radioactivity of unstable substances, or—as we will see later on—even the mechanism of nuclear fusion within the Sun and all other stars, work by virtue of the *tunnel effect*. For example, in the case of radioactive decays, particles strongly bound in atomic nuclei may have a small but finite probability of being outside the nucleus and therefore being emitted free, even with an appreciable energy. Also living organisms are affected by the *tunnel effect*, in relation to several biochemical reactions occurring in the cells. This has contributed to the opening of the new research field of quantum biology.

Another assumption made by quantum mechanics is that all the dynamic variables belonging to the system itself, such as energy or momentum, can assume quantized and well-defined values. In principle, before executing a specific measurement on a given quantum object, we cannot ever say with certainty what its energy or angular momentum is unless the quantum state that represents it were not a so-called eigenstate of energy or of angular momentum—i.e., a condition for which a precise value (eigenvalue) is perfectly defined. This is extended to all the other possible properties of our particles, as well as to the various physical quantities that characterize them. Thanks to the mathematical properties of the wave function, an arbitrary state of the electron may be expressed by a linear combination of various eigenstates of energy or other variables, through the coefficients that weigh the probability of each eigenstate-eigenvalue pair. We talk in this case of the quantum superposition principle. The values of the coefficients provide the so-called probability amplitudes for this to occur. Perhaps it would be better explained with an example. Let us imagine that we have a particle, an electron, which can assume only three values of energy (eigenvalues): E_1, E_2, and E_3, associated with three possible eigenstates, Ψ_1, Ψ_2, and Ψ_3, respectively. The three eigenstates are, as the jargon has it, mutually exclusive and orthogonal, exactly like the three Cartesian axes, x, y, and z. And just as every vector of geometric space possesses three components along each of the three Cartesian axes, so the comprehensive quantum state (wave function Ψ) that describes our particle will be decomposed into three orthogonal eigenstates of energy, according to the linear combination: $\Psi = \alpha\Psi_1 + \beta\Psi_2 + \gamma\Psi_3$, where α, β, and γ are three numerical coefficients.

Before an eventual measurement, before looking at the Moon as Einstein might say, the electron may have a priori any of the three energy values E_1, E_2, or E_3, actually, all of them *at the same time* and with different relative probabilities. But if we make the measurement of its energy, the laws of quantum mechanics tell us that the state Ψ, which globally describes the electron, immediately *collapses* into one of the three possible eigenstates Ψ_1, Ψ_2, or Ψ_3. Let's suppose that it is Ψ_2, with its corresponding eigenvalue E_2. The probability of obtaining the eigenstate Ψ_2, therefore, will be equal to β^2, given that $\Psi^2 = \beta^2 \Psi_2^2$. It is implicit in this discussion that the sum of $\alpha^2 + \beta^2 + \gamma^2$ is equal to 1, to guarantee that the total probability of having any of the three eigenvalues is itself also equal to 1 (100 percent). Hence, for example, if the probability amplitude for the state will be 30 percent, then the probability for the same state will be (30 percent)^2 = 9 percent.

A third salient aspect of quantum mechanics is that every classical physical quantity or variable (which is observable), has a corresponding mathematical operator that acts on the wave function. We can think of operators as an extension of the normal algebraic operations, such as addition or finding square roots, which when applied to real numbers generate other numbers. The application of operators to the wave function provides the eigenstates and eigenvalues of the given operator (if they exist). These are determined by solving the equation that describes the evolution of a given quantum system, the Schrödinger equation:

$$i\frac{h}{2\pi}\frac{\partial}{\partial t}\Psi(x, y, z, t) = H\Psi(x, y, z, t).$$

To understand the equation, or even to use it, is difficult and goes well beyond our aims here. But we can say, however, that we are dealing with a complex differential equation for which the domain of application to the microcosm is guaranteed by the presence of the Planck constant. The equation is the quantum equivalent of the Newtonian $F=ma$; its solutions are the eigenstates and the eigenvalues of the wave function of the particle subject to a given potential, in turn generated by a specific interaction (force), as well as the relative probability amplitudes, the previously discussed coefficients. Let's consider, for example, the classic case of the gravitational potential determined by the Newtonian interaction acting on a particle with mass. The interaction to which our particle is subject is included in the Hamiltonian operator H (energy), which describes it completely.

A classic example of the application of Schrödinger's equation is that of the hydrogen atom, a relatively simple quantum system consisting of two particles, an electron and a proton. One starts with the wave functions for the single, free electron that take the form of plane waves to indicate that in the absence of any force, an electron extends itself throughout spacetime (at least in principle). But due to the electromagnetic force between proton and electron, there exists a Coulomb potential energy described by the specific form of the operator H. The solutions of Schrödinger's equation are a series of eigenstates, each one with a specific value of the energy level of the electron and extending over a zone of space around the central proton. Our single electron will spend its time randomly between the various eigenstates with different eigenvalues or probability amplitudes. We thus reencounter the results of Bohr's model, but in a more elegant and convincing way: the electron can be found in a stable condition in particular regions around the atomic nucleus, each one defined by a specific value of energy and other physical quantities. In practice, we will never know where the electron actually is; we will only be able to define the relative probabilities (a priori) between the various spatial positions respective to each eigenstate. Equivalently, as we said, we can say that the electron will simultaneously be in all the positions allowed by the solutions of the equation. It is thereby understood that following an experimental measurement of the electron (its wave function), it will collapse, as we have stated repeatedly, into one of the possible eigenstates, with its corresponding eigenvalue.

To illustrate this, the orbitals of the electron of a hydrogen atom are represented in figure 4.8, which are calculated by solving Schrödinger's equation. Each one of these is characterized by specific values of the so-called quantum numbers. These are the numerical values, quantized, of the dynamic quantities that describe the quantum system (e.g., energy, momentum, etc.). In an analogous way, Schrödinger's equation allows us to calculate the energy levels of the particles inside the atomic nucleus (namely, protons and neutrons). In this case as well, although the calculations are more complex, quantum mechanics shows its power with regard

FIGURE 4.8
Visualization of the different orbitals of the electron of a hydrogen atom, corresponding to the various solutions (eigenstates) of the Schrödinger equation. Each orbital represents a volume of space around the central proton that may be occupied by the electron. With each orbital, there is an associated probability and a specific value of the binding energy between the electron and the proton (eigenvalues).

to explaining and interpreting the structure of the microcosm, achieving a fantastic concordance between theoretical predictions and experimental data. And, as previously noted, chemistry (and biochemistry!) itself—the ties between different elements and the various chemical compounds—is described and understood thanks to quantum mechanics.

Another strange, exquisitely quantum property of the microcosm and of the elementary particles that populate it is *spin*, a truly special quantum number. From classical physics, we know that the motion of a body that turns on itself or circles around another—a spinning top, the Moon orbiting the Earth, or a seat on a merry-go-round, for example—is defined by a vector physical quantity, the angular momentum. For an orbiting body that can be reasonably assumed to be pointlike (figure 4.9) the magnitude of the angular momentum (i.e., its numerical value) is equal to the mass of the body multiplied by the rotation radius and the revolution speed. The direction of the vector is orthogonal to the rotation plane, and its direction is upward or downward, depending on whether it is spinning clockwise or counterclockwise, respectively (figure 4.9). For bodies with an arbitrary mass distribution that are not pointlike, the angular momentum has a more complex form—we have to consider the so-called moment of inertia—but for our elementary

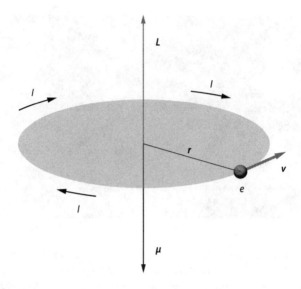

FIGURE 4.9
Classical definition of the angular momentum *L* of an electron bound to an atomic nucleus. *L* is equal to the product of *v*, *r*, and *m* (the mass of the electron). The vector that describes *L* is oriented perpendicularly to the rotation plane of the electron. The charge of the electron in motion generates a current *I*, and consequently a magnetic moment *μ*, oriented in the opposite direction to that of *L*.

particles this is not relevant. As is the case for energy and momentum, angular momentum is a quantity that obeys a conservation principle in physics.

Angular momentum can also be defined for our elementary particles, such as for the electrons distributed around an atomic nucleus—I have deliberately not said *orbiting*, in order to emphasize the aleatory nature of quantum motion. We speak in this case of an orbital angular momentum, which also is associated with a quantum number. And as is expected for quantum quantities, the angular momentum is a quantized variable, as if the electron could rotate only at specific speeds, producing as a result discrete and quantized values of the angular momentum. This behavior was discovered in a famous experiment conducted by Otto Stern and Walther Gerlach in 1922. The measurement is illustrated in figure 4.10. In the experiment, a heated metal emits silver atoms. They are electrically neutral but possess an angular momentum that is quantized according to the laws of quantum mechanics. Such an angular momentum produces a dipole magnetic moment, a small magnet, just as is classically obtained by making a

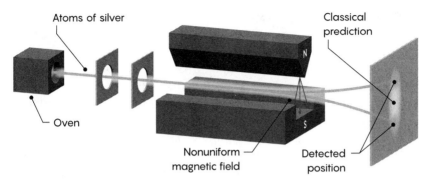

FIGURE 4.10
Description of the Stern and Gerlach experiment.

current circulate in a coil (figure 4.9). If this dipole passes through a uniform magnetic field, it will have no reason to be attracted toward a positive or negative pole, and the resulting force exerted on it will be zero because, at the two poles, a magnetic force will be exercised that is equal and opposite. In the case of a nonuniform field, such as in the experiment by Stern and Gerlach, the result will instead be that the total force is *not* null. If the atom was a classical object, depending on the random orientation of the atom-dipole, its trajectory would also be deflected in a random way, as shown in figure 4.10 (classical prediction). The atom is a quantum entity, however, and the two German physicists wanted to highlight this specific nature of it. They believed that the atoms of silver were in a quantum state corresponding to a total angular momentum with three possible eigenvalues: $-h/4\pi$, 0, and $h/4\pi$. If this had been the case, Stern and Gerlach would have had to observe on the screen three points of arrival of the atoms, not a continuous distribution: a spot corresponding to an absence of any deflections (eigenvalue of 0) at the center; another shifted above this, and a third symmetrically below. But this is not what occurred. The atoms were indeed in a state having a total angular momentum of just zero, without the expected values $-h/4\pi$ and $h/4\pi$. But the pair did not know this … Consequently, whether in a classical or quantum sense, their experiment *should have produced* a null result, registering no deflections at all. In reality, as shown in the diagram, there were two impact points formed on the screen and not three as they expected—but they were not even distributed in a diffuse way, as would have been compatible with classical mechanics. To a certain degree, Stern and Gerlach showed the discrete nature of angular momentum, but they did not obtain the quantitative result that they were expecting.

The explanation of the experiment, and of the enigma, did not come until 1925, when it was provided by the Dutch scientists Samuel Goudsmit and George Uhlenbeck. They observed that it is true that the atoms of silver possess no angular momentum, but that there also exists an intrinsic momentum, called *spin*, which is added to the orbital one, and which particles in purely rectilinear motion also possess. In simple terms, it is as if the electron or the proton were rapidly rotating around an axis as they propagate, acquiring angular momentum from spinning. The spin, a quantized variable, can have values equal to half-integer (1/2, 3/2, etc.) or integer (0, 1, 2, etc.) multiples of $h/2\pi$. The action of the magnetic field of the experiment has the same effect as a quantum measurement on any other physical variable. It forces the atom state to fall (i.e., collapse) in one of the spin eigenstates. In this way, the so-called quantization axis is defined; in the act of the measurement, the preexisting symmetry made up of all of the infinite spin orientations is broken. From the laws of quantum mechanics, it turns out that for a spin 1/2 particle the possible spin components along the quantization axis will be +1/2 and −1/2; a spin 1 particle, instead, will have three possible values for the spin components: +1, −1 and also 0 (for simplicity's sake, from now on I will omit the factor $h/2\pi$). It is worth stressing, as for any other quantized variables, that before the measurement our electron is simultaneously in the two states of +1/2 and −1/2 spin, as if it could rotate at the same time clockwise and counterclockwise: the magic of quantum mechanics!

The importance of spin goes well beyond the introduction of another quantum number, albeit specific to the microcosm. In 1925, the Austrian Wolfgang Pauli, the author of some exceptional theoretical contributions to physics, articulated the so-called exclusion principle that was subsequently named after him. Pauli showed that two identical particles with a half-integer value of the "newborn" spin cannot occupy the same quantum state, as defined by having all quantum numbers coincide. At least the direction of the spin vector will have to be different. One of the two half-integer spin particles, for instance, can have spin components of 1/2 and the other of −1/2. The reader might conclude that we are now splitting hairs, but this is far from the case. The repercussions of this—for the world of particles, for astrophysics, ultimately for the entire universe—are enormous. The Pauli principle can be expressed in a formal manner as follows: the overall state of a system of identical particles with half-integer spin (let's say two particles, for simplicity's sake) must be described by a total wave function that is antisymmetric for the exchange of the particles. In simple terms, if the first particle is described by the quantum state Ψ_1 and the second by Ψ_2, then the overall state of the two particles will

FIGURE 4.11
Enrico Fermi.

be represented by $(\Psi_1\ \Psi_2)$. Applying the antisymmetry condition for the exchange, we find that $(\Psi_1\ \Psi_2) = -(\Psi_2\ \Psi_1)$. This condition implies that if the two particles have quantum numbers, including those of spin, that are exactly the same, we will get: $(\Psi_1\ \Psi_1) = -(\Psi_1\ \Psi_1)$. The only way of satisfying such equivalence is if the total function $(\Psi_1\ \Psi_1) = 0$, and therefore the state does not exist! An important aspect of this argument is the requirement that the particles involved should be identical, indistinguishable one from the other—much more so even than identical twins, between whom many minor differences can always be found. The two electrons, in effect, are totally defined by their own mass, by their own electric charge, and by their own spin—quantities that are exactly the same for all the electrons in the universe. The same obviously applies for the atoms, the molecules, the photons, and all the particles that we will encounter in the course of our journey. It remains understood that the exchange of which we speak is only a virtual mathematical operation, and therefore, it does not correspond to a real displacement in space of our identical particles.

Parallel to the discovery of the exclusion principle, notable advances were made in the formulation of the so-called statistics of elementary particles. In 1926, Enrico Fermi (figure 4.11) and, independently, the British theoretical physicist Paul Adrien Maurice Dirac, obtained a relation that describes the energy distribution of a statistical sample of particles with half-integer spin as a function of the temperature, including the Pauli exclusion principle within the calculations. By *statistical system*, I mean something similar to a gas constituted by an arbitrarily large number of identical entities such as molecules. In these systems, the motion of each element is virtually nonobservable and noncalculable, while some combined (statistical) properties of the whole can be measured, such as temperature and pressure, as in the case of a normal gas. The theoretical formalism of the Fermi-Dirac statistics, in particular, is capable of quantitatively explaining physical processes that involve electrons in thermodynamics and in solid-state physics. As a result, particles with half-integer spin are now called *fermions*.

An analogous statistic was elaborated in the same year by the Indian physicist Satyendra Bose for particles with integer spin. The theory, which was subsequently developed by Einstein, describes the behavior of these particles— called *bosons* in honor of the Indian scientist—which are not subject to the Pauli principle: more bosons, in fact, may be found in the same quantum state, without any limitation in principle, just as if they could overlap in space one with another without coming to blows. This behavior really doesn't have any macroscopic equivalent! The preceding considerations on the exchange of particles are also applicable to bosons. In this case, the total wave function of the system must not be antisymmetric, as it is for fermions, but it is symmetric for the exchange of identical particles: $(\Psi_1 \Psi_2) = +(\Psi_2 \Psi_1)$; therefore, two or more particles may well have all the same quantum numbers, given that the relation $(\Psi_1 \Psi_1) = +(\Psi_1 \Psi_1)$ is always satisfied. As far as the elementary particles known at the time were concerned, the photon is a boson with spin 1, whereas the electron and the proton are fermions with spin 1/2. As we shall see, for the fermions, a conservation principle holds, but not for the number of bosons. The latter may be created and destroyed in nature, so long as their eventual masses are correctly transformed into energy, or vice versa: energy needs be conserved, without any ifs or buts.

The combination of the Pauli principle with the statistics of Fermi-Dirac and Bose-Einstein represented an enrichment of the formalism of nascent quantum mechanics, as well as a formidable instrument for understanding the microcosm, with important applications to the macroscopic world as well. For example, the stability of matter when we consider statistical combinations of particles is due to the exclusion principle. Electrons from external atomic levels (of higher energy) cannot fall into states of lower energy closer to the nucleus because they are already occupied by other electrons. This is equivalent to saying that forcing into the same state particles that are identical, with all the same quantum numbers, is to encounter a repulsive force—called of *exchange*—that tends to keep them apart. This phenomenon causes it to be the case that as the number of electrons increases, the volume of the atom grows correspondingly. Hence atoms cannot be compacted, and this is what renders solid matter rigid and fixes its dimensions, despite the fact that the atom is made of empty space, and therefore is potentially subject to being occupied by particles. Consequently, the impossibility of condensing beyond certain limits ordinary matter that's made up of electrons, protons, and neutrons (which are all fermions) is due to the repulsive exchange force generated by the Pauli principle. For the bosons, the absence of such limitations allows an arbitrarily large number of particles to be found in the same identical place.

A further ingredient, clinching the entire apparatus of quantum mechanics, is yet another principle arrived at during those heroic years of development: the Heisenberg uncertainty principle, proposed by the German physicist Werner Heisenberg in 1927. This principle governs the world of the microcosm, and its importance for the physics of elementary particles is such that we must discuss it here in some detail. The principle derives from the mathematical formalism of Heisenberg's quantum mechanics of matrices and has a profound significance. In practice, it asserts that it is impossible to measure specific pairs of physical quantities simultaneously with the same arbitrary precision. This applies particularly to the measurement of position x and momentum p. This limitation is expressed in the following celebrated mathematical formula:

$$\Delta x \Delta p \geq \frac{h}{4\pi}.$$

The formula reads as follows: the product of the uncertainty of the measurement of position (Δx) by the uncertainty of the measurement of momentum (Δp) is larger than (or at least equal to) the Planck constant divided by four times the value of π. Such a relation implies that:

$$\Delta x \geq \frac{h}{4\pi \Delta p}.$$

And from this second formula, it follows that if the result of the measurement of the position of our electron has a small experimental uncertainty (small Δx), then the uncertainty about its momentum will be large (large Δp), and vice versa. And it should be noted that all of this has nothing to do with our technical ability to achieve the measurement with arbitrary precision: it is actually an intrinsic knot, impossible to eliminate, inscribed in the laws of nature. Obviously, nothing prevents one of the two quantities from being measured by itself with infinite precision, but not both of them at the same time. Let's look at an example that I hope will clarify this, by way of an ideal experiment. Let's imagine that I wish to measure, in a classical sense, the position and momentum of a car traveling down a motorway at 100 km/h. Its position may be determined, for example, by taking a photograph at the precise moment t, as defined by my watch. This implies that photons emitted by the Sun hit the surface of the car and are reflected toward my digital camera. The momentum, on the other hand, can be measured by multiplying the weight of the car (say, 1,000 kg) by the speed recorded at the same instant t by a police radar gun. This is all well and good, because it is obvious that any interference caused by the observer (i.e., induced by my instruments) is entirely negligible. The photons that collide with the car produce no discernible effect on it, or on its motion—and the same is true in the case of the electromagnetic waves from the radar. Conclusion: in the macroscopic world, just as we expected, Heisenberg's uncertainty principle is completely irrelevant.

Now, however, let's envision another ideal experiment in which, instead of a vehicle weighing 1 ton and traveling at 100 km/h, I have an electron weighing 9.1×10^{-28} g, or approximately 500 keV/c^2, which propagates at a tenth of c. In a purely hypothetical manner, I decide to measure the position of the electron at time t. As usual, I request a photon to hit it and to bounce back toward my camera or in the direction of my eyes. In this case, however, after being hit, the electron, being itself a quantum object comparable to the photon, will be disturbed in some way, perhaps recoiling after that collision and acquiring a speed (momentum) in an arbitrary direction. This momentum will be added in an unpredictable way to the amount originally possessed by the electron. Consequently, at the same time that I wanted to measure its momentum, I would have an uncertainty, purely due to the action of measuring its position. To make the measurement of the electron position more precise, I would need to take a photograph at higher resolution, thereby sending more energetic photons to their target (the reason for this will be understood very shortly). Consequently, I would have a higher momentum transferred to the electron and an even larger uncertainty about the measurement of this variable. Improving the measurement of the position inevitably leads to a worsening of the measurement of the momentum! Such considerations are illustrated diagrammatically in figure 4.12.

With similar arguments, even when not completely rigorous, it is possible to understand that the orbits of electrons around an atomic nucleus, or rather their motion in volumes defined by quantum trajectories, are stable

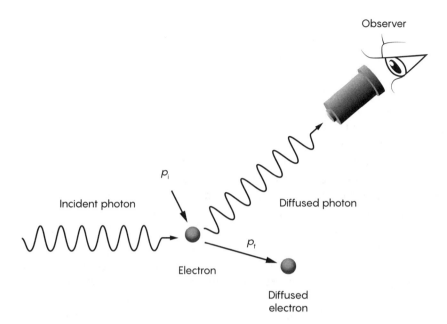

FIGURE 4.12
The Heisenberg uncertainty principle. If the photon is very energetic, the resolution of the "photograph" will be good (small Δx), but the electron will receive an extra high momentum that will add up in an unpredictable way to the original one (large Δp). On the other hand, for a photon of low energy, the picture will be of low spatial resolution (large Δx), but the electron's momentum will undergo little disturbance (small Δp).

and associated with solutions predicted by Schrödinger's equation. Imagine that you allow the single electron of the hydrogen atom to classically radiate photons, as it is accelerated, and fall rapidly on the nucleus (proton)—something that, as we have seen, does not happen for the quantum atom. In its catastrophic approach to the center of the atom, the position of the electron becomes ever more well defined, with only a small indetermination. In this way, however, the uncertainty regarding its momentum (energy or speed, in practice) increases in accordance with Heisenberg's principle, and hence high values for the momentum (speed) become possible. These tend to make the electrons shoot away, far from the nucleus. The concomitant and contrary pushes to fall and escape impose on the electron intermediate positions of equilibrium, which precisely coincide with those of our orbitals. Behold the power of quantum mechanics! As a corollary, this phenomenon explains why subatomic particles tend to reach extremely high speeds if confined. The reason for this is to be found, once again, in the uncertainty principle; the particles can acquire high speed simply by virtue of a reduction of Δx: the proverbial caged animal.

The substantial uncertainty regarding the position of quantum particles is reprised in the same classical conception regarding its trajectory, which doesn't make much sense when the distances at stake are minimal. The particles, in principle, do not follow well-defined trajectories like classical objects: now here and, a little later, further on ... Nevertheless, in the case of particles with a relatively high momentum (keV/c and more), which we can observe with our detectors, the uncertainty Δx regarding the position turns out in any case to be negligible compared to the typical experimental resolutions. These particles behave, in fact, like macroscopic objects and exhibit classical trajectories, while it remains virtually impossible to define the trajectory of an electron orbiting around an atomic nucleus; in this case, as in others, the uncertainty principle substantially influences the motion of the particles. The effects of the Heisenberg principle are combined with those produced by Pauli's exclusion principle. We have shown that in an atomic system with more electrons, an electron with higher energy and in an external level cannot radiate and therefore fall into a lower energy state because the spaces are already occupied by other electrons. This confirms why the quantum atom is stable, contrary to what should occur in a simple, classical atomic model, in which an electron accelerated by the centripetal force generated by the electromagnetic attraction with the nucleus is "forced" to emit photons, and hence progressively lose energy.

Another formulation of the Heisenberg principle concerns the simultaneous measurement of energy and time intervals: $\Delta E \Delta t \geq h/4\pi$. In this case as well, it can be demonstrated that it is impossible to measure the two variables simultaneously with the same level of accuracy. This form of the uncertainty principle, as we shall see, has a notable impact in defining the lifetime of unstable particles.

To summarize what has been covered so far: quantum mechanics describes and explains the motion, its evolution in time, and the properties of the particles that make up the microcosm, including the ways in which they interact with each other. We shall see that the theory, appropriately modified to take into account the quantum nature of these same fundamental forces (the quantization of the fields), is a robust instrument for calculating reactions, predicting events, studying the behavior of particles, and understanding their specific role in the big picture. For these same objectives, special relativity provides another necessary instrument for the interpretation and understanding of the microcosm. Quantum mechanics and relativity combined will guide our next steps.

We are now finally in a position to give a working definition to our subject of interest: the elementary particles. As already emphasized, to be elementary is to be devoid of internal structure, or pointlike. For a particle that does not appear pointlike, we can always hypothesize the possibility of dividing it in some way, thereby revealing its constituent parts. The notion of appearance in this context is, of course, quite vague. But in physics, all quantities and definitions must be objective, quantitative, and operational. Let's start, therefore, to define the relevant notions in the macroscopic world. To "see" an object means to detect the reflected light from a source. The light hits the retina of the eye, or an accessory or alternative instrument such as the photographic camera in our example, or a microscope. For this to happen, it is necessary that the wavelength of the incoming light is appreciably smaller than the dimensions of the object to be "seen." This is the reason why with visible light, for which λ goes from around 0.4 to 0.7 μm, we can observe bacteria and some viruses, but we cannot manage to see a molecule or an atom, and while a virus may have dimensions in the order of a few tenths of a micrometer, atoms turn out to be completely invisible on account of having a diameter that is typically less than one-thousandth of the wavelength of visible light.

Hence, we define the resolving power of a given radiation of wavelength λ as the minimum length Δr that is observable due to it. This length is proportional to λ (figure 4.13), or in mathematical notation: $\propto \lambda$. Remembering the relation $\lambda = h/p$ and $\Delta r \propto \lambda$, we then have $\Delta r \propto h/p$. From this formula, it follows that the observability of an object implies the use of a radiation with high momentum, or in practice of high energy. The resolving power of our observational instrument increases then with the energy of the radiation. We can extend such concepts immediately to material quantum particles and to their de Broglie wavelength. An electron with sufficient momentum will be able to "see" or not see an atom, a proton, or another electron.

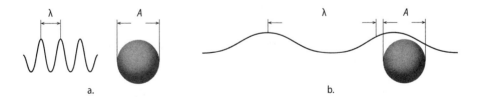

FIGURE 4.13
An electromagnetic wave with wavelength λ, or similarly, a particle with a de Broglie wavelength λ, is capable of probing (a), or (less well) (b) a particle of dimension A.

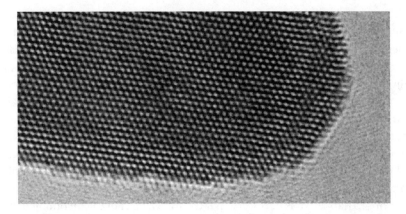

FIGURE 4.14
A photograph of a lattice of gold atoms obtained with an electron microscope. The atoms become individually visible, like rows of beads lined up next to each other.

Let's take, for example, the well-known case of the electron microscope, in which a beam of electrons takes the place of the light source of the optical one. The wavelength associated with its electrons is tens of thousands of times smaller than that of the photons of visible light. The results of this are spectacular. We even manage to see, in the literal sense of the term, the atoms as shown in figure 4.14. It's a really stunning result, don't you think, coming as it does after so many centuries of debate about the existence or nonexistence of atoms?

We can now understand why we are building ever more powerful particle accelerators: we desire to probe ever smaller structures and components of the microcosm. It is important to point out that the concept of seeing here should be understood as having some latitude, especially when we speak of projectiles of high energy and distances that are really microscopic. What we usually do in our labs is to send out electromagnetic probes (namely, photons) or particle ones (electrons, protons, etc.) off the structures that we wish to observe and, to identify the products of this interaction, we measure the kinematic features of the projectile and target after the collision. Remember Rutherford's experiment? By obtaining the angular distribution of the α particles deflected by the gold atoms, the great New Zealander discovered the atomic nucleus, or if you prefer, he "observed" it for the first time. A consequence of these considerations is that if I use a probe to see an atom and its momentum is not sufficiently high, it will appear elementary to me, and I will not be in a position to identify its internal structure. We can call this an operative definition, for it's certainly not an absolute one. Let's provide some data. If the projectile particle possesses, let's say, a momentum of 10 billion eV/c (10 GeV/c), which is typical of the interactions that we study at CERN, we will be able to observe structures of approximately 10^{-14} cm, equivalent to a tenth of the diameter of a proton. With such a resolving power, we have proven that the proton is not elementary, whereas the electron … still is, even at the level of 10^{-16} cm. The reader may understand that perhaps one day, with a superpowerful accelerator at our disposal, we will be able to show that even the electron is not as pointlike as it to us today. Never say never!

There is a nice collateral effect to the use of high-energy particles, caused by Einstein's relation $E = mc^2$. With the increase in energy, not only are we able to identify ever smaller structures, but we also have at our disposal a larger energy budget with which to create new particles, some of these perhaps as-yet unknown. And this is the profound significance of the synergy between undulatory mechanics and special relativity for the study of the microcosm. On the other hand, the laws of probability that form the basis of elementary particle mechanics impose an intrinsic uncertainty for all processes and reactions. As opposed to what happens in the macrocosmic world, governed by the deterministic laws of classical physics, only rarely will we be certain of what the outcome of a given reaction will be. We can only calculate the probability (say, 1 percent) that such a process will occur. Naturally, on account of the fluctuations implicit in the stochastic processes, the value of 1 percent will be verified only over a large number of attempts, or of events we produced and studied. The number 57 must come up in the lottery with a probability of 1/90—but nothing prevents it, in ninety goes, from coming up zero, one, two, three, or more times. Only after 1,000, 10,000, or more attempts do we see that the expected probability will be reasonably verified. These arguments hold true for collisions between particles, as well as for their decays. It is therefore understandable why, in studying a theory that can potentially predict certain kinds of (average and statistical) behavior of elementary particles, there must be a large number of collisions (events) produced—for example, with an accelerator—in order to then study them with the detectors of the experimental apparatus. In this way, by reducing the statistical error associated with the measurements, it will be possible to verify whether the theory quantitatively explains the actual behavior of nature. Then, the result of a measurement, following the accumulation of a given number of events, will always have to be compared with the *experimental background*. This consists of events with similar characteristics to those of the signal we are seeking, such as the same kind of particles produced after a particular collision, but due to processes of a different nature that are capable of making us erroneously believe that we have found precisely what we were looking for. Such a background always needs to be measured, or at the very least estimated by the experimentalist. Because these events are also subject to statistical fluctuations, it can happen that in repeating the experiment, we will obtain a notably larger or smaller number. Imagine, for example, that the estimated background for a given process of interest is equal to 50 events, and that we reveal 60 in total. Will we then be able to claim that we have observed 10 signal events with certainty? Definitely not! Given that the result of the measurement can fluctuate statistically, it is possible that in a subsequent experiment, the background may easily reach a number near to 60 (say, 59), therefore nullifying the evidence of our signal. It would be quite different if, instead of 60 events, we observed a much higher number like 750. It is clear that in this case, we would be entitled to claim to have measured "more or less" 700 signal events (750–50). Common sense also suggests that the probability that the number 50 fluctuates enough to become 750 in another experiment is really minimal—if not to say altogether nil! Needless to say, when confronted with a known signal process, we could also estimate the amount of signal events that we will eventually gather. If for the previous example this number is, say, 400, compared to the observation of 750 events and the expected background of 50 events, the measured (large) difference between 700 and 350 would reasonably be the indication of another (unknown) background source.

To avoid misunderstandings, scientists have imposed upon themselves cast-iron rules when it comes to defining the results of their measurements. Even if the probability of a background fluctuation is very small indeed (say, around 0.1 percent), we can only assert that there is an *indication* of the presence of the signal we are searching for. In order to speak of an *observation*, or of a *discovery*, the rule is that the probability that the background can fluctuate to simulate the hypothetical signal is no more than one part in 3 million! This may seem to be inordinately cautious—but as we know, the devil is in the details ... In reality, the mathematical calculations that form the basis of the data analysis in particle physics experiments are far more complex than our discussion here has suggested. Nevertheless, it is my hope that a clear message has emerged from it: the microcosm is governed in a fundamental way by the laws of probability and of statistics.

But now that we have set up the basic, necessary theoretical instruments on the table of our virtual laboratory, let's take up the thread of our narrative again. We have reached the beginning of the third decade of the twentieth century ...

5 THE FAMILY GROWS

The professors ... are wholly useless, most of them.

—John Major

Dear Radioactive Ladies and Gentlemen,

As the bearer of these lines, to whom I graciously ask you to listen, will explain to you in more detail, because of the "wrong" statistics of the N and ^{6}Li nuclei and the continuous beta spectrum, I have hit upon a desperate remedy to save the "exchange theorem" of statistics and the law of conservation of energy. Namely, the possibility that there could exist in the nuclei electrically neutral particles, that I wish to call neutrons, which have spin 1/2 and obey the exclusion principle and which further differ from light quanta in that they do not travel with the velocity of light. The mass of the neutrons should be of the same order of magnitude as the electron mass, and in any event not larger than 0.01 proton masses. The continuous beta spectrum would then become understandable by the assumption that in beta decay a neutron is emitted in addition to the electron such that the sum of the energies of the neutron and the electron is constant ...

I agree that my remedy might seem incredible because we should have seen those neutrons much earlier if they really exist. But only those who dare can win, and the difficult situation, due to the continuous structure of the beta spectrum, is lighted by a remark of my honored predecessor, Mr. Debye, who told me recently in Brussels: "Oh, It's much better not to think about this at all, like new taxes." From now on, every solution to the issue must be discussed. Thus, dear radioactive people, look and judge. Unfortunately, I cannot appear in Tübingen personally since I am indispensable at a ball here in Zurich on the night of 6/7 December. With my best regards to you, and also to Mr. Back.

Your humble servant W. Pauli

With this letter, dated December 4, 1930, one of the most famous letters in the history of physics, Wolfgang Pauli (figure 5.1) proposed with his "desperate remedy" the solution to the problem of the energy spectrum of beta (β) decay. Pauli had been invited to a conference on atomic physics at Tübingen. A prior engagement so important that it could not be foregone—namely, an evening's dancing with a girlfriend—caused him not to attend, and instead to address this letter to the delegates, in particular to Lise Meitner and colleagues. The problem was the apparent violation of the principle of the conservation of energy in the process that transforms a radioactive atom (nucleus) into another, accompanied by a Thomson electron of high energy. As we have seen, and as had been observed by Meitner, the energy available in the decay is not distributed equally between the two particles that are supposed to be part of the final state of the reaction (figure 5.2). The "desperate remedy" consisted of postulating the existence of a new particle that would solve everything, and which Pauli called the "neutron."

FIGURE 5.1
Wolfgang Pauli, the author of the neutrino hypothesis.

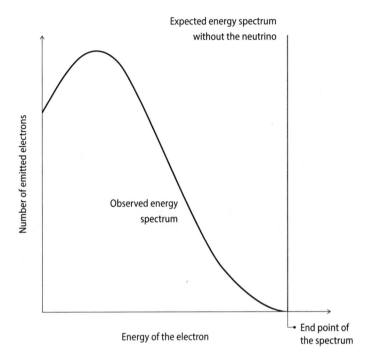

FIGURE 5.2

Energy spectrum of β electrons produced by a generic radioactive nucleus. Before the Pauli hypothesis, it was assumed that the process involved a two-body decay: Nucleus 1 → Nucleus 2 + electron, with Nucleus 2 and the electron dividing the available energy between them, with this always occurring in the same way, and therefore always with the same energy value for the electron. The observed spectrum, which is continuous, is an indication of either a problem with the conservation of energy in the reaction or the existence of a third particle (what we now call the neutrino) that takes a fraction of the total available energy and not always the same: the typical situation of a particle decay involving three bodies in the final state.

The available energy would now be divided between three partners rather than two, with a wide spectrum of possible "rations of energy" for each of the involved particles. But how can we explain the fact that there was so much reluctance to accept this hypothesis?

In line with what we have related previously, and in a manner that is undoubtedly extreme, various physicists (including Niels Bohr) had arrived at the drastic conclusion that in order to explain the measured spectrum of β electrons, it was necessary to renounce the "sacred" law regarding the conservation of energy, even though it was one of the pillars of science, amply verified today, in all experiments conducted in various research fields. To attempt to get rid of this law shows how iconoclastic scientists can be—men and women guided more by doubt than by the kind of certainty common among the modern exegetes of our society, who are sometimes too Manichean and inclined to simplify. In any case, Bohr's hypothesis seemed to many to be too extreme. To assume that the principle of energy conservation was not valid for β decay, despite being widely proved for a multitude of other processes, seemed to be an ad hoc solution, and one that therefore lacked generality and predictive power. In effect, with hindsight and with our current understanding of the world of elementary particles, Pauli's theory was completely reasonable and not in the least desperate. Is the electron's spectrum continuous? Well, if so, then it means that the cake of available energy is divided between three (or more) actors and not only two. In this way, in accordance with the probabilistic laws of quantum mechanics, the three particles originating from such sources as the radioactive decay of cobalt—nickel atom, electron, and "third wheel"—can each have attributed to them some energy limited only by the condition that the sum of their energies is the same as it was before the decay. So what's wrong, in the end, with introducing a new elementary particle (the neutrino, namely Pauli's neutron) to preserve the principle of the conservation of energy? Pauli, though, had sufficient reason to be less than wholly satisfied with his hypothesis. Above all, at the time, the number of elementary particles in the Rutherford-Bohr atomic model was really small: a nucleus made of protons, the total electric charge of which is balanced by an analogous number of electrons—the real neutron had yet to be discovered. That was it: two particles in total, in the whole of nature, and three when we add the photon, a corpuscle of energy without mass. Introducing another particle that had never been observed before, and was justified only by the need to save a conservation principle that always must be subject to scrutiny and judgment by the experimental results, seemed reckless. Furthermore, this new particle needed to be devoid of electric charge and to have a very small mass, in order to be compatible with the kinematical analysis of the β decay.

This is a good time to explain the neutron-neutrino enigma. When Pauli first put forward his hypothesis, it was debated as to whether, inside the nucleus, there could be a particle without electric charge and with a mass similar to that of the proton: what we know today to be the *neutron*. Ernest Rutherford, in fact, had intuited that in order to explain the actual values of the masses of atomic nuclei, it was necessary to assume the existence of heavy particles within them, which obviously had no electric charge in order to guarantee the neutrality of the atoms. Rutherford's idea, minimalist inasmuch as it did not propose the introduction of new particles, was that there were bound proton-electron states present in the nucleus that were neutral as a whole and of the right mass—that is, the *neutron system*. However, this hypothesis conflicted with the emergent theory of particles with spin. Furthermore, the model did not agree with the results of several experiments. The behavior of the nuclei of nitrogen and lithium, to which Pauli refers in the famous letter, represented a very clear stumbling block to the hypothesis of Rutherford's "bound particles." Ultimately, the application of Heisenberg's uncertainty principle to the hypothetical nuclear electrons led to a contradiction. As we have already seen, if in the relation $\Delta x \Delta p \geq h/4\pi$, we make Δx equal to the small nuclear dimensions, forcing the electrons into a much more confined space, then the uncertainty on p (Δp) will be much higher—specifically, it will be sufficient to allow the electron to acquire a momentum as high as 100 MeV/c, which is appreciably larger than that measured for the β electrons (of the order of 100 times lower).

While still unable to resolve the problem of the hypothetical neutral nuclear particle on account of it being too light, Pauli's neutrino made it possible to hit the proverbial two birds with one stone: saving both the principle of the conservation of energy and the statistics of particles endowed with spin. The solution to the dilemma arrived very quickly, thanks to an experiment devised by James Chadwick, another of Rutherford's students, who in February 1932 discovered the actual neutron, which was very massive and a little heavier than the proton. For this achievement, Chadwick received the Nobel Prize in 1935. The real nature of this particle, a sort of neutral proton, had already been well understood beforehand by the great Ettore Majorana (figure 5.3), about whom a vast amount has been said and written, not least on account of his mysterious disappearance when still a young man. I like to recall the opinion of him expressed by his "boss," Enrico Fermi. On March 29, 1938, Antonio Carrelli, the director of the Physics Institute of the University of Naples, where Majorana worked, phoned Fermi to inform him of the physicist's disappearance. At the other end of the line, Giuseppe Cocconi, a young researcher on a visit to Rome, was also present, and witnessed and later described Fermi's astonishment and the words he uttered in tribute:

> There are various categories of scientists in the world; people of second or third rank, for instance, who do their best and don't get very far. There are also people of the first rank who make discoveries of great significance, fundamental to the development of science. But then there are the geniuses, like Galileo and Newton. Well, Ettore was one of those. He had something that no-one else in the world has.

FIGURE 5.3
Ettore Majorana.

It was understood that Fermi was implicitly including himself in the second category, whereas he placed Majorana in the first, an immeasurably superior one. A few months later, Fermi wrote to Italian prime minister Benito Mussolini, pleading with him to do everything in his power to try and find Ettore Majorana. In the letter, he again expressed his great admiration for the Sicilian physicist, describing him as follows:

> capable at the same time of developing adventurous hypotheses and of acutely critiquing his own work and that of others; a calculator of outstanding expertise, a profound mathematician who never lost sight behind the veil of numbers and algorithms of the real essence of the physical problem, Ettore Majorana has to the highest degree that rare combination of attitudes that shapes the most exceptional theoretical physicists. And in truth, in the few years up till now during which he has pursued his activities, he has known how to command the attention of world experts, who have recognized in his work the imprint of one of the most powerful minds of our time—and the promise of further achievements.

Unfortunately, Majorana was never found, and to this day no one knows what happened to him.

FIGURE 5.4
Members of the "Via Panisperna" physics group in 1934. From left to right: Oscar D'Agostino, Emilio Segrè, Edoardo Amaldi, Franco Rasetti, and Enrico Fermi. Bruno Pontecorvo is missing (he was probably taking the photo).

Enrico Fermi, the "Pope," as he was called by his colleagues of the Via Panisperna group in Rome (figure 5.4), promptly supported Pauli's hypothesis, framing it within his theory of β decay. Fermi also changed the name of the particle, baptizing it with the Italian word *neutrino*, with a clear allusion to the smallness of its mass (in Italian the suffix *ino* indicates a diminutive). On the other hand, the neutron present in the atomic nuclei was really the culprit responsible for the β decay of the unstable (and therefore radioactive) nuclei, according to the reaction *neutron* → *proton* + *electron* + *neutrino*, or in abbreviated form, $n \rightarrow p + e + \nu$ (figure 5.5). In a specific sense, a cobalt nucleus contains 27 protons. When one of the neutrons of the nucleus decays according to the reaction described here, producing an electron, a neutrino, and a proton, the latter is added to that number of protons, thus creating a nucleus of nickel, which is precisely composed of 28 protons. Incidentally, Pauli and Fermi's neutrino is in reality an antineutrino $\bar{\nu}$ (i.e., the antiparticle of the neutrino), but this doesn't alter the

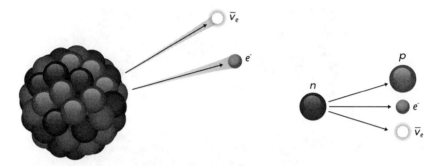

FIGURE 5.5
Diagrammatic representation of nuclear β decay, together with the reaction that forms its basis; that is, the transmutation of a neutron into a proton with the emission of an antineutrino that accompanies the beta electron.

substance of this argument. We can now anticipate that antiparticles are in every respect similar to particles except for a few opposite features, such as the electric charge (if present), and they are conventionally represented with a bar above the letter that identifies them.

In the introduction to his fundamental scientific article, Fermi wrote:

> What is proposed is a quantitative theory of the emission of β rays, in which the existence of the neutrino is assumed and the emission of electrons and neutrinos from a nucleus in the act of β disintegration is treated with a procedure similar to that followed in the theory of radiation to describe the emission of one quantum of light from an excited atom. Formulas are deduced for the lifetime and for the form of the continuous spectrum of β rays, and they are compared to experimental data.

With this paper, Fermi was introducing, in effect, a new force of nature—the *weak nuclear force*—which we will discuss, along with its three siblings, in the following chapters. It is interesting to note that the scientific article in which Fermi developed his theory of β decay was originally submitted for publication to the prestigious journal *Nature*. But its editors rejected the manuscript. Though disappointed, Fermi went on to publish in 1934 his work in the *Nuovo Cimento*, under the title "An Attempt at a Theory of Beta Rays," and simultaneously in the German journal *Zeitschrift für Physik*. Laura Fermi, Enrico's wife, included this instructive anecdote in her book *Atoms in the Family*, published after the death of the great physicist.

It is also important to note that in Fermi's model, the neutrino is not a preexisting particle in the atomic nucleus; rather, it is *created* in the act of transmutation of the neutron into a proton. This idea, which was revolutionary for the time, is fundamental to current quantum theories, which as we shall see, predict the creation and destruction of elementary particles in the process of their interaction. Fermi's hypothesis was rapidly accepted by the scientific community—and rightly so, since even today the theory in its original form is capable of explaining the results of many experiments. At the Solvay Congress in Brussels in 1933, Wolfgang Pauli said while speaking of his neutrinos:

> ... their mass cannot be very much greater than the mass of the electron. In order to distinguish them from heavy neutrons, Mr Fermi has proposed that they should be assigned the name *neutrinos*. It is possible that the mass of the neutrinos is equal to zero ..., and it seems to me plausible that the neutrinos have a 1/2 spin. ... We know nothing about the interaction of neutrinos with other particles of matter and the photons: the hypothesis that they have a magnetic moment seems to me to be completely unfounded.

With the discovery of the neutron and the hypothesis of the neutrino—twenty-five years would pass before its experimental identification—the menagerie of elementary particles was enriched with new specimens. Parallel to this, Rutherford's atomic/nuclear model becomes more credible: the mass of the atom and the positive charge reside in the nucleus; protons and neutrons are bound together by a force of a nuclear type, the strong force, which is more intense than the Coulomb electromagnetic one, which by itself would tend to cause the like-charge protons to repel each other.

The hypothesis that there existed a *nuclear exchange force*, analogous to that which existed for atomic electrons, was proposed by Werner Heisenberg. This possibility too probably had been already conjectured by Majorana, although his reluctance to publish his scientific results makes it difficult to attribute this idea to him. Heisenberg considered the neutron and the proton to be different quantum states of the same particle, the *nucleon*, with the two being distinguishable from the value of a new physical quantity, the *nuclear isospin*. In an atom, the number of protons, each having a positive electric charge, is balanced by negative electrons distributed around the nucleus, making the atom electrically neutral. Each element, with its specific chemical properties, is defined by the number of atomic electrons. Various versions of a given element exist (the isotopes), according to how many neutrons are present in the nucleus: just as their chemical properties are generally equal, determined by the number of electrons, so the nuclear structure is different. In truth, this applies rigorously to heavy elements, for which adding or subtracting a neutron does not produce any appreciable effect on the energy levels of the electrons, and hence on the chemistry they generate. But if we take a hydrogen atom instead (a proton and an electron) and add to it a neutron, we obtain deuterium D; hence, the aggregation significantly disturbs the quantum levels of the single electron. So, what are the consequences? Everyone knows that the water molecule, the chemical formula of which is H_2O, contains an atom of oxygen bound to two of hydrogen. If, in place of the hydrogen, we were to use atoms of deuterium, we would obtain what is commonly called *heavy water* (D_2O). This liquid is transparent, odorless, and tasteless like normal water—but toxic, precisely due to the modified chemical bonds to which life is so sensitive. This and many other analogous effects must make us reflect on the fact that the existence of life is really tied to a very fine thread, woven over billions of years of Darwinian evolution.

Let's then define the atomic number Z as the number of protons (or electrons) of the atom of a given element. The mass number A is instead equal to the total number of protons and neutrons. And let's take as an example the atom of carbon. It occurs naturally in three isotopes: carbon-12, carbon-13, and carbon-14. The atomic number of carbon is $Z=6$. Hence, the three isotopes have 6, 7, and 8 neutrons, respectively, for values of A equal to 12, 13, and 14. Carbon-12 is by far the most common isotope in nature, accounting for 98.93 percent of the total. The stability of a particular isotope is a function of the ratio of the number of protons and neutrons. An unstable isotope may transform itself spontaneously into another isotope, or into a different element, by means of an α, β, or γ radioactive decay, going on to occupy a quantum nuclear state corresponding to a lower energy level. For completeness, we define ions as those atoms with a number of electrons higher or lower than Z. Ions are created, for example, when, following a collision with an ionizing radiation, some electrons are kicked off from an atom, making it no longer electrically neutral.

The new nuclear physics developed greatly in the first decades of the twentieth century. So many exceptional physicists contributed to this development with fundamental theoretical and experimental works: Lise Meitner, Maria and Pierre Curie, the members of Fermi's group in Rome, and many more. Among these advances, we must mention at least the discovery of artificial nuclear fission, the discovery of the transuranic elements, and Fermi's realization of the atomic pile. It seemed like an already scripted story. And it proved altogether natural that from the suffocating atmosphere of a Europe subject to totalitarianism, race laws, and the drums of war, pioneering physics progressively transferred itself to the United States, with the Manhattan Project for the production of an atomic bomb representing the dramatic finish line for so many scientific and technological efforts. And, once again, the scalpel that served to remove a tumor was turned into a dagger—an instrument of death. In nuclear fission, which led to the first atomic weapon, and subsequently to nuclear reactors and controlled fission, the nucleus of a heavy element is split into smaller fragments—i.e., into the nuclei of atoms with a lower atomic number. An extremely small mass of the initial material is transformed into energy because the sum of the masses of the fission products is slightly lower than that of the initial nucleus. From the energy-mass equivalence relation, we know that even a small difference in mass is sufficient to generate enormous quantities of energy.

In a comparable way, the process of nuclear fusion formed the basis for the creation of the first hydrogen or thermonuclear bomb, by means of a reaction capable of producing even higher energies and more devastating effects than those of the fission bomb. Nuclear fusion occurs when nuclei of light atoms are forced together with very little distance between them, "melting" in the process (also thanks to the mentioned *tunnel effect*) and thereby overcoming the strong Coulomb repulsion barrier that acts between their many protons due to the positive charge of the nuclei. It is thus possible to create nuclei of increased atomic weights, passing (for instance) from hydrogen to helium. In this reaction, the total mass of the initial nuclei is also slightly larger than that of the nuclei after the fusion takes place, and this difference is transformed into energy in this instance as well. It is worth noting that normally, heavy nuclei tend to undergo fission, while light nuclei can fuse. The reason for this is that it is rather difficult to bring heavy nuclei close enough together to fuse them due to strong Coulomb repulsion, while, as we will see later on, the strong force that keeps nucleons together has a very short range. In the same way, small nuclei do not undergo fission due to the intense attractive nuclear force: their binding energy corresponds to the energy that was emitted when their constituent nucleons originally fused together to build the nucleus.

As we shall see, the process of nuclear fusion is the basis of the functioning of the Sun, and of all the other stars in our and in the other galaxies. Huge numbers of hydrogen nuclei (protons) fuse together to form helium nuclei, with the emission of energy in the form of light and heat (and neutrinos!), via a slow and stable process. Inside stars, the creation of progressively heavier nuclei due to fusion continues until it becomes energetically disfavored. Those elements that we find here on Earth are produced in this way. Controlled nuclear fusion is a kind of holy grail for physicists, and obtaining it is an ambitious objective. This process would allow us to produce infinite quantities of clean and cheap energy—in effect solving the energy needs of human society forever. However, the scientific and technological problems that need to be solved before this can happen remain numerous and complex, though not so much so as to cause us to despair of finding a solution!

Returning to our subject, the period in question was truly a golden age for modern physics, with breakthrough after breakthrough coming at a dizzying pace. In 1933, Francis Perrin, the son of the already mentioned Jean Baptiste Perrin, showed that the mass of the newly proposed neutrino is much smaller than even that of the electron, which in turn is about 1/2,000 of that of the proton and neutron. Today, we know that the neutrino mass is even less than about 1 eV/c^2; lower than 1/500,000, that is, of that of the electron. In 1932, the American physicist Carl Anderson discovered the first particle of antimatter, the positive electron or positron, which had been recently postulated by the English physicist Paul Dirac (figure 5.6). The latter's idea was to construct a quantum-relativistic equation that would

FIGURE 5.6
Paul Adrien Maurice (P. A. M.) Dirac.

describe the motion of elementary particles, given that the Schrödinger equation does not include special relativity. This might seem like an excess of zeal, since we know that electrons in atoms have energies of the order of tens or hundreds of electronvolts, and that, as demonstrated in the table shown in figure 3.5, their speeds are approximately equal to only a few percent of that of light. But a quantum equation that took relativity into account was needed in order to have command over much wider applications. Afterward the idea turned out to be a correct one, surprisingly capable of predicting new phenomena.

In 1928, Dirac rapidly arrived at an equation, named after him, that was not just aesthetically beautiful, but comprehensively relativistic; it also included spin in the description of quantum electrons—a big step forward from Schrödinger's equation. The small (though alas not insignificant) collateral effect is that Dirac's equation foresees solutions of negative energy for a free electron. Now, the reader may well know that the kinetic energy of a particle, in the absence of a potential produced by an external force, is classically defined as being positive—or at least of zero value—while a negative value for the sum of kinetic energy plus potential energy may manifest itself where there is a bound system—e.g., in the case of a satellite orbiting a planet. For a relativistic free particle, however, precisely the aforementioned negative solutions of energy appear instead, together with those of positive energy: $E^2 = p^2c^2 + m^2c^2$, and hence $E = \pm\sqrt{p^2c^2 + m^2c^4}$. These solutions are infinite in number, each corresponding to one of the possible values of the electron momentum, including the two limit cases where $p=0$, in which E takes the two values of mc^2 and $-mc^2$. Dirac's proposal, worked out in 1931 in order to take into account the embarrassing negative solutions, was very astute—but also very similar to a new desperate remedy ...

From a purely mathematical point of view, for a given absolute value of E, the following hypotheses are all equivalent: (1) particles that propagate forward in time with a positive energy; (2) particles that come from the future with a negative energy; and (3) antiparticles that proceed forward in time with positive energy. Cases (1) and (3) allow us to eliminate at the same time science fiction and negative energy, at the cost of introducing antimatter—in specific terms, an electron with a positive electric charge. A price worth paying, perhaps, given that the particles of antimatter are basically identical to those of ordinary matter except that they are endowed with an opposite charge, as well as having opposite values for some of the other quantum variables (the quantum numbers) that characterize them. The fact that popular literature and science fiction filmmakers subsequently abused the idea of antimatter (witness Dan Brown's *Angels and Demons*) is not a problem that concerns physicists. It is important to note that Ettore Majorana, not satisfied with the implications of Dirac's theory, had occasion to propose a key alternative reading of the quantum treatment of electrons and positrons—one that introduced, among other things, fermions that were electrically neutral (i.e., neutrinos and antineutrinos). The argument, a further testimony to the ingenuity of the Sicilian scientist, is complex, and we shall return to it briefly at the end of this book when we are dealing with the so-called Majorana neutrinos.

Anderson's discovery of the first particle of antimatter, in addition to proving Dirac's hypothesis pertaining to positrons to be correct and thereby extending the set of elementary particles, represented a milestone in physics for two other reasons. The first is that no one had yet felt the need for the positron, because if the particles known until then had contributed to our understanding of matter, of the atom and of the nucleus—sometimes unstable and radioactive—what role did the positron ultimately have to play? The second reason, connected to the first, is that the positron was not discovered by observing the matter that surrounds us. It was found by studying the flux of particles by which we are continuously bombarded from space (the so-called cosmic rays), even if, as a partial correction of this, in the case of positive β decay, the unstable nucleus emits a positron instead of an electron. The essence of the argument is unaltered: positive β decay was detected experimentally, two years after Anderson's discovery, by Frédéric Joliot-Curie and his wife, Irène Joliot-Curie, the daughter of Maria Skłodowska Curie.

Cosmic rays had already been discovered in 1912 by Victor Hess, thanks to his adventurous flights in aerostatic balloons (figure 5.7), daringly loaded with scientific equipment, including the so-called leaf electroscopes, which as students of physics will know are instruments for measuring electric charges. They consist of two minuscule metallic leaves placed inside an evacuated ampoule, where they collect the electric charge from an external electrode. The leaves tend to repel each other when electrical contact occurs, both becoming charged with the same sign: the larger the gap, the greater the charge. Thanks to a previous pioneering experiment by the Jesuit priest and physicist Theodor Wulf, it was known that the electroscopes lose their charge quite rapidly, presumably neutralized by the interaction with particles of an opposite charge produced by the radioactive elements present on Earth (or at least in the buildings in which the laboratories were housed). Wulf went so far as to climb the Eiffel Tower with his electroscopes, in order to place himself as far as possible from all sources of radioactivity. But his efforts were to no avail. The electroscopes lost their charge even more rapidly.

FIGURE 5.7
Victor Hess (second from left), before one of his flights into the atmosphere in an aerostatic balloon.

With Hess's experiments at high altitude, it became clear that outer space was itself an intense source of radioactivity. His results led to the discovery of cosmic rays. Truth be told, outstanding contributions to this discovery were made by the Italian physicist Domenico Pacini, who executed measurements that complemented those made by Hess, determining the attenuation of cosmic rays with detectors positioned deep underwater. The reduction of the radiation flux that was observed excluded the possibility that its source was terrestrial; instead, it was wholly compatible with the absorption of cosmic rays by the amount of water above the detectors. Today, we are familiar with this radiation, and by studying it we have learned a great deal about elementary particles and the astrophysical sources in deep space that are responsible for it: photons of high energy, protons, electrons, neutrinos, and light nuclei. In short, we have always been bombarded by particles. Approximately every minute, every square centimeter of the Earth's surface—humans included—is hit by an electrically charged cosmic ray. I would draw attention to the word *charged* because in the case of neutrinos, which are electrically neutral, the flux is immensely higher: the Sun alone causes humans and all living species to be doused and pierced through by 60 billion neutrinos per square centimeter … per second! Here is another incredible characteristic associated with the phantasmal Pauli particle. Thanks to a kind of anthropic principle, the probability of the interaction of neutrinos with matter, including of a biological type, is negligible: good news for human life, but not such good news for physicists wanting to study neutrinos. On the other hand, the other cosmic particles to which human beings have been exposed since their first emergence on Earth have contributed to genetic mutations and, ultimately, to natural selection. Given that we have successfully evolved from some kind of primordial soup in the vicinity of an underwater volcano, we should consider that on the whole, the effect has not been negative ...

Returning to Anderson (figure 5.8), his positrons were generated precisely by interactions of cosmic origin, a discovery that opened the way for the exploration of the microcosm of elementary particles as conducted by the fledgling physics of cosmic rays, in parallel to what eventually would be achieved with particle accelerators, in the form of a synergy that remains fruitful today and likely to develop even further in the future. But in the

FIGURE 5.8
Carl Anderson, the discoverer of the positron and of the muon.

meantime, the Pandora's box of which we spoke earlier in this book was starting to reveal new particles (elementary ones?). This complicated the lives of physicists by indicating a complexity of nature and of the universe that had seemed unthinkable until then. As a result, scientists were compelled to study the intimate structure of matter in ever greater depth.

The muon (μ) was discovered in 1936, also by Anderson and a student of his, Seth Neddermeyer, once again while analyzing the flux of cosmic rays. The new particle was called the *mesotron* or *μ meson*, in reference to the fact that its mass (106 MeV/c^2) was intermediate between that of the electron and that of the proton or neutron: approximately 200 times larger than the mass of the electron and 10 times smaller than that of the proton. Today, we know that the muon is not a meson, but a lepton—these are particle families that we will define later—and it exists in nature as a particle with either a positive or a negative charge—although many senior physicists remain attached to its original nomenclature. The particle is unstable, and on average, it decays in around 2.2 microseconds in an electron of the same sign; and, as we now anticipate, in two undetected neutrinos. The muons that are produced following the interaction of the high-energy primary cosmic particles with the Earth's atmosphere constitute, at sea level, the predominant component of the aforementioned flux of electrically charged cosmic rays.

A couple of years before the discovery of the muon, the Japanese physicist Hideki Yukawa (figure 5.9) had elaborated a theory capable of explaining the interaction that keeps protons and neutrons bound together in the interior of the atomic nucleus, developing the original Heisenberg hypothesis. From his calculations, there emerged the necessity of a mediator of the force—one that would behave in an altogether similar way to a photon that mediates the interaction between two electric charges. We will have occasion later to discuss this salient aspect of elementary particles—how they interact and how they can exchange "virtual" particles that, thanks to quantum laws, can materialize from nowhere and live for extremely short time intervals. Yukawa observed that the short range of the nuclear force, coinciding in practice with the dimensions R of the atomic nucleus (around 10^{-12}–10^{-13} cm), required this mediator to have a considerable mass, equivalent to about 100 MeV/c^2. It is instructive to follow the corresponding calculations that lead to this conclusion:

$$\Delta E \Delta t \geq \frac{h}{4\pi}; \quad \Delta E \sim mc^2; \quad \Delta t = R/c; \rightarrow mc^2 \sim 100 \text{ MeV}/c^2. \tag{3}$$

Therefore, it seemed natural to hypothesize that the Yukawa particle was indeed the newly born mesotron (muon). This appeared all the more

FIGURE 5.9
Hideki Yukawa.

reasonable when taking into account the fact that—apart from the electron, the positron, the proton, the neutron, and the proposed neutrino—the muon was the only particle that was known about at the time, and it was only logical to assume that the newcomer must serve *some* purpose! In fact, as soon as he heard of its discovery, the American physicist Isidor Rabi (who would go on to win a Nobel) remarked with mock surprise: "And who ordered that?" Furthermore, the observed decay of the mesotron into an electron made it possible to explain the nuclear β decay: some of the muons that are present in the nucleus would give way as they decayed into β electrons, accompanied by the Pauli neutrinos. But this was not actually the case.

The hypothesis about the muon remained valid for many years to come. But then there came an experiment carried out by three young Italian physicists: Marcello Conversi, Ettore Pancini, and Oreste Piccioni (figure 5.10). The trio began to study the muons of cosmic rays in the very midst of World War II, in Rome, sometimes even working during bombing raids. The aim of their experiment was to demonstrate the existence of a difference between positive and negative muons when they are stopped by a very dense material. If the muon had indeed been the Yukawa particle, the negative muons would have been captured by the atomic nucleus, unlike

FIGURE 5.10
Three special young men: from left to right, Oreste Piccioni, Ettore Pancini, and Marcello Conversi.

the positive ones, which would be repelled by the barrier of the same electric sign generated by the protons of the nucleus. These three brilliant (if not to say heroic) physicists demonstrated experimentally that even the negative muons were not absorbed by the nucleus and had sufficient time to calmly … die naturally—that is, to decay. This was a truly unexpected result. The muon confirmed that it did not play any specific role in the atomic structure, and ultimately in the construction of the matter of the universe. The Nobel winner Luis Alvarez, in his lecture during the 1968 ceremony, said: "Modern particle physics was born during the last days of the Second World War, when a group of young Italians, Conversi, Pancini, and Piccioni began a remarkable experiment." And another American Nobel laureate, Murray Gell-Mann, added: "The muon was like a child abandoned in a churchyard—it represented the end of the age of innocence." The real Yukawa particle was discovered a little later—in Bristol, in 1947—by Cecil Powell (figure 5.11), César Lattes, and Giuseppe (Beppo) Occhialini. It was the pion, which was detected with both positive and negative charge, with a mass similar to that predicted by Yukawa (around 140 MeV/c^2) and an average lifetime of 26 nanoseconds—much shorter than that of the muon. Once again, the particle was found in cosmic rays by using detectors positioned at high altitudes, on the top of mountains. Proper recognition of the discovery made by Conversi, Pancini, and Piccioni was unfortunately belated—perhaps due to the role that Italy had played in the war, not to mention its disastrous defeat—but today, physicists are in agreement as to the enormous significance of their findings.

FIGURE 5.11
Cecil Powell, the discoverer of the charged pion, the Yukawa particle.

But now the time has come to satisfy the reader's legitimate curiosity. How do we actually "observe" the elementary particles? Seeing them with the naked eye is obviously out of the question. We have learned that their dimensions are much, much smaller than the resolving power of visible light. I have also pointed out that, for example, in the study of the interaction of an electron of high momentum with a nucleus or another particle, we can derive information about the structure of the target, just as Rutherford did. What remains to be understood is how we reveal the scatterings of the projectile and the target, as well as of the various other particles that, as a result of the energy-mass equivalence relation, may be created during the collision. It is no accident that we speak currently of the "detection" of particles, indicating a kind of indirect observation of them. This is apt, and yet at the same time somewhat misleading if the word *indirect* suggests any lingering doubts about the real measurement of the particles' properties. There should be none. On the other hand, precisely because we have no hope of observing particles optically, we are obliged to measure their actions (interactions or decays) and thereby to arrive at an unambiguous way of proving their existence. This is a demanding enough task, no doubt, but this circumstantial analysis is necessary largely because for a physicist, it is not enough to be certain that a particle had been located. It is also vital to measure all the properties that characterize it: speed, mass, electric charge, energy, orbital angular momentum, spin, and so on.

To detect (observe) a particle implies evidencing the effects of its interactions by means of particular instruments called *detectors*. These, after such interactions, provide an electrical, acoustic, optical, or otherwise identifiable signal for the experimentalist conducting the measurements. For example, a particle that is electrically charged and energetic enough to travel through a medium produces ionizations—extracting electrons from atoms of the material and rendering them available for identification. Technically, our particle supplies to the electrons of some of the atoms found along its trajectory an energy higher than the binding energy that holds them in place around the nucleus. On the other side, an atom that has had at least one electron removed becomes a negative ion, with a resulting charge that's positive overall. The counting of these ionization electrons provides a representative number for the energy of the primary particle, while their position gives an indication of the trajectory that has been followed. This process is not unlike when the jet stream of a plane "measures" its trajectory, even if the plane itself is not visible from a distance. A substantial complication is encountered in the case of neutral or weakly interacting particles, such as neutrinos. These leave no track in their wake and produce an effect (seemingly on a whim) only when they decay in the process of dying, or when they interact with other particles through collisions. In this instance, the opportunity of identifying them further is completely lost. Finally, we have the case of particles that are so extremely unstable that even with the push of relativistic time dilation $T=\gamma T_0$—remember the formulas (2) in chapter 3—their lifetime is so short that we can reveal them only by studying the products of their decay, almost like reconstructing a vase by piecing it together after it has broken.

The discoveries of the positron, of the muon, and of the pion are useful examples illustrating the first detection methods used during the golden age of particle physics at the beginning of the twentieth century. We will talk more extensively about particle accelerators (chapter 6) and detectors (chapter 15), but let's start here with Anderson's first experiment. For the first positron detection, the American physicist used a cloud chamber. This device, much used in the first half of the twentieth century, can be schematized as a container of supersaturated water or alcohol vapor. Given that it involves a thermodynamically unstable system, the passage of an ionizing particle through the volume of the receptacle causes the vapor to condense in microbubbles of liquid (a cloud, to be precise) along the track of the particle. The greater the ionization, the denser the track, which

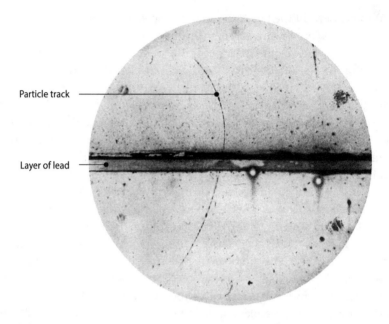

FIGURE 5.12
One of the events produced by positrons in Anderson's cloud chamber (see the text for an explanation).

can be photographed through a transparent wall in the box containing the vapor. It therefore amounts to a visual detector in which, as in the example of the jet plane, the trajectory of the particle appears optically visible. In figure 5.12, the event in which one of the first positrons was detected by Anderson is shown. The positron track is about 5 cm long and pierces a slab of lead that is 5 mm thick. A magnetic field applies a force to the particle, bending its trajectory in the plane perpendicular to it. The curvature of the track becomes larger as the module of its momentum mv decreases. Because the particle loses energy and momentum in its passage through the lead, the magnetic force increasingly bends its trajectory after it has come out the other side. This indicates that the particle originates from the bottom of the figure. Consequently, by knowing the direction of the magnetic field, one determines that its electric charge is positive (i.e., a given magnetic field bends the positive and negative particles in opposite directions). In other words, the event would have seemed identical to that generated by a negative electron coming from the top of the figure, but this is not possible because the particle has a higher speed in the lower part of the photograph. Finally, by measuring the total length of the track (5 cm), we gather that its mass must be compatible with that of an electron. In fact, a proton with equal momentum—and hence with equal curvature of its track—would have covered only 5 mm before stopping and losing all its kinetic energy. The reason for this will soon be clear.

In the experiments discussed here, the energy loss of a charged particle due to ionization along a certain length of its trajectory, whether through the volume of a cloud chamber or the thickness of a lead plate, decreases as its speed increases. It is almost as if a slow particle had more time to strip electrons in its collisions with the medium. As the speed (or, if one prefers, the momentum) increases, the loss of energy reaches a minimum and from then on increases only by little. The amount of energy lost due to ionization depends on the type of both the material and the particle, therefore contributing to its identification. Given equal momentum mv (e.g., determined by measuring the curvature of the track in the magnetic field), the mass m of the proton is around 2,000 times larger than that of the electron, and this implies a lower value of the module of the speed v for the proton; in turn, this causes the latter to lose energy more rapidly through ionization, and in the end, all things considered, means that it can cover only 5 mm, and not 5 cm, before stopping. Always tied to this effect, a reader looking carefully at figure 5.12 will notice that toward the upper end of the track, it becomes thicker, thus indicating that the particle is slowing down, losing speed, and therefore undergoing more ionization: a fact that supports Anderson's conclusions on the direction of origin of the particle.

A technological evolution of the cloud chamber was made with the bubble chamber, invented by Donald Glaser in 1952 (figure 5.13). It consists of a detector of the kind that has nowadays become extinct, having been superseded by hyperfast electronic ones. The bubble chamber is basically a cylinder coupled with a piston; the cylinder contains a liquefied gas (such as Freon) in unstable thermodynamic conditions, just as in a cloud chamber. If an ionizing particle crosses through the space containing the liquid or interacts with it, it loses part of its energy, producing pairs of negative electrons and positive ions. After a rapid expansion produced by the piston, the liquid begins to boil along the particles' trajectories, in correspondence with the electrons and the ions. It's something comparable to the bubbles that form in a saucepan of very hot water when salt is thrown in for the pasta. The microbubbles in the cylinder are immortalized by various photographic cameras pointing at the volume filled with the liquid. From the analysis of the photographs, it is possible to reconstruct the three-dimensional topology and the kinematics of the event. With this detector, many successful experiments have been realized. A typical event obtained

FIGURE 5.13
The largest bubble chamber ever built: BEBC (Big European Bubble Chamber), now a part of the open-air museum at CERN. Important experiments were undertaken with BEBC in the 1970s and 1980s. Courtesy of CERN.

with a bubble chamber, highlighting the excellent imaging features of the detector, is illustrated in figure 5.14.

The discovery of the pion, the particle predicted by Yukawa, was made by the Bristol group using another type of detector—the photographic emulsions—that in one way or another has left its mark on the history of the physics of elementary particles. Of all the particle detectors, the emulsions, sometimes also called *nuclear emulsions*, offer the highest resolution in the measurement of the position of the tracks left by ionizing particles, even today. Having emerged at the beginning of the twentieth century, this

FIGURE 5.14
An event in a bubble chamber. A γ-ray photon enters at the top of this image. Because it is electrically neutral, it leaves no track until it interacts with the constituent particles of the atoms of the liquid in the chamber. We recognize an electron and a positron that spiral in opposite directions due to the presence of a magnetic field. The long linear track is that of another electron produced in the interaction. As its trajectory is also bent by the magnetic force, the electron is accelerated and radiates a γ-ray along the tangent to the curve. The latter is invisible until it interacts with a nucleus, producing another electron–positron pair—the upside-down V at the bottom of the picture.

detector has a long history and has contributed to a number of fundamental discoveries in physics. An emulsion is made up of a large number of small crystals of silver bromide, with linear dimensions ranging from 0.1 to 1 micrometer, dispersed in a gelatin matrix. By interacting with the emulsion, light or charged particles cause some of the grains of silver bromide to be modified and to be subject to molecular excitation. The effect of this process is the formation of a so-called latent image. This image remains impressed in the emulsion, just as with an ordinary photographic film, up to the moment in which the successive development and fixing bath the image finally makes visible the microscopic grains of silver, which are dark and therefore identifiable. The main function of the gelatin is to provide a three-dimensional structure to help locate the small crystals of silver bromide and to stop them migrating during the process of developing and fixing, risking the loss of the impressed image. The emulsions used for the identification of charged particles are substantially different from those normally used in photography, due to the higher quantity of silver and a thickness that is between 10 and 100 times larger. Furthermore, in the case of emulsions, the silver grains that are visible after development are smaller and more uniform. The number of such submicron grains present along the passage of a high-energy particle is of the order of a few hundred per millimeter! They are made visible only thanks to optical microscopes that make it possible, by means of the procedure known as *scanning*, to identify and measure the microtracks left behind by the particles.

At the beginning of the 1970s, the emulsion technique was superseded by modern electronic position detectors before having a kind of renaissance in the 1990s. More recently, important scientific results have been obtained with a new generation of detectors, which use emulsions combined with modern systems of automatic image analysis, especially in neutrino physics and for the study of short lifetime particles. A curious application of this technique that deserves mention here is one that we have carried out in Switzerland to study Alpine glaciers. Given that the muons of cosmic rays are a kind of heavy light that continuously illuminates the Earth, it is possible to think of creating with these particles some radiographs that would be able to show the contrast between ice and the underlying rock, or so-called bedrock. Our emulsions are then equivalent to the film of a radiologist, which, illuminated by X-rays, allows us to see broken bones or the inside of a chest cavity. In our case, what are radiographed are structures of huge size, by means of the exposure of emulsion films placed in different positions around the glacier to construct a three-dimensional image. The muons, in fact, are absorbed and diffused differently by structures that are more or less dense (ice or rock), just as happens with X-rays and the human body. The study of the structure and the morphology of glaciers' bedrocks is of great interest to geologists and may well contribute to the understanding of the effects of global warming on glaciers. Similar studies have been carried out on volcanoes, for example at the universities of Tokyo and Nagoya in Japan, to analyze the structure of the internal lava ducts.

Returning to our elementary particles, one of the first events corresponding to the production and decay of a pion in an emulsion is shown in figure 5.15. The charged pion is unstable and decays into a muon with the

FIGURE 5.15
The discovery of the pion (π). The particle, originating from cosmic rays, travels just a few micrometers in the photographic emulsion before decaying into an invisible neutrino and a muon. The latter, after 600 micrometers, decays in turn into an electron and two neutrinos. These neutrinos are also not detectable because, being electrically neutral, they do not leave any tracks of ionization electrons.

same electric charge and a neutrino. The latter, as we have already pointed out, is electrically neutral and leaves no track. The negative muon, after a journey of approximately 600 micrometers, decays in turn into an electron and two neutrinos. Later in this book, we will discuss the features and decays of the particles involved in this event. For now, though, it is interesting to remark the extremely high spatial resolution of the emulsions: in a fraction of a millimeter, it is possible to reveal the birth, the brief life, and the death of each of our particles. As was the case with the discovery of the positron, so we also note in figure 5.15 the increase in ionization on the part of the pion and the muon when they are slowed down and at their points of stopping and decaying; the electron that is the product of the muon decay ionizes very little because the mass of the parent muon when at rest (106 MeV/c^2) transforms into the small mass of the electron (0.511 MeV/c^2, precisely), thus leaving a large amount of kinetic energy to the two neutrinos and to the very same electron, which turns out to be fast and hence only weakly ionizing. I have devoted so much space to these experiments precisely to emphasize the deductive effort that lies behind the analysis of event data in particle physics. This is a task worthy of Sherlock Holmes, where every detail counts toward arriving at a satisfactory solution to a case.

In 1949, the third member of the family of pions was discovered: the electrically neutral π^o, which is different from its charged relatives on account of having a slightly smaller mass (135 MeV/c^2), a much shorter lifetime (8.5×10^{-17} seconds, or approximately ten millionths of a nanosecond!), and a particular way of decaying into two γ-rays of high momentum, equal to 67.5 MeV/c for each of two photons, in the case where the π^o decays at rest. The notable differences between charged and neutral pions, as we shall see, mirror the different interactions involved in the decay processes. A similar case is that of the kaon (K), another particle discovered through the study of cosmic rays, that already mentioned powerful instrument for understanding the microcosm. Once again, an unstable particle is involved which, as we now know, appears in the guise of three possible states of electric charge—K^+, K^-, and K^0, with a mass equal to approximately half that of the proton (about 500 MeV/c^2). The discovery of kaons was not painless for physicists. Their behavior was so bizarre and unexpected that it obliged scientists to create for them a whole new chapter in the great book of elementary particles—the one of the so-called strange particles.

The neutral kaon was the first to be discovered, by George Rochester and Clifford Butler, in an experiment in 1947 using a cloud chamber. They observed that two neutral and unstable particles originating from cosmic rays were always produced together. This was an inexplicable fact, which in 1952 led Abraham Pais to propose the model of associated production. One of the two particles was our neutral kaon, while the other could be either a Λ^0 (lambda zero) particle or a Ξ^0 (xi zero), both heavier than the proton. In the detector, the final state particles appeared like a V (remember figure 5.14?), characteristic of the decay of a charged particle into two daughter particles of positive and negative charge. The properties of these pairs are indeed strange, to say the least. The Vs are produced frequently and abundantly in the interactions of cosmic rays, while their decay appears to be very slow—slow, that is, compared to that of other unstable particles known at that time. An example of this V is the bubble chamber event shown in figure 5.16, together with its diagrammatic reconstruction. The charged kaon was discovered by Powell in 1949 with an experiment using emulsions. He dubbed the particle τ (tau), and only later was it understood that it was nothing other than the charged counterpart of Rochester and Butler's particle. We will revisit our discussion of the so-called Powell's tau, and of his contribution to the creation of another intriguing puzzle, in chapter 9.

The first interpretation of the associated production capable of successfully explaining all the reactions involving the new particles was provided in 1953 by Murray Gell-Mann and Kazuhiko Nishijima. The two scientists proposed the existence of a new quantum number: the strangeness S, which is added to the electric charge, mass, and spin and has a value of zero for particles already known at the time such as the proton, the neutron or electron, and values of 1 or –1 for the particles and antiparticles termed strange, such as the kaon and the lambda. The total strangeness is a quantum number conserved in the production of particles (via strong interaction) but violated in the slow (weak) decay reactions—so much so that, with reference to figure 5.16, the sum of the strangeness of the pion and

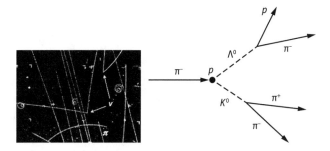

FIGURE 5.16
The associated strangeness production. The image on the left shows
the detection of an event in a bubble chamber. On the right, there is the reconstruction of the reaction: a negative pion that is part of the flux of cosmic rays interacts with a proton of an atom of the liquid filling the chamber. Two neutral particles are created, K^0 and Λ^0, which leave no ionization tracks. The first decays into two pions of opposite charge, and the second into a negative pion and a proton.

proton is zero, equal to that of the kaon (–1) plus the lambda (+1), which is still zero. In the decay process, the strangeness of both the kaon and the lambda disappears, thereby reestablishing a total strangeness $S=0$.

Nevertheless, the theory of Gell-Mann and Nishijima was just a phenomenological model that both explained and at the same time did not explain, although it did make it possible to predict which reactions could take place and how. Furthermore, the need for a global and unifying interpretative framework was becoming more and more urgent as new particles were being discovered throughout the 1950s and 1960s at an increasingly faster pace, thanks in part to the advent of the new accelerator machines. In this way, the situation of the periodic table of elements was being repeated, a zoological classification that could only be fully understood and justified with recourse to the atomic-nuclear model and quantum mechanics. At the start of the 1960s, besides Λ, Σ, Ξ, ρ, ω, ϕ, η, and others, the related antiparticles and the various excited states of higher mass, the antiproton (in 1955) and finally the neutrino (1956), had been discovered. Altogether, there were dozens and dozens of others that, as was by now quite clear, could not all be elementary. It was a family that proved to be too numerous, and ultimately rather chaotic.

6 ENERGIZED PARTICLES

He travels the fastest who travels alone.

—Rudyard Kipling

With the proliferation of discoveries of new elementary particles that started in the 1950s thanks to the advent of particle accelerators, physicists began to catalog them diligently into families, groups, and multiplets, according to the values of the quantum numbers that characterized them. In reality, identifying 100 alleged elementary particles, as was managed toward the end of the 1960s, was not really the most conducive for speaking about elementarity. Nevertheless, given the role played by the first accelerators in the discovery of many particles that were evidently not elementary, and the significance that such machines have in present-day physics, I think it is opportune to open a parenthetical discussion here in order to describe these fundamental instruments of discovery in more detail.

Accelerators are machines that nowadays are very complex and sometimes gigantic, capable of furnishing energy to particles so as to increase their ability to probe ever-smaller distances. In parallel to this, it enables us to have at our disposal an energy budget to create the mass of new particles that come in the wake of collisions. We have developed these concepts in the preceding chapters, and we have learned that as the energy grows, the speed of the particles approaches the speed of light, without ever being able to match it—let alone go beyond it—by virtue of the fact that they have mass—an aspect well illustrated in figure 3.4 in chapter 3. In a sense, as Ugo Amaldi observes, we should talk about energizers rather than accelerators. In effect, the particles in the accelerator machines soon reach speeds approaching that of light, at which point their speed hardly varies, while their energy continues to grow.

The first question that we must answer is: where do we get the particles that are injected into the accelerators? In the case of protons, for example, we start with a standard bottle of hydrogen gas, whose atom is made of a proton and an electron. The gas passes through devices that, thanks to intense electric fields, strip the single atomic electron from its nucleus, thus liberating the proton, ready for the successive acceleration. You should imagine that with a small gas bottle, one can supply protons to the CERN Large Hadron Collider (LHC) for … several billion years of operation! On the other hand, we have seen that electrons can be easily generated by a cathode tube and then sent straight to a downstream accelerator.

Acceleration (and energization) of an electrically charged particle can proceed through the application of electric fields. In very schematic terms, a planar capacitor such as the one shown in figure 6.1 constitutes an elementary particle accelerator. Although conceptually correct, this method of energization can be utilized only with great difficulty, because it may require unrealistically high voltages to increase the energy of our corpuscles. In figure 6.2, a more complex and efficient system is illustrated, one that entails a series of consecutive accelerations. It is something like a missile, with multiple stages rather than just one. A series of units, in practice similar to that shown in figure 6.1, are placed along the axis of the motion of the particles; these units have an aperture aligned with the particle direction to permit them to enter one acceleration unit after the other. The polarity of the electrodes varies at high frequency in a manner that is synchronized with the passage of the beam. After a positively charged particle has been attracted by the first negative electrode, the latter is rapidly made to become positive, which entails applying a new repulsive force to the particle that is thereby pushed in the same direction—and so on. In this way, we construct a battery of accelerating elements, each of which contributes some energy.

The main technological challenge is that of trying to rapidly change the polarity of the electrodes, a condition that is obtained with an alternating electric voltage, as shown in figure 6.3. In this example, there are two improvements over the previous scheme. The first consists of the fact that the capacitors have been replaced by hollow metal tubes. When the particles pass through the tubes, they do not feel any force, given that the electric field is only present at the interface between two consecutive elements where particles are accelerated. The second is that the length of the tubes increases proportionately with the particle's increase in speed, in order to make constant the time spent traversing each unit. This ingenuity is necessary in order to accelerate protons, whereas for the electrons, it is unnecessary because they soon reach a constant velocity that in practice coincides with that of light

Energized Particles

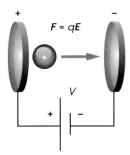

FIGURE 6.1
A positively charged particle (q), such as a proton, feels the electric field E established between the electrodes of a capacitor set at a voltage V. A force F is generated that accelerates the particle by pushing it from the positive to the negative electrode.

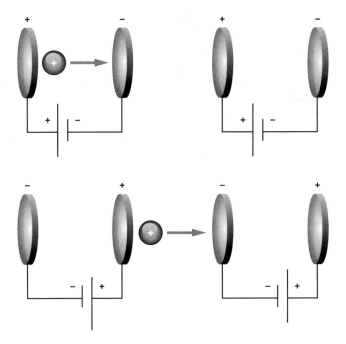

FIGURE 6.2
A positive particle is accelerated by the electric field present in the first unit. It is attracted by the first negative electrode (upper image). When the particle reaches it, the polarity of the electrode is reversed (lower image), the particle is pushed toward the first electrode of the second unit (now negative), and so on until the desired energy is attained.

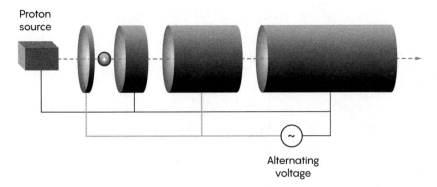

FIGURE 6.3
Conceptual plan of a linear accelerator with electrodes of increasing length and alternating accelerating voltage.

(figure 3.5). The device shown in figure 6.3 was proposed in 1928 by the Norwegian physicist Rolf Widerøe and the Hungarian Leó Szilárd, and immediately applied as the first particle linear accelerator. The next step, something of an intuitive one, was to develop the particle's trajectory in a vacuum to avoid losing energy in collisions with atoms of the air. The real limitation of this acceleration method is the size of the apparatus needed to reach sufficiently high energies. The typical values of the electric fields within a linear accelerator consist of millions of volts per meter, and this effectively limits the energy of the outgoing protons to a few tens of megaelectronvolts (MeV).

In reality, Ernest Rutherford had already hypothesized that the use of machines devised to increase the energy of electrons and atoms would open new and unexpected research avenues. One of his students in the Cavendish Laboratory in Cambridge, John Cockcroft, together with his colleague, Ernest Walton (figure 6.4), consequently engaged in studying the design for an electrostatic accelerating machine. In 1932, the pair created the proton accelerator that still bears their name. The instrument was based on a method for multiplying a relatively low voltage arbitrarily until very high voltages were reached. Cockcroft and Walton succeeded in supplying an energy of 200,000 eV—modest by today's standards, but capable at the time of producing the first nuclear reaction artificially induced by an accelerator. A well-deserved Nobel Prize was duly presented to them in 1951. However, I should point out here that my pre-pre ... predecessor to the chair of physics at the University of Berne, the Swiss Heinrich Greinacher, had independently invented the voltage multiplier several years before the English duo.

A significant advance in the accelerator technology was achieved when, starting in 1930, the American nuclear scientist Ernest Lawrence thought of

FIGURE 6.4
Ernest Rutherford, flanked by his two brilliant students, John Cockcroft (on his left) and Ernest Walton.

developing Widerøe's original idea further by adding the effect of a magnetic field to that of the electric one; the result was a machine in which the particles follow a circular trajectory and has smaller dimensions than those of a long, linear accelerator. I'll explain with the aid of figure 6.5. The cyclotron, as it was originally conceived, consisted of two D-shaped semicylinders, which are hollow inside and connected to a generator of high-frequency alternating current, exactly as in the case of the linear accelerator. Between the two Ds, there is a proton source, provided by ionized hydrogen gas—if we use hydrogen atoms, we electrostatically strip the single electron so that the only thing left is its nucleus, consisting of just one proton. Under the influence of the electric field, the protons move toward the electrode of negative polarity. A constant magnetic field **B** perpendicular to the plane of the semicylinders generates a force that bends the trajectory of the protons, causing them to rotate inside the cylinder. When the protons reach the edge of one semicylinder, they are confronted with an electrode that has now become negative, and hence they are newly attracted. This process is repeated thousands of times. At every turn, however, the radius of the trajectory augments slightly, due to the increased speed (momentum) of the particles, until the protons are extracted from the accelerator and emitted at maximum velocity and energy. With such a cyclotron, it is possible to accelerate protons up to a speed of around 80 percent of c.

These machines quickly became very important instruments in the hands of physicists. Many research institutes immediately constructed large and powerful cyclotrons, and the results were not long in coming.

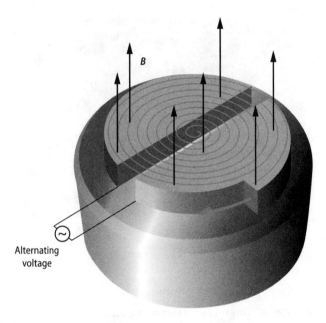

FIGURE 6.5
Operating principle of the cyclotron. In the real implementation, the circular trajectories of the particles toward the end of the accelerating process (with longer circumferences) becomes increasingly dense, contrary to what is just conceptually shown here.

Until today, the cyclotron remains a type of accelerator that is frequently employed for various applications to technology and medical science; however, having a maximum attainable energy equal to only a few hundred megaelectronvolts, it is not used in particle physics experiments anymore. Currently, there are worldwide about 1,000 cyclotrons dedicated above all to the production of radioisotopes for medical diagnostics and therapy. A *radioisotope* is a radioactive isotope of an element usually artificially generated by bombarding a suitable target with protons that have been accelerated in a cyclotron. The radioisotope can be injected into the body of a patient in a solution with a glucose base. The sugar, and with it the radioactive substance, attaches itself to a possible tumor, thus making it possible to identify the lesion through the detection of the radioactive decay products, and by measuring their point of departure. An example is positron emission tomography (PET). Figure 6.6 shows the medical cyclotron in operation at the University Hospital in Berne.

A further advance in the accelerator technology was achieved with the development of the first synchrotrons, which were conceived independently

FIGURE 6.6
Photograph of the interior of the cyclotron of the University Hospital in Berne, used for the production of radioisotopes for medical diagnostics and therapy. Modern cyclotrons have evolved a more complex structure than the original one devised by Lawrence.

around 1945 by the Russian Vladimir Veksler and the American Edwin McMillan. I like to point out that the two of them, at the end of a global conflict and the beginning of the Cold War, started an intense scientific correspondence, becoming good friends in the process of engaging in fruitful discussions about nuclear physics—surely a shining example of the role that scientific collaboration can play in promoting international peace. Their idea was a conceptually simple one. In a cyclotron, there is an alternating electric field that accelerates and a constant magnetic field that causes the curvature of the particle trajectories. This is the reason why, as the energy and speed of the particles grow, the radius of curvature of their trajectories progressively increases (figure 6.5). The synchrotron (figure 6.7), on the other hand, consists of an evacuated doughnut-shaped ring into which the protons are injected as they are emitted from a linear nuclear preaccelerator. The particles follow a circular orbit in the vacuum tube. This requires that with the increase in energy, the intensity of the magnetic field also gradually grows to maintain the particles along their established trajectories. This is achieved by increasing the current in the coils of the electromagnets of the device. Given that the accelerated particles turn at increasing speeds, the frequency of the oscillating acceleration voltage must increase proportionally, in parallel and in synchrony (the factor from which the machine gets its name) in order to always intercept the particles at the right moment

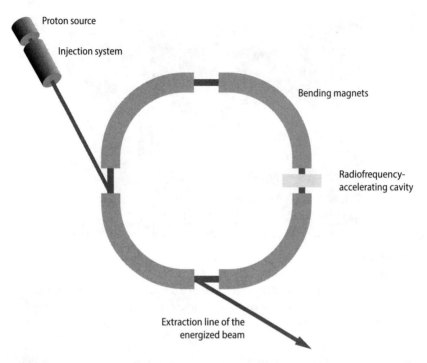

FIGURE 6.7
Plan (seen from above) of a synchrotron and its constituent elements.

and provide them with the necessary push. Consequently, whereas in a cyclotron the trajectory is varying and the magnetic field constant, in a synchrotron the trajectory is constant, but the magnetic field and the voltage frequency are variable. With equal energy being furnished to the particles, the synchrotron requires a weight of magnets, and therefore a cost, significantly lower than that needed by the cyclotron. It is possible to express in a compact form the formula that links the bending power of the magnets—that is, the intensity of the associated magnetic field—and the maximum momentum that is attainable: $p = Br$, where r is the radius of curvature of the particle trajectory. This relation implies that:

$$p\left[\frac{\text{GeV}}{c}\right] \approx 0.38 \times B[T]\, r[m].$$

From this, it follows, for example, that if we have at our disposal magnets that produce a magnetic field of 1 tesla (T) and a radius of curvature of the doughnut tube of 20 m, the momentum obtainable for our particles will be about $0.3 \times 1 \times 20 = 6$ GeV/c.

Another important aspect of this matter is linked to the type of particles that we are accelerating. In the case of electrons, there are intrinsic limitations that impinge on the performance of the machine. As will be discussed in the following chapters, an accelerated (charged) particle emits a radiation made up of photons—an emission that in fact has been given the name *synchrotron radiation* (or synchrotron light). The corresponding amount of energy lost into photons emitted along the tangent of the circular trajectory increases very rapidly as the energy of the particle increases, and is inversely proportional to the radius of curvature; furthermore (and this is an important aspect), it diminishes greatly as the mass of the accelerated particle grows. In other words, this effect is very relevant for accelerated electrons and negligible for protons, due to the much larger mass of the latter. We arrive at the case in which, even continuing to supply energy to the electron, its total energy does not grow on account of the loss through radiation. To efficiently increase the energy of the electrons until they reach the desired value, it is necessary to reduce their centripetal acceleration, increasing the radius of the vacuum tube (and with it the size of the machine). No problem occurs with the protons, on the other hand, which are 2,000 times heavier than the electrons. In this case, the limitation is a technological one: creating magnets powerful enough to bend the trajectories of high-momentum particles.

Worthy of particular mention here is the substantial improvement of the acceleration system that was being developed, thanks to the ingenuity of the American physicist Luis Alvarez, a method that's still used today in modern accelerators. In place of the capacitors or hollow tubes, and of the relatively high oscillating current, the "motor" of the synchrotron consists of a radiofrequency unit; that is, a sealed metal cavity in which energy is injected in the form of electromagnetic radiation of radio frequency, with a wavelength befitting the cavity dimensions. In this way, within the cavity, stationary sinusoidal waves are generated that make it possible to apply an electric force synchronized with the passage of the particles. This mechanism is illustrated in figure 6.8. A large number of particles (let's assume they are protons) are injected into the synchrotron in an almost continuous manner. Some of these will arrive on time, in phased order to the appointment with the accelerating electric field that's present in the cavity (the "synchronous particle" in the diagram). Others will arrive early and will have a more intense boost (higher value of V); others still will arrive late, being subject to a lesser accelerating force. The early particles will tend to escape along the tangent of the accelerator, following a slightly longer circumference, and will reach their next appointment slightly late. On the

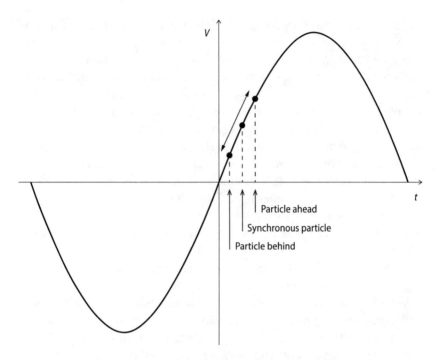

FIGURE 6.8
The principle of phase stability. The particles that are either early or late relative to the synchronous ones oscillate around them, thus creating "packets" that are globally accelerated by the electric field of the radiofrequency cavity.

contrary, the particles arriving late will be less accelerated and will travel a more internal trajectory, reaching the next cavity first, and therefore being subject to a more intense force. The reader will have grasped that this mechanism tends to create, naturally and dynamically, clusters of particles that oscillate forward and backward around the so-called ideal (synchronous) particles. This is the phase stability principle discovered by Veksler and McMillan, which in some way makes less stringent the demands for constructing and operating particle accelerators. The small instances of geometric unevenness and the uncertainties in the values of the accelerator parameters are corrected by the compacting phase stability mechanism. After some turns, everything inside the machine settles almost automatically …

In reality, the oscillations of particles in the accelerator happen both longitudinally and transversely, and increase in amplitude with the energy growth. This causes the clusters to acquire a spatial extension of some significance, of the order of tens of centimeters. As a consequence, it is necessary that the vacuum tube should have a diameter sufficient to contain the particles, and this in turn requires the use of increasingly larger and heavier magnets as the maximum energy of the accelerator grows. For this reason, the first synchrotrons had significant bulk and sometimes required thousands of tons of iron to create their bending magnets. A staggering example is the synchrophasotron (figure 6.9) of the (at that time) Soviet laboratory in Dubna, designed by Veksler. In 1957, this machine achieved a record energy of 10 GeV per proton, but at a cost of a 200-m circumference and a weight of 36,000 tons! To obviate such frustrating limitations to the construction of ever-more-powerful accelerators, in 1950 the ingenious discovery of *strong focusing* was made by Nicholas Christofilos (figure 6.10), and independently by Ernie Courant, Stan Livingston, and Hartland Snyder. The idea behind this discovery is an analogy with the method of focalization of a light beam, which can be made very thin in both transverse directions, thanks to the combined action of convergent and divergent lenses. The magnets used to bend the trajectories of the particles in the synchrotrons are dipoles, as shown in figure 6.11. A quadrupole instead has two positive and two negative poles. Its effect is to cause the trajectories of the bunch of particles to converge in a transverse direction and to diverge

FIGURE 6.9
Detail of the Synchrophasotron in Dubna, Russia. Courtesy of JINR.

FIGURE 6.10
Nicholas Christofilos, who together with Ernie Courant, Stan Livingston, and Hartland Snyder, was the discoverer of strong focusing in particle accelerators.

in the other (figure 6.12). With surprise, it was observed that, by interspersing bending dipoles with quadrupoles, alternately focusing and defocusing in the two directions orthogonal to that of motion, the beam narrowed in its entirety. The need for a large vacuum cylinder and hence of heavy and costly magnets was in this way drastically reduced. The schematics of a synchrotron exploiting strong focusing, with the use of dipole and quadrupole magnets, is illustrated in figure 6.13.

Allow me now to make a brief digression before addressing the following development in accelerator technology. Once they have been raised to maximum energy, the particles in a cyclotron or synchrotron are extracted from the machine and directed toward a target, where they produce collisions with the atoms and nuclei of the medium and eventually generate new particles from the materialization of part of their energy. The detectors positioned around the target make it possible to reconstruct the events: by studying them, physicists gather precious data that might perhaps lead to

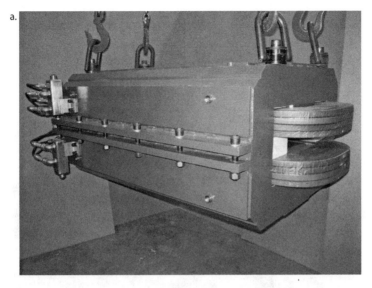

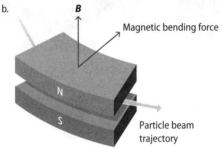

FIGURE 6.11

(a) Example of a dipole used in a particle accelerator; (b) conceptual plan of a dipole. The magnetic field B between the two poles of the magnet generates a bending force. Courtesy of CERN.

the discovery of new particles, and also to the understanding of the mechanisms behind their interaction. Everything is quite clear so far. With some applications, using the formulas in chapter 3, we can demonstrate that in the collision of a particle accelerated to an energy E, with a target particle (therefore stationary) of mass M (a proton, a neutron, or a whole nucleus), the maximum available energy E_{avail} for the creation of the mass of new particles is equal to about $c\sqrt{2ME}$. Let's take an example. Assume that on escaping our synchrotron, we have protons each with a total energy of 10 GeV, in keeping with the relation $E^2 = p^2c^2 + m^2c^4$. It is possible to show that the de Broglie wavelength, which is associated with them, is sufficient to

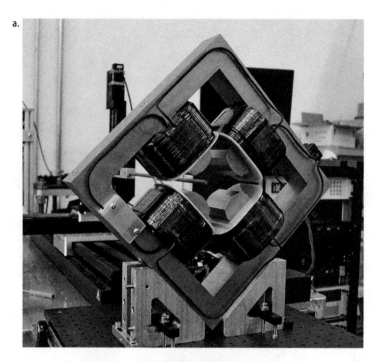

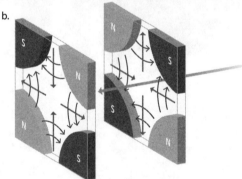

FIGURE 6.12
(a) Example of a quadrupole used in a particle accelerator; (b) diagram of a succession of two quadrupoles for respectively focusing and defocusing the beam of particles in opposite directions.

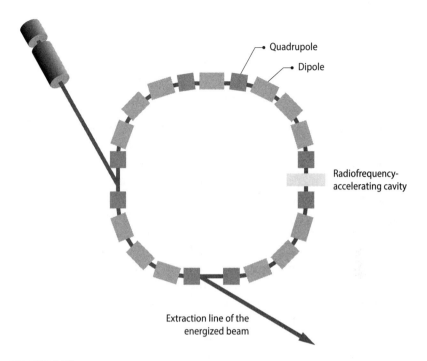

FIGURE 6.13
Diagram of a synchrotron with strong focusing (seen from above). The number of magnets (dipoles and quadrupoles) increases with respect to the case of weak focusing, but their dimensions diminish for equal power of the machine.

probe dimensions comparable to those of another proton or neutron. Let's then hypothesize that the collision occurs between an accelerated proton and another proton of the nucleus of an atom of which the target is composed. The mass of the proton is nearly 1 GeV/c^2, for which the E_{avail} ≈ $c\sqrt{2ME}$, which is to say E_{avail} ≈ 4.5 GeV. Consequently, the usable energy is substantially less than that of my accelerated particle. With equally simple calculations, it is possible to show that if the two particles with equal mass (two protons, for example), each of energy E and coming from opposite directions, collide with each other head on, the energy which may be used to create new particles is $E_{avail} = E + E$. Reprising the numerical values of the previous example, and assuming that E is still equal to 10 GeV, we have $E_{avail} = 20$ GeV, which is considerably larger than 4.5 GeV. Also, in classical physics, the damage suffered by a car traveling at 50 km/h when it hits a wall is definitely less than it would be if two cars traveling at 50 km/h were to collide head on, with a resulting speed of 100 km/h! The relative speed of relativistic particles cannot be so readily added up, but nevertheless, the relation $E_{avail} = E + E$ is rigorously valid for high-energy collisions.

From this brief discussion, it's clear that we have an experimental advantage in producing head-on collisions "in the center of mass," as it is expressed in the jargon, with particles coming from opposite directions rather than "on a fixed target." Well then, let's take two accelerators and arrange it so that the respective extracted beams collide with each other! This is easier said than done, though. Moreover, the smallness of the particles, the transverse dimensions of the particle packets, and the typical number of particles in each beam would make such a collision highly improbable. It would be smarter to employ the same accelerator to make the particles circulate, accumulate, and accelerate in opposite directions, and finally, to make them collide at specific points throughout the machine, with suitable detectors placed around the collision point. Then, if the particles have the same mass and an opposite electric charge, as in the case of matter and antimatter—such as electrons against positrons, or protons against antiprotons—it will be possible to use the same radiofrequency cavities and the same magnets, as well as the same vacuum tube. In fact, from an electrical perspective, the positive particle that crosses an element of the accelerator in a clockwise direction undergoes the same effect—accelerating, curving, or focalizing—as a particle with opposite charge proceeding in a counterclockwise direction. We have just invented the "collider." In reality, the original idea for the collider belonged, once again, to Widerøe, though many years were needed to develop really functional accelerators. One of the main problems was managing to inject and to accumulate in the machine, in a stable way, a large enough number of particles and antiparticles to produce a reasonable probability of producing collisions at the points anticipated for them. Moreover, during the phases of injection, accumulation, and collision, a stability of operation for many hours (or even days) is required.

Colliders today represent the frontier in accelerator technology. Beginning in the 1950s, great physicists have been contributing to the development of this technique—one that has produced scientific results of outstanding importance. One name may serve here to represent the many who have worked in this area. Bruno Touschek (figure 6.14), the genius Austrian physicist who escaped from a Nazi concentration camp, choosing to live in Italy, created at the beginning of the 1960s, in the INFN laboratory in Frascati, a groundbreaking machine, AdA, which was a ring for accumulating electrons and positrons. AdA was followed in 1968 by the arrival of a much more powerful machine, ADONE, which was capable of causing electrons and positrons to collide at 1.5 GeV, producing a total energy of

FIGURE 6.14
Bruno Touschek and his Frascatan creature: AdA (*Anello di Accumulazione*, i.e., Accumulating ring). © INFN Frascati National Laboratories.

3 GeV. With the success of ADONE, the way was opened for the coming of modern accelerators, and for ... a missed Nobel Prize that I will tell you about later in this book.

To conclude this chapter, I would like to remind the reader that nowadays, researchers use accelerators for physics experiments employing both the fixed-target and the collider method, depending on the applications and scientific objectives they are pursuing. In the second case the energies of the particles are higher, but the probability of obtaining collisions is lower. Technical limitations play a crucial role here, especially for the accumulation of the required number of particles in a restricted space, as well as in ensuring that the opposing beams from which they are coming can effectively meet and generate collisions. In the first case, the lower available energy is balanced by a much larger interaction probability. In technical terms, we speak of the *luminosity* of the accelerating machine—a quantity proportional to the number of collisions that occur per unit of time for a given physical process.

In recent decades, we have constructed many accelerators, of various types, power, and complexity; machines capable of accelerating nuclei, protons, and electrons together with their antiparticles; and even muons.

FIGURE 6.15
Internal view of the underground tunnel of the LHC at CERN: the accelerator behind all the records. The long tubes are the accelerator's powerful bending magnets (dipoles). Courtesy of CERN.

The state of the art is represented now by the gigantic LHC (figure 6.15) at CERN, the largest machine ever built by humans: thousands of superconducting magnets installed around a vacuum tube with a circumference of 27 km, placed 100 m below ground in the beautiful countryside between Lake Leman and the Jura mountains in Switzerland. It is a machine that has broken every record and is capable of accelerating protons up to 7,000 GeV for each of the colliding beams, giving a combined total of 14,000 GeV. This is the machine that has made it possible to discover the famous Higgs boson, dubbed "the God particle," in an inspired editorial turn of phrase.

But now we must return to the 1960s, to when a young physicist named Murray Gell-Mann was reading James Joyce ...

7 THREE QUARKS FOR MUSTER MARK!

> The universe is full of magical things patiently waiting for our wits to grow sharper.
> —Eden Phillpotts

In 1964, the time was ripe for yet another revolution in the emerging world of particle physics. Independent of each other, two American physicists—Murray Gell-Mann, who we have already seen introducing the quantum number of strangeness, and George Zweig (figure 7.1)—proposed the solution for the evident nonelementary nature of the plethora of particles that would have chaotically crowded the books of "particle zoology," if such books had actually existed: hundreds of particles that had appeared from the study of emulsion exposed to cosmic rays, or in bubble chambers hit by beams of high-energy protons accelerated by cyclotrons and synchrotrons. The Gell-Mann/Zweig hypothesis elegantly resolved this problem, as we shall see, introducing the handful of elementary building blocks (just three or four) that make it possible to build many of the presumed elementary particles (notably the so-called hadrons). It's a system that is in some respects similar to the atomic one, which with opportune combinations of protons, neutrons, and electrons makes it possible to catalog all of the elements of the periodic table existing in nature.

The word *quark* comes from a literary work, *Finnegans Wake* by James Joyce, which Gell-Mann was reading at the time when he was developing his model. "Three quarks for Muster Mark," wrote Joyce, and the phrase—difficult to decipher—at least had the merit of pointing to a good number, three, which was ideal for returning the multiplicity of the world of particles to a more simple and unified system. As Gell-Mann himself had occasion to admit, the quark model was nothing but a mathematical-geometrical tool to organize particles and explain their multiplicity, but it was not yet founded on a dynamic model that could frame their existence in a theory based on the laws of nature. Zweig's theory, quite similar and in some respects

FIGURE 7.1
Murray Gell-Mann and George Zweig, the inventors of the quark model.

dynamic, given that the physicist took real particles into account, envisaged constituent entities that he termed "aces," after those in a pack of playing cards. However, there are four aces in a pack of cards, not three like the quarks of Muster Mark ...

Before tackling the interactions between particles and the forces that attract, bind, or repel them (i.e., the fundamental interactions) in the next chapter, it is perhaps worth anticipating now that a large number of the very many "creatures" populating the particle zoo in the 1960s are subject to strong interaction, one of the four forces of nature. Consequently, they were called *hadrons*, from the Greek word ἁδρός (*hadros*), meaning "strong." The quark model referred to hadrons and not to electrons, muons, and neutrinos—the particles we now call *leptons*, from the Greek word λεπτός (*leptos*), meaning "light"—which by then were well known and incapable of feeling the strong force. The hadrons, in turn, were grouped into two families: the first—the *baryons* from the Greek βαρύς (*barys*), meaning "heavy"—included particles with mass higher than or equal to that of the proton; the second comprised the lighter ones, the *mesons*, such as pions and kaons. In reality, the historic subdivision between mesons and baryons no longer corresponds to the original meaning, because over the years we have discovered mesons that were much heavier than many of the baryons. With the quark model, as we shall now see, the difference between baryons and mesons becomes logical and comprehensible.

The original quark theory predicts that every baryon is made of a suitable combination of three really elementary pointlike entities: the quarks *up, down,* and *strange*. These are fermions with spin 1/2, that possess specific values of the various other quantum numbers, and have an electric charge that is fractional, in units of the elementary charge of the electron, $e = 1.6 \times 10^{-19}$ coulombs (by definition equal to $-1e$), while protons, neutrons, pions, and others have a charge of $+1, -1$, or 0, omitting for simplicity's sake the term e. Instead, while the *up* quark (u) has an electric charge of $+2/3$, the *down* (d) and the *strange* (s) ones have a charge of $-1/3$. This was a genuine revolution that was algebraically necessary to take into account all the possible combinations of quarks, and to understand the phenomenology of the many hadrons existing in nature. Let me explain. Let's take the proton, a particle with spin equal to 1/2 (with the two orientations $+1/2$ and $-1/2$) and electric charge of $+1$. This will be made of two *up* quarks and one *down*: uud. The electric charge of the proton is now found to be $+2/3 + 2/3 - 1/3 = 1$. Moreover, two of the three quarks will need to have the spin component along the quantization axis equal to $+1/2$ and one at $-1/2$ (here too, we overlook the factor $h/2\pi$), so as to guarantee a total value of $+1/2$ for the proton. Similarly, we can obtain the value of $-1/2$ for the spin of the proton. Perfect! And what about the neutron with null electric charge? Simple—it will consist of the quark combination ddu with total charge: $-1/3 - 1/3 + 2/3 = 0$ (figure 7.2). And what about the antiparticles, such as the antiproton? Nothing could be easier: there are also antiquarks, with the same quantum numbers, but with the signs reversed. So then the antiproton is made of two *up* antiquarks (also indicated by $\bar{u}$) and by a *down* antiquark, with a total electric charge of $-2/3 - 2/3 + 1/3 = -1$. So it's simple and logical, notwithstanding the introduction of fractional charges never revealed before and, not to keep the reader in suspense, *still not observed, even today*!

This aspect in particular used to vex the parents of the quark model—so much so that Gell-Mann had occasion to remark: "The research into stable quarks of charge $-1/3$ or $+2/3$ at the highest possible energy of the accelerators will help to reassure us of the nonexistence of quarks as actual particles." The model was in effect being considered as a pure theoretical artifice, without a real equivalent in nature. Nevertheless, the theory seemed solid, and as if by magic, all the baryons took their place in one of the pigeonholes allowed by the model. On the other hand, it's assumed

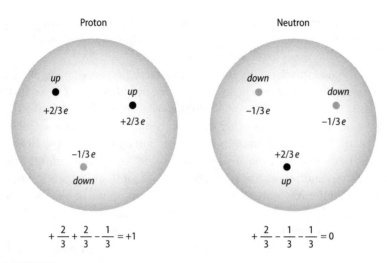

FIGURE 7.2
The composition of protons and neutrons in terms of quarks and the values of their total electric charge, in units of the elementary charge *e*.

that all mesons are composed of a quark and an antiquark. Let's take, for example, the negative pion π^-. It can be constructed by combining an *up* antiquark and a *down* quark, $\bar{u}d$, with an effectively negative electric charge, as follows: $-2/3 - 1/3 = -1$, with the two spins of the quarks aligned in opposite ways (+1/2 and −1/2) to guarantee the null total spin of the particle. But the success of the model does not stop here. Remember the *strange* particles? Well, they possess in "their interior" a *strange* quark (or an antiquark) that takes the corresponding quantum number (strangeness *S*) of value −1 (+1 for the antiquark). Strangeness is a typical example of the so-called flavor quantum numbers. We will find out more about this issue later ... For the moment, though, let's continue with our LEGO brick construction. What about the K^- strange meson, for instance? Simple—it's composed of a *strange* quark and an *up* antiquark, $s\bar{u}$, which, needless to say, have a total electric charge of −1. Analogously, the Λ^0 particle will be a bound state of three quarks: *uds*. I think we can conclude that the pessimism felt by Gell-Mann before this resounding success at simplification does not seem justified. The model works too well to be merely an artifice, a game that, by sheer chance, seemed to explain how nature manifests itself.

Let's get our breath back and look at figure 7.3, where the properties and composition of some of the hadronic particles are listed in terms of quarks. It is important to note that while the strange quark possesses its own specific quantum number, in order to take into account the phenomenology

Particle	Quark composition	Mass (MeV/c²)	Electric charge	Spin	Isospin	Isospin projection	Strangeness
p	uud	938.3	+1	1/2	1/2	+1/2	0
n	ddu	939.6	0	1/2	1/2	−1/2	0
Λ⁰	uds	1116	0	1/2	0	0	−1
Σ⁰	uds	1192	0	1/2	1	0	−1
Σ⁺	uus	1189	+1	1/2	1	+1	−1
Σ⁻	dds	1197	−1	1/2	1	−1	−1
Ξ⁰	ssu	1315	0	1/2	1/2	+1/2	−2
Ξ⁻	ssd	1321	−1	1/2	1/2	−1/2	−2
π⁺	u$\bar{d}$	139.6	+1	0	1	+1	0
π⁻	$\bar{u}$d	139.6	−1	0	1	−1	0
π⁰	(u$\bar{u}$ − d$\bar{d}$)/√2	135.0	0	0	1	0	0
ρ⁺	u$\bar{d}$	775.4	+1	1	1	+1	0
ρ⁻	$\bar{u}$d	775.4	−1	1	1	−1	0

FIGURE 7.3
Properties of some of the hadrons (baryons and mesons).

of strange particles, the *up* and *down* quarks are characterized instead by the value of the isospin. Remember Heisenberg's theory? He explained the proton and neutron as different states of the same particle, the nucleon, with a value of the quantum number of isospin (this is also a vector) equal to 1/2 and with two possible orientations: +1/2 for the proton and −1/2 for the neutron. It is only natural, therefore, that with the quark model, the value of quantum numbers, including isospin and strangeness, more properly should be referred to the constituent quarks.

The table in figure 7.3 is instructive for evidencing other characteristics of the quark model. Look at the composition, in terms of quarks, of the three pions. When we indicate the negative pion synthetically with $\bar{u}d$, we mean, according to the rules of quantum mechanics, that its overall wave function is the convolution of the two wave functions attributed to the quarks $\bar{u}$ and d, respectively. In the case of the π°, we have two independent ways of constructing it; that is, with either a $u\bar{u}$ pair or a $d\bar{d}$ one. The combination of the two possible ways is shown in the diagram with an appropriate normalization factor (1/√2), and this is predicted by the quark model in order to account for all the characteristics that are experimentally observed for the particle. Another notable aspect is that, given that the model was based on only three quarks (u, d, and s), there was a number of combinations that were not sufficient to construct all the particles known at that time. This is reflected, for instance, in the fact that both Λ^0 and Σ^0 have the same composition of quark (uds), although they are different particles and of similar (but not equal) mass. The explanation is that other quantum numbers (e.g., the isospin) may be different for identical trios of quarks. In the specific case of Λ^0 and Σ^0, they are differentiated precisely by the value of the isospin, which is in turn determined by how the isospins of the constituent quarks u and d are arranged. With the quark theory, it becomes evident and understandable that many of the particles considered independent are nothing more than higher-energy states of a basic particle, of smaller mass, just like what happens at the energy levels of the hydrogen atom's electron compared to the so-called ground level. We talk in this case of the spectroscopy of a given particle. I can fully understand that these considerations may seem a little abstruse, but they are useful for understanding the ingenuity and complexity that forms the basis of the quark model of hadrons. Extending these arguments further, the classification was soon completed, framing the various particles in so-called multiplets, a schematic and summary way of cataloging hadrons on the basis of the values of their quantum numbers, including flavor, as imposed by the quark model. Hoping not to scare the reader excessively, in figure 7.4, I show one of these multiplets, which is known as the *decuplet of baryons of smaller mass with spin equal to 3/2* ... OK, I'll explain.

On the X axis, we have the value of the isospin with its sign, in technical terms the projection of the isospin vector along the direction of quantization (I_z). On the Y axis, the value of the strangeness S is represented. Let's now take the baryons of lower mass, characterized by having the orbital angular momentum equal to 0 and the total spin value equal to 3/2; the three quarks which they consist of have their individual spins all oriented in the same way: +1/2, +1/2, and +1/2, with a total of 3/2. In the figure, there is a third (inclined) axis representing the electric charge Q of the particles. Let's start to pigeonhole the hadrons. The first particle is the Δ^{++}, which we place in the top right corner. Discovered by Fermi in 1952, this particle has an electric charge equal to +2, being composed of three u quarks and consequently devoid of strangeness ($S=0$). The three isospins of its quarks will all be oriented in the same direction (let's say, positive), for a total value of $I_z = 3/2$. Proceeding toward the left, we insert the Δ^+ that contains a quark d and two u, with an electric charge +1 and $I_z = 1/2$.

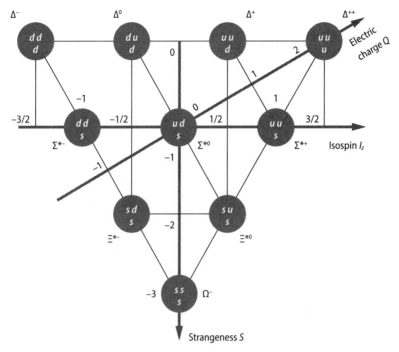

FIGURE 7.4
A special baryon decuplet.

The scheme is beautiful—simply and geometrically organized—and the reader, while sparing me from continuing further, should nevertheless know that our cataloging task ends by inserting the Ω^-, a particle composed of three s ($S=3$) quarks with $I_z=0$ and an electric charge of –1, precisely at the bottom corner of the decuplet. Pretty good, no? Well, the remarkable thing is that Ω^- had not even been discovered when Gell-Mann, whilst working on his model, had hypothesized its existence in 1962, precisely in order to fulfill his decuplet symmetry! The discovery of the particle, in a bubble chamber experiment made in 1964 with the Brookhaven synchrotron on Long Island, New York, was game-changing. The Ω^- was something like the proverbial missing link, needed to definitively confirm the quark model. It was not simply a mathematical artifice, but a theory that was also predictive. The beautiful event that provided the discovery is shown in figure 7.5. Its complex kinematical analysis leads to the unequivocal identification of the small track at the bottom, as indeed generated by the missing particle! For the sake of completeness, we should observe that the quark

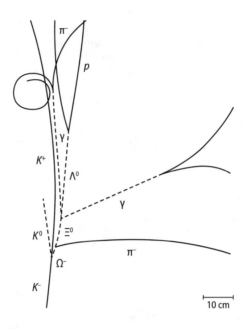

FIGURE 7.5
The discovery of the missing link, the Ω^-, in 1964. From the complex reconstruction of the event and from the measurement of the kinematical quantities relative to the various particle tracks, it is established that the short track at the bottom is actually left by the new particle.

model is based on mathematical and geometrical arguments pertaining to the mathematical group theory, which in turn is governed by very general symmetry principles. The discussion now becomes abstract and complex, and once again, it would require a deviation from our course. Nevertheless, the mathematical aspect of the matter is quite clear and intuitive. I have at my disposal six elementary building blocks (three quarks and three antiquarks) with which to construct all the possible hadrons (baryons, mesons, and their antiparticles). Each of these bricks has specific properties—the quantum numbers, which in turn assume discrete values. Consequently, it is only natural that multiplets and "periodic tables" appear, in order to show the possible quark permutations and combinations. A more profound matter arises from the fact that from the study of particles, it was soon understood that the laws of geometrical symmetry are intimately bound to the structure of matter and to fundamental interactions, and that they even govern the mathematical aspects of the corresponding physics equations. This is one of the most fascinating and intriguing characteristics of the microcosm, and one that is in many respects still not properly understood. Then, the connection that exists between symmetry and its manifestation in physics leads us to speak of an expectation of a kind of geometrical beauty in nature at its deepest level, almost akin to a justification of why the universe is put together in a particular way rather than otherwise. The laws of symmetry permeate the world of elementary particles to such an extent that physicists are often guided by them when choosing between one theory and another. We will touch on precisely this aspect when confronting some of these theories that are not only beautiful, but also capable of interpreting reality.

The quark model did not exhaust its role in cataloging the hadrons and making the hypothesis about their substructures sounder. The theory made it possible to elegantly predict which reactions were possible, and it also provided a scheme with which to interpret them. Let's take as an example the event shown in figure 5.16 in chapter 5; now we can read it as it is symbolically represented in figure 7.6, assuming the correct composition of the particles involved in terms of quarks. At the same time, the principle of the conservation of strangeness becomes more comprehensible. In the diagram, we identify three moments in the process; the pion and the proton before the collision, the interaction between the constituent quarks, and their rearrangement into the final state, made up of two strange particles—a K meson (or kaon) and a Λ baryon. In the interaction, the *up* antiquark ($\bar{u}$) of the pion and the *up* quark of the proton (u) annihilate each other (we will have occasion to revisit this process)—matter against antimatter—with

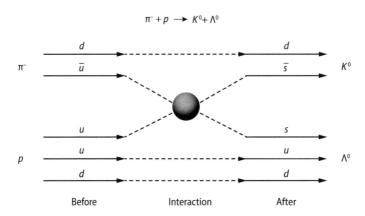

FIGURE 7.6
Example of a collision between hadrons, interpreted using the quark model. The initial quark configuration belonging to the colliding hadrons changes after the interaction to form the hadrons in the final state.

the production of pure energy. According to the laws of special relativity, this available energy can be reconverted into matter, creating from nothing a *s* quark-antiquark pair. The other quarks somehow remain identical and recombine with the two *s* quarks to build the kaon and the lambda. Obviously, this is only one of the possible reactions, with its specific probability of happening; what is most interesting is that the quark model explains it, if not in a dynamic sense (we go on to see its significance), then certainly at the level of possibility, and accurately takes into account the various conservation principles.

With a little effort, the reader may be able to verify that both the conservation principles of the electric charge (before and after the collision) and that of the strangeness—remember that this quantity is conserved in the hadron production reactions—are holding in the process illustrated in figure 7.6. Finally, given that the sum of the masses of the pion and the proton (1,080 MeV) is less than that of the reaction products, kaon plus lambda (1,610 MeV), it is necessary that the colliding pion has sufficient kinetic energy to be transformed into the mass of the final particles. All very reasonable, is it not? In reality, the original theory of Gell-Mann and Zweig included a very weak spot—almost the equivalent of a stone guest. At the time, no one had yet observed (sorry, *detected*) a free quark. This matter is not insignificant, given the fact that even today, despite significant experimental attempts to do so, the direct identification of a quark has not been made. So what? It is glaring proof of the fact that, despite being successful and aesthetically pleasing (Dirac *docet*), the model is entirely geometrical and mathematical rather than being based on dynamic considerations deriving from the interaction at the root of the processes. To predict the motion of the planets and the occurrence of eclipses without knowing the role of gravity would be similarly unsatisfactory. In chapter 11, we will indeed discover what is *The True Theory of Quarks and Their Interactions*: a powerful and solid theoretical framework that includes the quark model in a scheme that is more comprehensive, and that also explains why we will never be able observe free quarks! But we will have to be patient for a while before getting there.

Despite the lack of free quark detection, the quark hypothesis was quickly accepted by the scientific community. On the experimental side, in addition to (unsuccessful) attempts to find particles of fractional charge, the main effort was immediately concentrated on realizing experiments that were conceptually similar to Rutherford's, which in its time had enabled the discovery of the nucleus. In effect, in the mid-1950s, the American physicist Robert Hofstadter and his collaborators demonstrated, using such experiments, that both the proton and the neutron are not pointlike particles, but rather that they have a size of about 10^{-13} cm—10 times smaller than what is typical for a nucleus. To accomplish these measurements, the physicists had studied the angular diffusion or so-called *scattering* of high-energy electrons, which were available thanks to the linear accelerator at the Stanford Linear Accelerator Center (SLAC) laboratory at Stanford, California. Notwithstanding this, the majority of physicists did not consider that the result necessarily implied the existence of other particles inside the nucleons. Be that as it may, in 1961, the discovery earned Hofstadter the Nobel Prize. The method that he used precisely recalled the original methodologies of Rutherford. As described by the relation $\Delta r \propto h/p$, the resolving power of a projectile increases with its momentum. It is, therefore, necessary that the de Broglie wavelength should have dimensions comparable to those of the objects one is aiming to investigate. Let's take an electron as our projectile: it will or not be capable of probing a proton. With a few simple calculations, we derive that the electron momentum required to observe an individual proton in the nucleus must be in the order of 1 GeV/c; with at least 10 GeV/c, we will be able to reasonably show evidence for its internal constituents, such as the quarks of Gell-Mann and Zweig. To be precise, the parameter that determines the dimensions that I can explore with my collisions is the fraction of the momentum of the projectile effectively transferred to the target in the collision. But for our purposes, this is a mere detail.

A series of conclusive experiments were again carried out at SLAC, shortly after the formulation of the quark hypothesis. Between 1968 and 1969, Jerome Friedman, Henry Kendall, and Richard Taylor (figure 7.7) obtained a result that pointed to the existence of pointlike scattering centers inside protons and neutrons. For this experiment, which was later considered to have achieved the discovery of quarks, they were the joint recipients of a well-deserved Nobel Prize in 1990. The experiment was significantly more complex than Rutherford's, but it was conceptually similar. A beam of 20 GeV electrons (an extremely high energy for the period) interacted with a target, and the diffusion angle of the electrons was measured by a series of electronic detectors with the ability to cover a broad angular interval around the collision point. The expected distribution for the electron scattering angles was computed on the assumption that the interaction (called *elastic diffusion* in this case) would happen on nucleons without any internal structure. The experiment revealed instead a number of events indicating quite large diffusion angles, incompatible with the theoretical assumptions, a fact that indicates the presence of pointlike objects within the nucleons (figure 7.8) that are capable of making head-on collisions with the electrons of the beam.

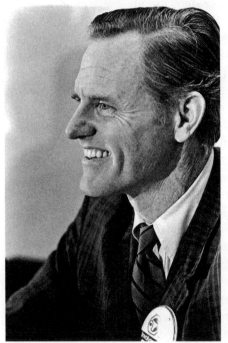

FIGURE 7.7
Taylor (top), Kendal (bottom left), and Friedman (bottom right), the discoverers of quarks.

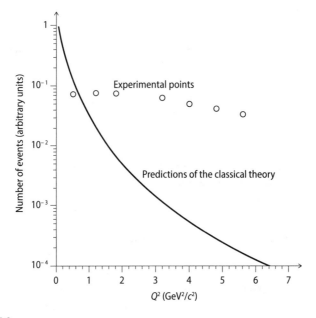

FIGURE 7.8
The result of the SLAC experiment that proved the existence of scattering centers (the quarks) within the proton and the neutron. In abscissa, there is a variable function of the scattering angle (i.e., the square of the momentum exchanged in the collision); on the ordinate, a quantity that is proportional to the number of observed events. The classical theory (in the absence of quarks) predicts that the number of scatterings decreases very rapidly with increasing angle (the continuous curve), something that doesn't happen, as demonstrated by the open circles representing the experimental results. The events with large angles outnumber those expected, while there are fewer with small angles than predicted. The incompatibility between data and theory is very significant, indicating the existence of pointlike entities capable of scattering the electrons at very large angles *following hard head-on collisions.*

Once the SLAC results had been published, it became clear to everyone that the scattering centers within the proton, originally called *partons*, were precisely our quarks. The overall scenario that emerged is summarized schematically in figure 7.9. Just as in a nest of Russian dolls, within the interior of each intimate structure of matter there is another one, equally complex but on a much smaller scale. Atom, nucleus, nucleons, quarks ... To the best of our current knowledge, attained with experiments using very high-energy probes, the quarks still appear to be pointlike. They are as elementary

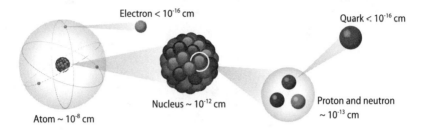

FIGURE 7.9
The "Russian doll" of the structure of matter. With each stage, the dimensions of the inner objects diminish significantly. Today, there exist only upper limits for the size of quarks and electrons, which therefore are assumed to be pointlike.

as the electrons and with a size at least 100 million times smaller than that of an atom (less than 10^{-16} cm)! You will recall that we have observed to what extent the diameter of the nucleus is derisory compared to that of the whole atom. Well, approximately the same relation exists between the proton and its constituent quarks. Just as in the case of the atom, the proton and the neutron are both essentially empty. All the visible matter in the universe is in fact concentrated in minuscule centers (namely, quarks and electrons). And all around, the quantum vacuum that we shall later see populated by phantom entities: the virtual particles. It's a mysterious and intriguing scenario that we are only just beginning to glimpse on our journey among the elementary particles. And it really is just the beginning … Soon we will expand our horizons, and after having understood the fundamental role that forces exert between the elementary particles, our course will lead to the discovery that physical reality is much more complex than we can conceive. Perhaps just three quarks are not enough for nature …

8 MAY THE FORCE BE WITH US!

The Force is what gives a Jedi his power.
It's an energy field created by all living things.
It surrounds us and penetrates us. It binds the galaxy together.

—Obi-Wan Kenobi

Now we must tackle an aspect of our subject that is undoubtedly even more important than the basic need to classify and to create a unified scheme for the elementary particles—namely, that of the forces (or what we can call the interactions). This is the very thing that is behind the dynamics of the lives of particles—and that ultimately justifies and explains them. To proceed in the study of the elementary particles without requiring this extra effort from the reader would be like classifying plants and animals and studying their morphological differences without knowing how a cell works, or what deoxyribonucleic acid (DNA) is.

We have already observed that, given the dimensions of its fundamental constituents, matter is essentially empty. The atom is empty for approximately 99.9999999999999 percent of its volume, and the proton and neutron are largely empty too, for 99.9999999 percent of theirs. Therefore, the entire universe, already itself very diluted, with its galaxies and stars extremely distant from each other, is made up for the most part of empty space. Imagine that its actual density is around 10^{-29} g/cm^3, including matter and energy! But from this, two questions automatically follow: What makes matter as rigid as we find it to be? Why can't I push my hand through a wall? The answer lies in the forces acting between the particles of which ordinary matter is constituted. In nature, as far as we know, there are four fundamental forces or interactions: the gravitational, the electromagnetic, the strong nuclear, and the weak nuclear. Of these interactions, the first to be known, the gravitational, is also the least relevant and most quantitatively negligible compared to the others, at any rate as far as the world of particles is concerned. This was not the case an infinitesimal fraction of time after the Big Bang, as we will see later, but for now, it is safe to assume—it's an excellent working hypothesis, in fact—that the gravitational force is irrelevant when studying the interaction between elementary particles. Its weakness, though, seems to contradict the very visible interactions between planets, stars, and galaxies, and the evident intensity of gravity, which pulls all material bodies toward the center of the Earth. Objects fall on account of their mass, sometimes in a precipitously spectacular manner; the Earth and Newton's apple reciprocally attract, and the Sun holds our planet on an orbit that is 365 days long, thanks to gravitational attraction. And yet gravity is the weakest force when we normalize it to a single particle. A kilogram of iron is heavy only because it contains around 10^{25} atoms, each one made up (for the most common isotope of iron) of 56 protons and neutrons and 26 electrons. The gravitational attractive force between two electrons is completely negligible compared to the Coulomb repulsive force generated by their electric charge. Without delving into complex considerations of general relativity, it is possible to say that the mass of the particles generates the gravitational force, just as the electric charge generates the electromagnetic interaction. But the intensity of the latter is around 10^{37} times greater (!) than the gravitational one at the scale of the typical distances of particle physics experiments. This explains why, in the overall balance, the electric repulsive force so completely overrides the feeble gravitational one.

However, the question of the actual rigidity of matter remains unanswered. Let's try to explain with an example. If I wanted to pass my hand through a wall, I would find myself exerting a pressure that would see the electrons of the atoms that make up that part of the surface of my hand encountering the electrons of the atoms of the cement. The latter are organized into an ordered and rigid structure, with strong links between them, in contrast to what happens with liquids and gases. The concomitant and coherent electric repulsive interaction between the myriad electrons involved exercises an overall, very intense force that impedes the interpenetration of hand and wall, even though both are essentially made of empty space. As introduced in chapter 4, the Pauli exclusion principle plays a highly important role in this mechanism. In fact, the pressure that is eventually exerted on the atom's peripheral electrons by an external force is transmitted to the other, more internal ones, and eventually to the electrons of the other atoms of hand and concrete. Electrons are prevented from cohabiting with their fellows in an atom beyond certain limits imposed by the principle, and oppose themselves to an ultimate compacting of the atom, therefore applying a contrasting force to the external ones. The same holds for the strong nuclear force. The latter binds protons with protons, neutrons with neutrons, and protons with neutrons, regardless and with the same intensity to build up the atomic nucleus. The strong force is also responsible for the interactions between the other, more exotic hadrons (baryon and mesons). The fact that it must be strong is readily understood since it has to compete with and triumph over the opposing force of repulsion due to the same electric charge of the protons, as shown in figure 8.1. If this weren't the case, electromagnetic repulsion would prevent the formation and the existence of stable nuclei, and therefore of ordinary matter ... including human beings. The strong force between two protons placed almost in contact is approximately 100 times more intense than the electromagnetic one at the same distance. The fourth interaction that exists in nature is the weak nuclear force. It is responsible in particular for β decay, as well as the functioning of the Sun and all the other stars of the universe. All fermions feel the weak force. That said, as the term itself suggests, its intensity is approximately 100,000 times less than that of the electromagnetic force—also in this case for typical distances comparable to the size of the proton. The weak interaction becomes dominant and measurable only when the other interactions are not active

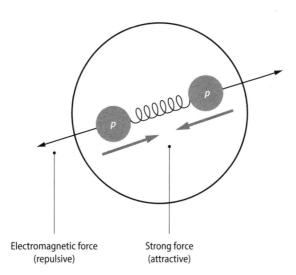

FIGURE 8.1
The electromagnetic repulsive force between two protons (e.g., in a nucleus) is outweighed by the concomitant strong nuclear force, which is much more intense. For this reason, protons and neutrons are bound together in the constitution of the atomic nucleus.

or are prevented by conservation principles. Neutrinos provide us with an example: they are not hadrons and thus are not receptive to strong interactions, and not being electrically charged, they are not even subject to the electromagnetic force. The weak force is the only way that they have to interact with other particles, and hence with ordinary matter—apart from a negligible sensitivity to the gravitational force, as we will see later on.

Figure 8.2 illustrates the four interactions in a schematic way, their respective fields of influence, and their relative strength in the abovementioned case of two protons coming into contact: the protons feel all four interactions and therefore provide a particularly good example. It is evident that at a macroscopic scale, and in relation to our everyday experience, all the actions and effects tied to forces can be attributed to gravity and electromagnetism, including bodies that fall (gravity), objects subjected to mechanical pushes and pressures (in the final analysis due to the electromagnetic interaction between superficial electrons), reactions and chemical processes, fire, explosions (once again caused by the electromagnetic

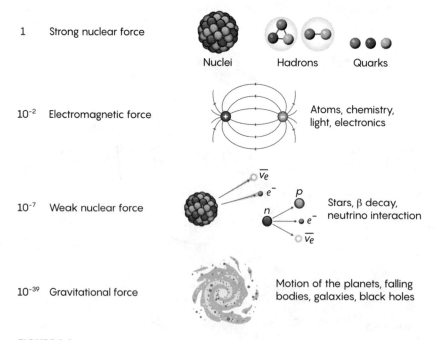

FIGURE 8.2

Illustration of the four fundamental interactions in nature, with examples of their fields of applications. On the left are shown the relative intensities of the forces in the case of two protons in contact with each other, assuming the intensity of the strong nuclear force equal to 1.

interaction), and so on. In all these cases, forces are applied to large agglomerates of particles, multiplying the effect of the elementary interaction with each one of them. If it was not for the exotic microcosm, and for the Sun and the other stars, we certainly would feel no need for the other two nuclear forces, the weak and the strong. The differences in range of the four interactions are also quite marked. While gravitational and electromagnetic forces are intense over small distances and formally extend up to infinite ones, strong interaction is felt only if the particles are closer together than 10^{-13} cm (roughly the nuclear dimensions). Weak interaction, moreover, virtually acts only if the distance is less than 10^{-16} cm. By varying the distance, the relative strengths of the forces also change. As we will see later, for distances that are extremely small, the weak force becomes more intense than the electromagnetic one, and at even closer distances the gravitational force also becomes extremely strong!

A classical version of the concept of interaction between particles is that of the field of force, an extension of the scheme used to explain gravitational and electromagnetic effects that we have referred to previously when discussing particle accelerators. Let's consider electrons and positrons—the former with an electric charge of –1, and the latter with one of +1 (in units of the fundamental charge e). The Coulomb electromagnetic force between two particles is proportional to the product of their respective charges divided by the square of the distance between them ($\propto e^2 / r^2$). The resulting force, as is well known, will be repulsive between the two positrons or two electrons—and attractive between an electron and a positron. We can interpret the interaction between particles indifferently, assuming that one of them creates an electric field around itself, which is to say a kind of disturbance of the space such that when another charged particle finds itself in the vicinity, a resultant force occurs, equal to the product of the intensity of the electric field at that point for the value of the charge of the second particle. Generally, we assume that the second charge is quantitatively much smaller than the first, in order not to notably perturb the system. This is the case for the electric field generated by an iron nucleus that, with its 26 protons, has an absolute value proportional to 26. If an electron with an electric charge of –1 is in the vicinity, given the large difference between 1 and 26, the resultant field lines will be very similar to those shown in the top left of figure 8.3. In the case of two electric charges equal to +1 or –1, on the other hand, which are identical in absolute value and equal or opposite in sign, the overall electric field is very different to that of a single particle, as is shown in the diagram for two specific cases.

Considerations of an altogether similar kind apply in the case of gravity, so long as one replaces the charge that generates the electric field with that

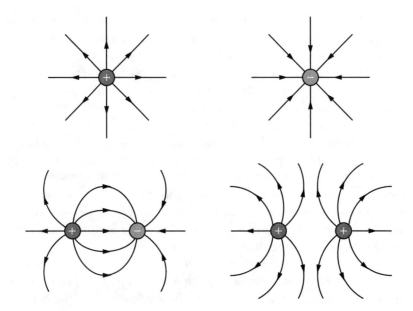

FIGURE 8.3
Top half: an illustration of the electric field generated by a positive and a negative charge, respectively. The field is more intense where the density of field lines is greater (in the vicinity of the charge that generates it). Lower half: the field lines that exist between two opposite charges, such as an electron and a positron, and between two of the same, as in the case of two positrons.

charge creating the gravitational one, which is to say the mass—which, indeed, only has a positive sign and therefore makes the force always attractive. At this point, nothing prevents us from postulating the existence of a strong and a weak charge to account for the other fundamental forces. We are gradually getting to the point ... however, before tackling this subject, let's consider an alternative approach, which is very practical for describing the interactions between elementary particles. Imagine two people in two boats that are floating, stationary, facing each other on the surface of a small lake. If they begin to throw a ball back and forth repeatedly between the two boats, as shown in figure 8.4a, the boats will slowly move apart—subjected to a force, or if you prefer, to a repulsive interaction. The physics reason for this is the principle of conservation of momentum, for the system constituted by both the ball and the boats. The boats, then, represent our particles, and the ball is the mediator of the force—another particle that causes the others to interact by transferring momentum between them. With a little imagination, this

FIGURE 8.4
The repulsive or attractive force between two particles (the boats) can be explained by way of the exchange between them of other particles called *mediators*—the ball (a) and the boomerang (b). The action of the exchange of mediators generates a repulsive or attractive force, just as happens with particles subject to the four fundamental interactions.

time involving the exchange of a boomerang as a mediator between our experimenters, we are also able to explain attractive forces (figure 8.4b). Remember Hideki Yukawa's idea? On the basis of the Japanese physicist's hypothesis, it was believed that the muon could be the mediator (the ball …) of the strong interaction between protons and neutrons within an atomic nucleus, though it was subsequently discovered that instead, this mediator needed to be Powell's pion.

The particle-mediated interpretation of the interactions is a very effective way of describing the four fundamental forces for carrying out calculations relating to particle collisions and decays. This is conceptually equivalent to envisaging them in terms of fields of force, although the latter scheme presents aspects that are not really obvious. For instance, in claiming that an electric charge creates a field around itself, we must not forget that it is not established instantaneously in all the space. The signal of the presence of the charge that generates it may propagate at most at the speed of light; a second particle placed at an arbitrary distance from the first will have to wait a certain time before knowing of the existence of its sister particle. With the particle-based approach, these concepts become more natural. If the mediator is a corpuscle without mass, the two interacting particles separated by a distance l will register their reciprocal presence only after time $t=l/c$. In the case of a mediator with mass, such time will in principle be a little longer. These arguments, however, are somewhat academic, given the minuscule distances involved and the very small time needed for the travel.

Continuing with the concept of the mediator, we can develop the case of the electromagnetic interaction between two electric charges, notably our two colliding electrons. It is only natural to assume that the mediator of their interaction is the photon. We have seen that the wave-particle dualism applies (de Broglie matter waves), and to an even greater extent to the electromagnetic waves that can be interpreted through the propagation of quanta of energy $E=h\nu$ and momentum $p=E/c=h/\lambda$ (precisely, the photons), where λ is the wavelength of the associated electromagnetic radiation. The two electrons, in approaching each other, are mutually affected and reciprocally exchange photons. The fact that these possess energy and momentum is found again almost exactly in the interpretation in terms of fields: the electric field, just like the fields of other forces, contains energy and momentum. It's quite simple, isn't it? Well, not really.

Let's analyze the process illustrated in figure 8.5. Electron 1 approaches electron 2, and at a certain point "decides" to emit a photon and recoils, just like the boat when the ball is thrown, to conserve the total momentum: as the photon moves in the direction of the electron carrying momentum away, so the electron recoils with equivalent momentum in the opposite direction. Similarly, electron 2, in absorbing the ball—sorry, the photon—acquires momentum, and its original trajectory is modified in the same way, this time in the opposite direction to the other one. But there is a detail that should not be overlooked. The analogy with the thrown ball is not complete. From studying particle accelerators, we have learned that an electron subjected to centripetal acceleration curves its trajectory and loses energy—part of that provided by the energizing electric fields—radiating photons along the tangent. On the contrary, it may be demonstrated (although we shall not do so here) that an electron that coasts at a constant speed or that is, at the limit, stationary (depending on the reference frame), cannot emit any photons—otherwise, in order to simultaneously conserve both momentum and energy after the emission, it would be forced to cannibalize its own mass. But the electron is precisely defined by its mass, just as it is identified by its electric charge, spin, and so forth. There does not

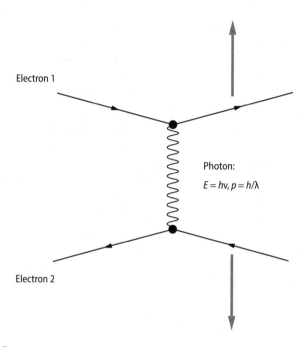

FIGURE 8.5
Electromagnetic interaction between two electrons mediated by a photon. The electromagnetic force, expressed in terms of the field, can be interpreted as being due to an arbitrarily high number of fundamental interactions of the type shown here.

exist an electron that weighs less than its peers: it is not possible for the photon to play the biblical role of Adam's rib! For this reason, electron 1 could not emit the photon needed to interact with its equals. So what then? Fortunately, the Heisenberg uncertainty principle comes to our aid with its energy-time formulation:

$$\Delta E \Delta t \geq \frac{h}{4\pi} \to \Delta t \geq \frac{h}{4\pi \Delta E}. \tag{4}$$

Let's assume that the two electrons are at a distance Δx when they exchange the photon. Traveling at the speed of light c, it will take a time $\Delta t = \Delta x/c$ to arrive at electron 2. Now the Heisenberg principle tells us that, not being able to measure time and energy simultaneously with the same precision, for (4) a small and well-defined value for Δt implies a large indetermination on the energy ΔE and vice versa. From this fact, it follows that if the energy of a particle is equal to zero, it can be worth $0 + \Delta E$, so long as this lasts for a time (correspondingly brief) equal to Δt. I know it's complex, but try to follow me: a photon of energy equal to 0, not having mass, also has $p=0$, and therefore simply does not exist. But given that Heisenberg allows it, a photon that is initially "nonexistent" can "appear" with an energy of up to $0+\Delta E$, so long as it lives for a time Δt related to ΔE by the uncertainty principle formula. At this point, the die is cast. The momentum of the photon emitted by the electron 1 is borrowed immediately from the energy bank of the "vacuum," but only for the time $\Delta t = \Delta x/c$ required to reach electron 2. In the end, nobody objects because the uncertainty principle allows me to have, for an extremely short interval, my account in the energy bank in the red—without paying any interest! Even if there is a small price to pay, actually. My photon is not real but virtual, and we will return to such strangeness in chapter 10, in order to discuss it extensively.

Everything outlined here is a pure mathematical artifice, if you like, to describe the interaction between electrons according to the concept of mediators. But it hardly matters. We could have explained everything in terms of fields, and this would have been no more satisfactory. In the end, no one observes what happens at the actual moment of the interaction, blurred as it is by the uncertainties of quantum mechanics. What we measure are our electrons before the interaction and the effect that the electromagnetic force has on them (deflections and accelerations). The attentive reader will have noticed, however, that there is a small flaw in these arguments. Electron 2 finds itself the recipient of some momentum and uses it to deflect its trajectory. Electron 1 instead borrows from the "bank" the same momentum, and on recoiling adopts a behavior comparable to that of its sibling. There's no problem here. For the same reason as before, in experimental terms we will never know which of the two electrons has emitted and which has received $\Delta p = \Delta E/c$. We can say that each of the two eventualities takes place with a probability of 50 percent. If we were to exhaust all of the credit granted by the bank, the energy of our photon would be at most $E = h/4\pi\Delta t \rightarrow E = hc/4\pi\Delta x$, and its momentum $p = h/\lambda = E/c \rightarrow p = h/4\pi\Delta x$. From these relations, it follows that as the distance between the two electrons grows, the energy and the momentum of the mediating photons diminishes proportionately. This in effect renders the electromagnetic force progressively weaker, as happens experimentally in nature, and we encounter again the aforementioned dependence according to the inverse of the square of the distance. The final point that needs to be discussed relates to the interaction between charges of opposite sign. In this case, our simple interpretative model does not work, so in order to explain the attractive force, we will have to consider quite complex arguments, which, once again, go beyond our objectives. Qualitatively, we can affirm that the indetermination on the value of the momentum exchanged between two electrons may also be extended to include the direction of the overall force applied to the electrons, as in the case of the boomerang. The virtual particles, such as our mediating photons, constitute one of the characteristic aspects of the theories that describe the forces between quantum particles. For now, it is important to repeat that they are, above all, a descriptive model of the interaction, or (if you prefer) a pure mathematical tool. Virtual particles are free to break the most banal rules of common sense. For these, the equality $E^2 = p^2c^2 + m^2c^4$ can be violated as well, and the virtual photons can have mass without paying any cost. The only requirement is that the intervals of time and energy must always be regulated by the Heisenberg principle.

An argument analogous to the one for virtual photons can be made, and I consider it to be very instructive with regard to the mediators of weak interaction. As we shall see in more detail in what follows, the mediators of the weak force are three very heavy particles: W^+, W^-, and Z^0. The mass of the first two is around 80 GeV/c^2, and that of the third 90 GeV/c^2. We are dealing with really significant masses, around 80–90 times that of a proton. The reason behind this strange choice of nature is still unknown to us, but to be honest, we also don't know why the proton has a mass of 1 GeV/c^2 and the electron a mere half MeV/c^2. Perhaps one day, we will manage to understand it by elaborating the so-called Theory of Everything, which has been pursued for so long by physicists, and which we have unfortunately yet to achieve. But let's return to our task: applying the same considerations used for the virtual photons. A mediator of 80 GeV/c^2 emitted by one of the two weakly interacting particles, perhaps itself having a mass much, much lower than 80 GeV/c^2, must perforce be virtual. Applying equation (4), and continuing with our banking analogy, the great loan of energy will have to last a truly derisory time, during which W will be able to traverse a distance that is correspondingly extremely small (the mentioned 10^{-16} cm), even assuming that it practically travels at the speed of light. This explains the very low intensity of the weak force: if the particle that absorbs the W is obliged to pass so close to that which emits it, it is quite easy to understand why this is an eventuality that is geometrically very rare. Such a improbability implies that the interaction does not happen frequently, which is synonymous with the weakness of the force. Similarly, the strong interaction has its mediator, responsible for keeping together the quarks in the interior of hadrons, forcing protons and neutrons to stay bound together in the constitution of the atomic nucleus, and making hadrons interact with other hadrons in the collisions that we create in our accelerators. The name of this powerful mediator is the *gluon,* from the word *glue.* Never has a more apt name been coined, given its role as a kind of quantum-superglue between particles. Powell and Yukawa's pion mediator at this point becomes a useful approximation only applicable to low-energy reactions.

The picture of the mediators is completed by the quantum of the gravitational interaction, the hypothetical graviton that we can do without well enough in the macroscopic world of gravitational forces, but which probably becomes necessary in the extreme microcosm. Here, the discussion is quite similar to that of electromagnetic force, which also has a range that is infinite. Gravity is too weak to merit being contemplated in our experiments and theories about elementary particles: consider that for the discovery of gravitational waves, the enormous cosmic collapse of two black holes has caused a wrinkle of spacetime that has produced in the Laser Interferometer Gravitational-Wave Observatory (LIGO) detector a shift of less than a thousandth of the diameter of a proton. Nevertheless, because the intensity of the gravitational force grows as the distance between interacting particles diminishes, we can hypothesize that it played an active role in the very earliest moments of time after the Big Bang (barely 10^{-43} seconds), when the distances between constituents of the then-microcosm were incredibly small and the density of energy infinitely high—and gravity even exceeded the other forces in intensity. It's a fascinating situation for physicists, which today is completely incomprehensible from both the experimental and theoretical point of view. Quantum mechanics and general relativity (Einstein's theory of gravity) conflict inexorably in this exotic realm of primordial matter, despite both being considered today to be extremely valid, in the microcosm and macrocosm, respectively. This represents a big problem for physics. Unfortunately, due to its considerable complexity, we will only be able to discuss it in a marginal way, at the very end of our journey among elementary particles.

Another important aspect, which follows from all that has been stated here, is the estimation of the frequency with which a collision or decay reaction occurs following the action of one of the fundamental forces. We have seen that to assert that an interaction is more or less strong is equivalent to affirming that the process in question can happen with more or less ease, and therefore with higher or lower probability. The reactions mediated by the strong force occur very frequently in the collisions between hadrons—those that involve weak force with much less frequency. With this objective, we define a variable that is much used in the study of elementary particles: the cross section. Let's return for an instant to using the representation of interactions in terms of fields of force. Imagine that our particle (a tennis ball) is traveling toward another particle (yours truly, wielding a racket). The force field generated by me (the target) will be all the more intense and have a greater range the bigger the racket is. If I could hold a massive racket with a head having a surface area of 1 m², I would be able to intercept the ball (the particle) without difficulty. In a synthetic manner, the interaction *ball–author of this book* would have a large cross section. If instead I had a miniature racket with a tiny head of only 1 cm², the chance of interacting with the projectile would no doubt be minimal—and in this case, we would have a small cross section. It is clear, therefore, that this quantity must be expressed in units of surface area—in square centimeters, for example. The same argument is illustrated in a more rigorous way in figure 8.6. A proton electromagnetically scattered by a nucleus will find itself confronting a collision surface appreciably larger than that of the target due to the effective range of the electromagnetic force. Here too, we are dealing with a surface. The target area is usually referred to with the Greek letter σ. Given the extremely small size of elementary particles, it is convenient to use minuscule fractions of square centimeters; the typical unit is the barn (b), which is equal to 10^{-24} cm².

Now let's try to illustrate how the cross section functions in our study of elementary particles. Let's consider hypothetically that we have a beam of protons of about 5 GeV/*c* momentum, which collides with a target made

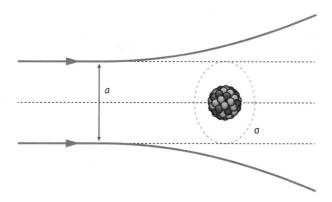

FIGURE 8.6

Illustration of the concept of cross section (surface σ of the circle of diameter *a*). A scattering center, such as an atomic nucleus, bends the trajectories of colliding particles (protons in our case), subsequent to an interaction. This happens if the incoming particles are within the shadow of the circle that defines the effective range of the force. In this example, the nucleus, in practice, does not move during the collision, given its much larger mass than that of the colliding proton.

of iron; given the sufficiently high energy of the beam, the protons will be able to "see" the constituent parts of the nucleus (remember de Broglie) and interact via the strong force with the protons and neutrons of which it is made. In the collision, other particles will be produced, arising from the energy to mass conversion. We may suppose, for example, that in the specific proton-proton collision, two protons and one neutral pion are produced: $p+p=p+p+\pi^0$; obviously, this is only one of the many possible reactions, with its specific value of the cross section quantifying the probability that it will occur. The flux Φ of projectile particles is defined as the number of impinging particles per unit surface area and time. It can be demonstrated (though we will not do so) that the number of events W, corresponding to a given reaction, that I will obtain in a certain interval of time (e.g., a second), is a function of the value of the cross section of the specific reaction, naturally normalized to the incoming flux: the more intense the beam, the larger the number of events generated per unit time: $W=\Phi\sigma$. If I double the number of incoming protons, I will consequently double the value of W. In the same way, the number of events that I will produce in the same time interval for a different reaction will be larger or smaller according to the corresponding value of the cross section σ. The latter may be calculated from the physical characteristics of the particular interaction that is involved. For every collision process, such as the reaction $p+p$ referred to previously, we will have to solve complex quantum mechanics equations in which operators appear that connect the wave function of the particles before and after the interaction has taken place. But fortunately, we are nevertheless in a position to predict and estimate some of the principal quantitative features of the reaction, even without performing complex calculations.

Arguments of a similar nature to those discussed here in relation to collisions also apply to the decay of unstable particles, once we have substituted for the concept of the cross section that of the mean lifetime of a particle. If this lifetime is short, decay is highly probable and will most likely be mediated by strong interaction. If instead, though unstable, the particle lives for an appreciably longer period, it will be a weak or electromagnetic interaction that will mediate the decay. In terms of numbers, a decay due to the strong force results in lifetimes in the order of merely 10^{-22}–10^{-24} seconds; in the case of the electromagnetic interaction, we are talking about 10^{-16}–10^{-21} seconds; and finally, for the weak force, the average decay times are extremely long (between 10^{-6} and 10^{-10} seconds). In the case of a free neutron, it even reaches about 900 seconds, which is virtually an eternity in this context! Remember the case of the muon, which decays in 2.2 microseconds? Well, we are obviously dealing there with a weak decay, given the almost biblical lifespan of the particle ... I invite the reader to notice that the relative difference between a lifetime of 10^{-20} and 10^{-10} seconds is enormous: it is in the same ratio as 1 hour compared to more than a million years!

The average life expectancy of a particle, the abovementioned mean lifetime τ, is one of the fundamental quantities that qualify and define it. By now, we know well that the mechanisms at the basis of the microcosm of elementary particles are determined by the aleatory nature of the physical processes, aptly described by the prescriptions of quantum mechanics. For a single, specific decay, we will never know with any degree of precision when it will occur. If we take as our starting point a large set of identical and unstable particles, each one will decay after a different amount of time—some sooner, some later. Our initial group will be reduced exponentially (figure 8.7), and in practice, all the particles will have decayed in the end.

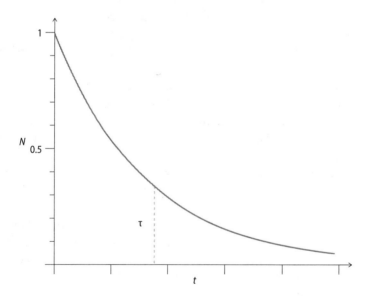

FIGURE 8.7

Let's take an initial set of identical and unstable N_0 particles (normalizing N_0 to 1); at a certain point in their lifetime, they will begin to decay, and their residual number N will be a function of the elapsed time. The decay follows an exponential function, determined by the fact that the number of particles that decay is proportional to the time we wait. Many particles will decay at the start, while a few will be relatively long-lived. The average value τ, by which the original sample is reduced to roughly a third, is defined as the mean lifetime of the particle in question.

However, we will be able to define the time after which the number of initial particles has been reduced to approximately one-third, and therefore obtain the mean lifetime of the particles in question. Connected to τ, and analogous to what has been discussed in chapter 4, we can also define the decay amplitude Γ (i.e., the particle decay probability per unit time). This quantity is inversely proportional to τ. In the very common case in which a particle can decay in a number of ways (e.g., in an electron-positron pair, in three pions, and so on), we will have different partial values Γ_i, such that their sum is equal to the total decay amplitude: $\Gamma_{tot} = \Sigma \Gamma_i$.

One last useful concept from the baggage of elementary particle physicists is that of resonance. Classically, we have resonances in mechanics and in electromagnetism. Given an oscillating system with its own frequency (e.g., a swing or an electric circuit with an alternating voltage), by applying to this system a force that is itself periodical, we can obtain a resonance when the frequency of this oscillation coincides with that belonging to the system, be it mechanical or electronic. In these conditions, the magnitude of the oscillation of the original system increases as a consequence, becoming able to diverge. There is a rather famous example of soldiers marching across a bridge with a particular pace (frequency), which may cause extremely vigorous oscillations of the structure, and may even end up destroying the bridge. Or consider the simple swing instead; if the person pushing it deploys his or her strength with the right frequency (that of the swing itself), the oscillations can diverge, at the peril of the person sitting in the swing ...

In particle physics, the mathematical description in terms of wave functions offers analogies with the classical formalism of resonances. The concept of resonance is particularly connected to the hadrons, to our particles composed of quarks and subject to the intense strong interaction. The unstable hadrons decay with extremely short lifetimes (10^{-24} s), which are so brief that even with a very high time dilation γ-factor (e.g., equal to 300), the space traveled by the particle before decaying does not exceed 10^{-15} cm—a distance definitely too small to be detected. Nevertheless, these extremely short-lived kinds of particles can be identified thanks to the mechanism of resonance, and this is what interests us. Let's look at a few examples. Let's assume we have a collision reaction between two hadrons, with the creation of three hadrons in the final state: $a+b \rightarrow c+d+e$. It should be noted that some of the five particles can also be the same: it is not necessary that they are of a different type. And let us further suppose that in addition to this process, there is another, concomitant one, by means of which an intermediate particle R is produced, which as soon as it is created

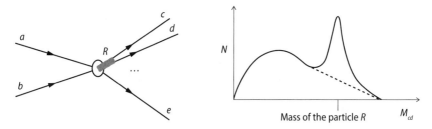

FIGURE 8.8
Diagram of the production of a hadronic resonance, identified by the intermediate particle R through the computation of the invariant mass of the two particles c and d.

decays through strong interaction in the two particles $c+d$ (as illustrated on the left in figure 8.8). We then have it that for some $a+b$ collisions, the process will be $a+b \rightarrow R+e \rightarrow c+d+e$. At this point, reprising the concepts of relativistic kinematics encountered in chapter 3, let's attempt to calculate the so-called invariant mass of a system of two particles, c and d in our case. This quantity coincides with the mass of a hypothetical intermediate particle that decays in $c+d$, just as for the intermediate state R. Starting from (1) we can demonstrate that:

$$M_{cd}c^2 = \sqrt{(E_c + E_d)^2 - (\mathbf{p}_c c + \mathbf{p}_d c)^2}.$$

This means that from the measurement of the energy and momentum of particles c and d, we can derive the mass of the parent particle R.

In the graph on the right of figure 8.8, we show the distribution of the invariant mass of particles c and d, obtained by calculating this quantity thanks to the relation $M_{cd}c^2 = \sqrt{(E_c + E_d)^2 - (\mathbf{p}_c c + \mathbf{p}_d c)^2}$ for many instances (events) of the collision between a and b. What happens? For those events in which the particle R has not been produced, we will obtain in practice any values of M_{cd}, that in the jargon are termed *combinatorial* because they are only determined by the accidental combination of the energies and momenta of the particles c and d. For the events where R is produced, however, the calculation of the invariant mass will give precisely the mass value of the particle R. Therefore, a resonance (a peak) will appear in the invariant mass distribution between c and d; for specific values of this quantity, the amplitude of the process increases just as the oscillation of the swing does, and some events are added to the underlying combinatorial distribution (the dotted line in the diagram), which are indicative of the creation of the particle R. Being produced through strong interaction, and decaying

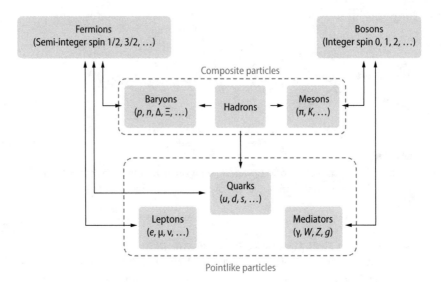

FIGURE 8.9
Classification of the elementary particles. The first two big families are those defined by the statistics: fermions and bosons. The former include both composite particles (baryons) and pointlike ones (quarks and leptons). The bosons include mesons and the mediators of forces. Baryons, mesons, and quarks are subject to the strong force. All the electrically charged particles feel the weak, as well as the electromagnetic force. The neutrinos only experience the weak interaction. In this diagram, we have not included gravity, although all the particles in the illustration are subject to it—including neutrinos, albeit to a minimal extent.

"strong" as well, the mean lifetime of R will be very short. Due to Heisenberg's uncertainty principle, $\Delta E \Delta t \geq h/4\pi$, the brevity of such a lifetime will imply a notable size in terms of energy (or mass) indetermination for the corresponding resonance. This is also obtained from the relation of inverse proportionality between τ and Γ_{tot}. In other words, the width of the peak in figure 8.8 increases as the lifetime of R shortens. Typically, we are talking of a few hundreds of MeV/c^2, corresponding to lifetimes of around 10^{-24} s, just as described in (3) in chapter 5. The identification of resonances allows us to identify (possibly new) particles with an extremely short lifetime, which otherwise would be impossible to detect. This method has proven to be very effective in the history of elementary particles, leading in some cases to remarkable discoveries.

Having exhausted these necessary premises, we can now turn our attention from the relationships between the various elementary particles to the interactions to which they are subject, with the objective of devising a coherent scheme for particles, mediators, and forces. Once more, wherever possible, we will take advantage of our historical path—above all, in order to underscore how the current interpretative scenario has not been given by nature all at once: rather, as you will have gathered by now, it has been extracted with difficulty by generations of physicists striving with ingenious experiments and complex theories. Figure 8.9 gives an overview that is useful for cataloging what has been discussed so far. Armed with this information, we are now ready to confront the aliens and the worlds hidden on the other side of the mirror ...

9 THE WORLD THROUGH THE LOOKING-GLASS

> The surest sign that intelligent life exists elsewhere in the universe is that it has never tried to contact us.
>
> —Bill Watterson

Toward the end of the 1950s, an important strand of research within the nascent physics of elementary particles caused a revolution with game-changing effects that had as its protagonist the neutrino of Wolfgang Pauli and Enrico Fermi: the proof of the violation of parity, constituting the collapse of one of the most solid myths of the physics of the time. This subject is intimately connected to a problem that may appear academic and of little real scientific relevance—though this is not the case—and that was brilliantly tackled by the American physicist Richard Feynman (figure 9.1) in his celebrated university lectures.

Let's imagine that we are communicating with a hypothetical friendly alien living on a distant planet, and let's further suppose, for simplicity's sake, that the communication can take place by radio, leaving aside the technical problems arising from the extremely long time that would intervene between transmission and reception of the messages. There are no televisions or transmission of images by cell phones. Given that our newfound friend is a scientist like ourselves, once we overcome the barriers to communication via the power and universality of scientific language, we will be in a position to explain the characteristics of our own worlds, evaluate our levels of cultural development, and talk about our respective art, politics, and society. At a certain point in our discussion, we would get around to how we are made, about our bodies and appearance: we humans with a head on our shoulders; they with their three heads, two arms, and four legs. We would quickly be able to explain and to understand that our heads are above, given that the concepts of *up* and *down* are easily grasped

FIGURE 9.1
Richard Feynman.

once we have established the direction of gravity on our respective planets. Our apples fall from up to down, as do the fluorescent fruits of their yellow-barked trees—we will understand the color from information as to the frequency of luminous radiation (which we will undoubtedly manage to interpret correctly). Continuing with the description of ourselves as biological organisms, we will have occasion to explain to our new friend that unlike his own two hearts, placed symmetrically halfway up his scale-covered body, our single heart is located in the left side of our chest.

"The left?"

"Yes, on the left …"

But how can we explain what *left* means, and the difference between *left* and *right*? In effect, we both will remain appalled and frustrated in the face of a problem that at first glance seems quite banal: how to distinguish right from left.

To simplify life for the reader, let's admit from the outset that, even hypothesizing complicated algorithms involving electric currents, magnetic or gravitational fields, ideal thought experiments (*Gedankenexperimenten*, in German, recalling those proposed, for example, by Albert Einstein to explain his special relativity), or logical-mathematical efforts, we have no way of giving an operational definition of right or left. Obviously, to say "right" or "left" is purely a matter of convention, but to say that the heart is located here or there in relation to a vertical direction is information that's anything but superfluous. Such an impossibility is the corollary of one of the principles considered to be so self-evident and acquired as to have virtually never been put into question until humanity had begun to concern itself with elementary particles.

The mathematical concept that lies at the basis of the impossibility of defining left and right is that of the conservation of parity. We are already familiar with the law pertaining to the conservation of energy, which applies to all natural processes, and therefore to the reactions that involve elementary particles. Given a reaction, the total energy (momentum and mass) of all the particles before the process must be the same (i.e., invariant) after the reaction has taken place. A similar rule, for example, is always verified for the total electric charge, as well. These conservation principles are, on the one hand, useful instruments for establishing if a reaction may or may not happen, while on the other hand, they are deeply connected to those laws of symmetry which mysteriously seem to govern the microcosm of particles.

Classically, the concept of parity is related to a spatial reflection, such as the one given by a mirror. An image in the mirror, in fact, appears to us to be reversed (the right appears left, and vice versa). More correctly, the application of a transformation of parity is what transposes the coordinates x, y, and z into $-x$, $-y$, and $-z$. Such a mechanism is much clearer at the quantum level, where each particle is represented by a wave function—a mathematical function of the variables x, y, z, and t that identify the position of the particle in spacetime. To apply the parity operation on a wave function technically implies the action of a quantum operator that transforms, as previously mentioned, x into $-x$, y into $-y$, and z into $-z$. The wave function that results from the operation may or may not still represent the same particle. We say that a particle has a positive parity (+1) when its wave function Ψ (x, y, z, t) remains identical after the operation, and negative (−1) if it becomes −Ψ. In all other cases, the particle (or the function Ψ that describes it) does not possess a definite parity. Let's consider the following example, not physically correct but useful to understanding the concept of parity. If the particle is described by the one-dimensional function

$\Psi(x) = cos(x)$, then the application of the operator of parity P would imply $Pcos(x) = cos(-x) = cos(x)$, and hence a positive parity for the function Ψ, as well as for the particle described by such a wave function. Instead, in the case of $\Psi(x) = sin(x)$, we will have $P sin(x) = sin(-x) = -sin(x)$, implying a negative parity for my particle. From what we just said, it is clear that the possible eigenvalues of the P (parity) operator are +1 and −1.

The parity values associated to the various particles contributed to their categorization from the outset. Furthermore, the fact that such quantities were conserved in all reactions (or so it was claimed …) made it possible to distinguish between processes that could potentially occur—and that therefore conserved parity—and others that could not (i.e., those that would violate its conservation). When Cecil Powell discovered the particle that he called "meson τ^+" in 1949, he observed that it decayed into three charged pions ($\pi^+ + \pi^+ + \pi^-$). Shortly afterward, however, another particle was discovered, the θ^+, with a mass and lifetime identical to that of the τ^+, and which decayed into two pions ($\pi^+ + \pi^0$). The parity of the two final states, composed by a system of three and two pions, respectively, were measured to be −1 and +1, and as such, had to be those of the two parent particles, τ^+ and θ^+. It was obvious that for the principle of conservation of parity, either the two apparently identical particles were in reality different from each other—or the parity conservation was violated in the decay of the θ^+/τ^+ particle. It is worth revealing immediately that θ^+ and τ^+ are really one and the same particle (i.e., the K^+ meson, or positive kaon), and that parity is not conserved in its decay, which proceeds by means of weak interaction and can happen therefore indifferently in two or three pions. Naturally, this thorny contradiction, aptly named at the time "the tau-theta puzzle," did not fail to trouble the sleep of many physicists, who were very reluctant to abandon a supposedly sacrosanct principle such as that of the conservation of parity. It was due to the patient and iconoclastic work of two brilliant young Chinese-American physicists, Tsung-Dao Lee—whom I had the great pleasure of knowing as a professor at the school of Subnuclear Physics in Erice in 1983—and Chen Ning Yang (figure 9.2) that the issue moved in 1956 toward a natural (but for all that, no less startling) conclusion. The two analyzed a large amount of experimental data, arriving at the incredible assertion that although it was evident that parity was solidly conserved in various processes of electromagnetic and strong interactions, the same could not be said for reactions produced by weak interactions. At that point, a decisive experiment became necessary in order to show incontrovertibly whether parity was indeed conserved or not by the weak interaction.

FIGURE 9.2
Tsung-Dao Lee (on the left) and Chen Ning Yang.

Lee and Yang were part of the Chinese scientific community in the United States, which included the brilliant Columbia University physicist Chien-Shiung Wu (figure 9.3), today known as Madame Wu. In discussion with her two theoretical colleagues, Madame Wu understood well enough the importance of carrying out a conclusive experiment. The idea was conceptually quite simple. It involved studying the classic β decay of a particular radioactive nucleus, cobalt-60, which is transformed into nickel-60 following the reaction $^{60}_{27}Co \rightarrow \, ^{60}_{28}Ni + e^- + \bar{\nu}$ (where $\bar{\nu}$ is an antineutrino) to search for a possible left-right asymmetry in the emission of the electrons. The 27 protons and the 33 neutrons that make up the cobalt nucleus are bound to each other by the strong force, and each of them settle in the "virtual pigeonholes" that are allowed by the Pauli exclusion principle, assuming the two possible spin values of +1/2 or −1/2. In the end, the total spin of the cobalt nucleus is comprised of 5 units. In Madame Wu's experimental apparatus, cobalt was arranged on a very thin metallic sheet so as not to absorb the β electrons once they were emitted. An intense magnetic field perpendicular to the plane of the sheet constituted the quantization

FIGURE 9.3
Chien-Shiung Wu (Madame Wu), discoverer of the violation of parity.

axis to align the spin of the cobalt nuclei participating in the experiment. To increase the degree of alignment (polarization) of the spins, Madame Wu was led to develop experimental conditions that were extreme for the period: intense magnetic fields, high vacuum, and, above all, extremely low temperatures of only a few millikelvins. Finally, thanks to the collaboration of various specialists, this complex experiment was successfully carried out.

If parity had not been violated in the weak decay of cobalt nuclei, Wu would have had to discover that the same number of electrons emitted on one side of the sheet were emitted on the other, making it impossible to distinguish right from left, and therefore verifying that parity was also conserved in weak interactions. On the contrary, however, the experiment clearly showed that the electrons tended to be emitted preferentially in the opposite direction to that of the polarization of the spins of the nuclei, aligned with the axis of the magnetic field. The experiment was concluded successfully on December 27, 1956—a date that speaks admirably of how Madame Wu and her colleagues spent their Christmas vacation that year ...

Obviously, as always happens in science, independent verifications were required. Leon Lederman decided to conduct another of the experiments proposed by Lee and Yang, involving the decay of pions and muons, the former produced thanks to the powerful cyclotron at Columbia University. The results came almost immediately, confirming those obtained by Madame Wu. On January 5, 1957, their university organized a press conference to announce the discovery—to my knowledge, the first time that such a means of communication was used to make known to the public a discovery in physics. Lee and Yang won a Nobel Prize—unlike Madame Wu, who in my opinion more than deserved one for her crucial, fundamental experiment. We could say much about what was almost certainly an example of gender discrimination, given that a similar situation arose with the first Nobel awarded to Maria Skłodowska Curie, who was at risk of seeing it given only to her husband, Pierre. But that's another story—one that unfortunately continues to repeat itself in the world of science, just as in society at large.

But let's return to the consequences of the cobalt-60 experiment. It became immediately clear that the results of the experiment pointed toward the neutrino—which had recently been discovered by Frederick Reines and Clyde Cowan—as the culprit behind the violation of parity. The explanation for this result is linked to the concept of the helicity of a particle. As we have seen, we can imagine that every elementary particle, with a great deal of simplification, revolves around one of its own axes of symmetry, assuming therefore an intrinsic (quantized) angular momentum, the spin. The rotation may be clockwise or counterclockwise, and is described by the spin component, the direction of which may be antiparallel or parallel relative to that of the momentum (figure 9.4). In the first case, we speak of a particle with left-handed helicity (or LH), and in the second of right-handed helicity (or RH). Now, the helicity of the particle (i.e., its left-handed or right-handed nature) may be inverted by a Lorentz transformation. In simple terms, let's imagine an example in which we are pursuing an LH particle and overtake it by slightly increasing our speed; its direction of motion, and hence the direction of the momentum vector p, will change by 180 degrees in relation to us, while obviously the sense of rotation (and hence the direction of its spin) will remain unaltered. The result of this transformation will be that the particle changes helicity. The situation is completely different for a particle such as the neutrino, which in the first approximation may well be assumed to be without mass. In this hypothesis, the particle will always travel at the speed of light in whatever reference frame, and for us, it will be impossible to overtake it, to make it change the direction of p, and therefore its helicity. In conclusion, we can affirm that all particles endowed with mass possess potentially both states of helicity (polarization) LH and RH (at the same time until a quantization axis is defined), while neutrinos

FIGURE 9.4
The definition of the helicity of neutrinos and antineutrinos. The neutrinos have the momentum and spin vectors oriented in opposite directions, unlike antineutrinos.

do not: neutrinos will always be LH and antineutrinos RH. As we have previously mentioned, we now know that neutrinos have an extremely small but nonetheless nonzero mass, and that as a result, there is a negligible probability that they exhibit the supposedly wrong helicity. In practice, however, this fact does not substantially alter our conclusions.

Wu's experiment demonstrated with certainty that the blocked helicity of the neutrinos made definite a priori the emission direction and helicity of the electrons, with a consequent parity violation. This is shown diagrammatically in figure 9.5. When the cobalt nucleus decays, its 5 units of spin are distributed between the nickel nucleus (4 units) and the electron-antineutrino pair (1 unit), with all the spins oriented in the same direction: $5 = 4 + 1$. We may then suppose that both the nucleus at the outset and the one on arrival are basically at rest, given their large mass, and hence have zero momentum. This must then be the same for the electron-antineutrino pair, in order to conserve the null value at the point of departure. This fact implies in turn that the electron and the antineutrino must be emitted in opposite directions, as shown in the illustration. From this consideration, the solution to the problem is derived. The electron is emitted in the direction opposite to that of the magnetic field, which is then the direction of the initial spin—the sum of the two spins of the electron and of the neutrino. If the electron had been produced toward the opposite direction, the antineutrino would have been forced to have LH helicity to conserve spin and momentum, but we know that antineutrinos are only RH, and therefore this cannot happen. Faced with the incompatibility of different conservation laws, nature has made its choice by privileging the conservation of energy and momentum, at the expense of parity. Figure 9.6 illustrates a corollary to this conclusion. If we imagine that we are "photographing" our LH neutrino (or RH antineutrino), together with its image reflected in a mirror, by looking at the two photographs (which at first glance appear

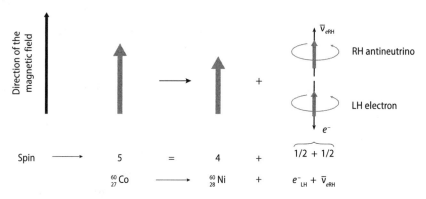

FIGURE 9.5
Description of Madame Wu's experiment on parity violation.

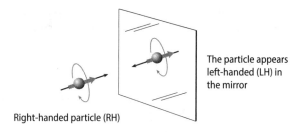

FIGURE 9.6
The mirror image of an existing right-handed antineutrino corresponds to a left-handed antineutrino that (in practice) does not exist in nature.

identical), we could affirm which one is the mirror image—obviously, the one that shows an RH neutrino or an LH antineutrino, both in essence nonexistent in nature. The world in the mirror is different (and distinguishable) from the real one, at least as far as weak interactions are concerned. It's an incredible result that caused shock among physicists. "I can hardly believe," Pauli joked, "that God is weakly left-handed."

So what about our extraterrestrial? Nothing could be easier. We simply have to describe to him Madame Wu's experiment, or an equivalent one involving neutrinos. With a little patience, thanks to parity violation in weak processes, we will be able to explain on which side our heart is located. In the general discussion about right and left, we will also have helped him to understand that when two Earthlings meet, they are accustomed to shake hands with their right hands, perfectly identifiable now in his eyes. The only problem, though—hardly an insignificant one—is that on the occasion of a notional meeting between us on a neutral planet, if our friend was to come toward us raising his *left* hand, we would do well to run away at full tilt: either he has made a mistake in executing Wu's experiment or—worse—his planet and his body are made up of antimatter. The result of our encounter would be a catastrophic reciprocal annihilation into photons. Luckily, nature has provided us with the means of solving this new problem. The solution is to be found in another violation: the violation of *CP* (in other words, of the product of the parity operation *P* for the "inversion of the electric charge" operation *C*). *CP* is also violated in nature, albeit to a much lesser degree—approximately one part over 1,000 in some weak reactions that involve quarks, a level that is nevertheless sufficient to design a hypothetical experiment to definitively resolve the left-right problem, even in the presence of the matter-antimatter ambiguity. Incidentally, it is thanks to *CP* violation that our universe is not simply a boring combination of photons originating from the original annihilation between the particles of matter and antimatter present in the same (and conspicuous) numbers immediately after the Big Bang. In effect, a small asymmetry between particles and antiparticles was likely caused by *CP* violation at the beginning of the life of the universe, and this has caused it to be the case that after the huge initial annihilation, a "small" amount of excess matter remained: the matter that today constitutes all the galaxies, stars, and planets that make up the current cosmos … including humankind. This is surely enough to make us eternally grateful to *CP* violation. And yet, not even the *CP* violation that has just been described, and which happens for quarks, appears to be quantitatively sufficient to justify the matter-antimatter asymmetry in the newborn universe. We need to look at another *CP* violation source, one that once again comes down to neutrinos … we will do that soon.

10 THE MOST BEAUTIFUL THEORY IN THE WORLD

> Antiworlds. They spend their time mostly looking forward to the past.
>
> —John Osborne

Everything that surrounds us, from the elementary particles to the stars and planets, is part of an enormous universe in which we are situated, probably generated in a random manner, by an immense, silent, and dark explosion produced by a hugely energetic, hot, dense, and (at least at the beginning) homogeneous system, the Big Bang. It is evident that the explosion was silent, given that there was certainly no elastic medium capable of transmitting any resultant sound waves (as the air on our planet), and above all, that it was dark because—as we shall see later—photons took a considerable time to decouple from the initial superheated matter and transmit their light. Our universe is so improbable in its specific cocktail of physical constants as to make us suppose that, in keeping with the anthropic principle, it was one among the innumerable possible universes that won the lottery and came into being, and us with it. As we have seen, chance, fluctuations, and probabilities are the bases of the laws that govern the microcosm—and ultimately the whole of nature, in all its complexity. A particularly intriguing aspect is the fact that the study of nature over time has revealed a deep link between the probabilistic laws that governed the birth of the universe and its subsequent evolution, and the extension of the concept of beauty, described in the language of mathematics and expressed in the principles of symmetry.

The English mathematician Godfrey Hardy said:

> The mathematician's patterns, like the painter's or the poet's, must be beautiful; the ideas, like the colors or the words, must fit together in a harmonious way. Beauty is the first test: there is no permanent place in this world for ugly mathematics.

In the same vein, Hermann Weyl affirmed:

> My work has always tried to unite the true with the beautiful, and when I had to choose one or the other, I usually chose the beautiful.

And, finally, Paul Adrien Maurice Dirac:

> The laws of physics must possess mathematical beauty.

It was Dirac again, who when speaking of Einstein's general relativity, said provocatively:

> Whoever appreciates the fundamental harmony that exists between the world in which Nature operates and some general mathematical principles cannot but feel that a theory of such beauty and elegance must be substantially correct ... independently of whether it corresponds to experimental data.

And then he added:

> The mathematician is engaged in a game in which he alone writes the rules, while the physicist plays by the rules provided by Nature. But with the passage of time, it becomes ever more evident that the rules that a mathematician finds interesting are the same ones chosen by Nature.

Giving myself license to add to this captivating line of thinking, I would say that on the whole, *there is no place in the world for ugly physics*. Sometimes, however, nature has dissented, rejecting symmetry and selecting the ugly (the breaking or violation of a symmetry) in exchange for existence. Moreover, we must not forget that the scientific method teaches us that the experimental measurement is sovereign, even when faced with an exquisitely beautiful theory. Every theory must be motivated by the necessity of interpreting the experimental evidence, and perhaps to predict phenomena not yet observed. The data that emerge from the observations are the words with which nature speaks to the scientist, and we do not have the right not to listen to them because they might seem discordant in relation to our way of thinking at the time. As I say to my students: *the ugliest experimental data are more important than the most beautiful theory*. An ugly bit of data, for example, is one that shows disagreement with a theory that possesses mathematical beauty, or that, in keeping with the principles of plausibility, we would not have expected. Nevertheless, physicists have been quick to appropriate the concepts of symmetry and beauty—initially attributes of art and of the contemplation of nature, and hence necessarily qualitative and subjective—developing laws, tangible factors, numbers, correlations, and principles of necessity. The word *symmetry* derives from the Greek συνμέτρω (*syn metro*)—that is, with measure. In Latin, this becomes *cum mensura*, which leads to *commensurable*, a concept that is operative in both mathematics and physics. *Harmony* derives from the Greek ἁρμόζω (*harmozo*), to connect, once again, a concept that is common to the cognitive and operative processes of science. The distance between such definitions and the methodologies of the scientist is short indeed. This etymological coincidence or connection leads us to the assertion of a very important principle: beauty—that is, for a physicist, *symmetry*—is a fundamental element of the laws that describe nature. Here's Weyl again:

> A thing is symmetrical if we can act on it in such a way that after our action, the thing appears exactly as it was before.

Therefore, a glass enjoys a rotational symmetry around an axis if, revolving it around such an axis, it appears the same to us as before the transformation (a rotation, in this case). Extending this concept to encompass the laws of physics, let's take the example of Galilean spatial translation. If I measure the electric charge of the electron with my experimental apparatus here in Berne, at the Physics Institute of the university where I work, I will obtain the same result as in any other place in the world, so long as I keep the experimental conditions under close, consistent control. This is a concept that develops and formalizes that which we commonly understand. The beauty of the universe as a whole, understood and observed well beyond the limits of our senses, appears to the scientist to be founded on rules of symmetry—such as the bilateral symmetry of mammals or the pentaradial symmetry of echinoderms, or the harmony of geometric or architectural structures, the golden section, etc., for instance—and on the interpretative models that follow from these: the physics theories. These, in turn, carry these very same symmetries, or others that are hidden or manifest.

In more technical terms, a symmetry reflects the existence of a non-observable quantity. The absence of an absolute origin of space generates the aforementioned invariance of spatial translation, which in turn implies the conservation of momentum in classical mechanics. The absence of an absolute origin for time instead generates invariance for time translation, which in turn implies the conservation of energy. In classical mechanics, consequently, a given symmetry causes invariance in relation to a certain class of operations. The invariance principles imply conservation laws of physical quantities, as the brilliant German mathematician Emmy Noether had already proved in 1918. Symmetry, then, as a source of objective and quantitative beauty? The physicist Eugene Wigner said, profoundly, that "symmetries are the laws that the laws of Nature must observe." The rules of mathematical symmetry bring discipline to the physics laws. The movement of the planets, for example, is governed by Newtonian mechanics, but the specific $1/r^2$ dependence of the force does not derive from the law of gravity, but rather from the symmetries of the three-dimensional space in which we find ourselves. This obviously all applies until a more extensive or general theory will clarify the interpretative scenario in the light of new experimental facts that are not explained by the less complete theory. Once again, an example is offered by Newton's mechanics, which is incorporated into Einstein's gravitation, with both based on symmetries of varying complexity.

All very interesting, you say, but how is this connected to the world of elementary particles? The connection exists, and it is a deep one. From our study of particles, we have understood that even the laws that regulate their interactions and their physical properties descend from recondite concepts of symmetry. Furthermore, the whole of particle physics is permeated by these concepts. Particles are described by quantum states represented by mathematical functions that also respect particular rules of mathematical symmetry. The relation between symmetries and conservation principles in classical mechanics extends to quantum mechanics, and to the elementary particles that obey its laws. To these are added other specific symmetries of the microcosm. This discussion is particularly relevant when we study the beginning of everything that occurred with the Big Bang. From its birth, the universe creates, by expanding, the time and the space (actually the space-time) in which it develops. The first particles to fill it copiously, created by energy-matter conversion, begin to interact among themselves. The physical laws that the universe obeys while expanding beyond measure are mathematically beautiful and respond to profound symmetry principles. In the earliest moments in the life of the universe, it exhibits an extremely high degree of symmetry. Afterward—and by *after*, I mean a derisory fraction of time subsequent to the Big Bang, when the density of energy substantially diminished—something happens … Let's recall the example in the previous chapter relating to parity violation. Nature has chosen to contravene a rule of symmetry, and hence a source of mathematical and physical beauty, in exchange for a world that is certainly more extravagant, but also more robust in evolutionary terms—and perhaps for an equilibrium and a harmony, therefore, that are founded on ugliness? Have we reached the end of our praise for symmetry? Not at all. The difference between a theory that works compared to one that does not describe anything often resides in mathematical beauty and its degree of symmetry. It seems, however, that to obtain real-life situations, with unpredictable developments such as the existence of particles endowed with mass, the creation of self-conscious life, and a universe that is more or less stable for billions of years, the breaking of the symmetries on which "real" theories are based is a necessary and functional element. Nevertheless, even after the breaking or the violation and the "disfigurement," the vestiges of original beauty remain evident; or even if hidden, are still present to guide us in our study of nature.

Let's turn to one of these beautiful theories—actually the most beautiful of all physics theories—and let's see how it is capable of elegantly describing the reality of our elementary particles. We are talking about Quantum Electrodynamics, the celebrated QED, which is precisely based on mathematical symmetry rules. The model was developed at the beginning of the 1950s with independent contributions by various theoretical physicists: the pioneering work of the Swiss Ernst Stückelberg, the important contribution of the Anglo-American Freeman Dyson, and the theoretical construction of the Japanese and American trio who shared the Nobel Prize for their work in 1965: Julian Schwinger, Sin-Itiro Tomonaga, and Richard Feynman (figure 10.1).

Before starting to talk about QED, I urgently need to give a warning to the reader ... to unfold their arms. The following pages, all the way to the end of this chapter, are densely packed with arguments that require more than the usual level of concentration and engagement. I apologize for this, but I am convinced that some of the themes with which we are dealing require a certain amount of study. The idea that forms the basis of QED is simple, and in some respects, we have already anticipated it. The theory of the interaction between particles through the concept of fields of force is classical, just as in the case of gravitation and of electromagnetism. But what happens if we reformulate the latter interaction to include special relativity and quantum mechanics, with its energy quanta and the wave

FIGURE 10.1
Feynman, Schwinger, and Tomonaga, winners of the Nobel Prize in 1965 for the formulation of quantum electrodynamics (QED).

functions that describe the particles? We develop QED, a theory that introduces the quantization of the electromagnetic field and that replaces the concept of action at a distance with that of the exchange of virtual photons, the mediating bosons of the interaction between electrically charged fermions (see figure 8.9 in chapter 8). Now it becomes necessary to introduce a formalism originally proposed by Feynman, which has been named after him: *Feynman diagrams* or *Feynman graphs*—a visual solution, and not just, for describing the interaction of quantum particles, which will allow us to understand quite intuitively the relevant processes of QED and the other theories that derive from it. Let's start with a case that we have already encountered in chapter 8: the electromagnetic interaction, and in particular the Coulomb scattering, of an electron by the electric field generated by the charge of the protons of a nucleus, as illustrated in figure 10.2.

The two axes in the illustration represent time (*x*-axis) and space (*y*-axis) respectively, in contrast to figure 8.5, where both axes describe spatial coordinates. An atomic nucleus comprising of Z protons with a total electric charge equal to Ze interacts electromagnetically with an electron coming from a certain direction, by means of the exchange of a *virtual* photon through which a force is applied, just as previously discussed. Both the electron and the nucleus travel in time (positive direction of the *x*-axis). The electron also moves in space and reaches the appointment with the virtual photon moving toward the bottom of the *y*-axis to be subsequently deflected backward, rebounding toward the top half of the graph. This means

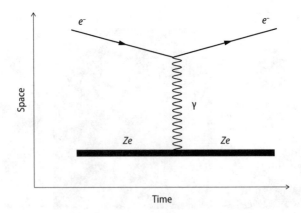

FIGURE 10.2
Representation of the electromagnetic interaction between an electron and an atomic nucleus, schematized through a Feynman diagram.

that after the interaction, the electron will move away from the nucleus. The latter does not move (to any substantial degree), always maintaining the same spatial coordinates because its mass is much greater than that of the electron—it's a bit like an elephant being hit by a ping-pong ball ... The wavy line represents the photon, the quantum of the electromagnetic interaction. The fact that the electromagnetic interaction between a negatively charged electron and a positive nucleus is attractive—contrary to what appears in the graph—is irrelevant; what matters is that after the photon exchange, the direction of motion of the electron changes.

To summarize, Feynman's diagrams describe the interaction of particles through the exchange of bosons, mediators of the forces. The graphic rules of the graphs are illustrated in figure 10.3, which also features two diagrams (g and h) describing two specific electromagnetic interactions—a useful introduction to the various conventions that will be used in the following discussion. The continuous segments identify the elementary fermions that are interacting (quarks or leptons), which in turn are described by appropriate wave functions. The point of encounter and contact between the two fermions is called a *vertex*. The particles that link two vertices are virtual. For these, as we have already said, the kinematic obligations of the real particles (e.g., respecting the relativistic invariant $E^2 = p^2c^2 + m^2c^4$), do not apply, only for a very short time and in a way that is compatible with Heisenberg's uncertainty principle. As shown in figure 10.3g, the electron, positron, muon, and antimuon are all real particles, whereas the photon is virtual. That said, the quantum numbers that describe the reaction, such as the electric charge, must be conserved at each vertex. The arrows on the fermion lines indicate their direction of propagation along the time axis (abscissa) and along that of space (ordinate). The rule is the opposite in the case of antiparticles. The electron comes from above and from before. The positron comes from below, and it too comes from before (well said!). The arrow of the positron, however, indicates that it is mathematically and physically equivalent to an electron coming from above and from afterward (i.e., from the future—remember the solutions of Dirac's equation?). But let me repeat: it is merely a convenient computational tool and has nothing to do with science fiction—agreed? The same argument applies for the two exiting muons, coming from the materialization (conversion) of the virtual photon. The negative one (the particle) has an outgoing arrow, while its positive partner (the antimuon) has the opposite direction, even though it is also an outgoing particle.

A fundamental feature of Feynman diagrams, which is necessary for interpreting the formulas of the interactions and for carrying out the

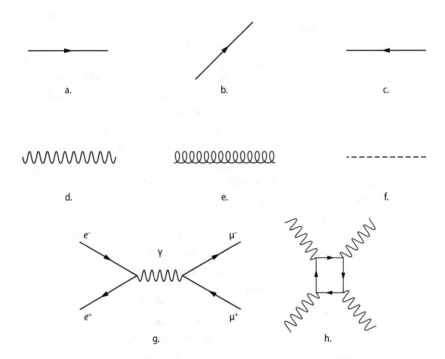

FIGURE 10.3
The graphic conventions of Feynman diagrams: (a) a fermion that propagates in time, but not in space (therefore stationary); (b) a fermion that propagates forward in time and space; (c) a fermion that comes from the future, equivalent to an antifermion that goes forward in time; (d) representation of the photon, the mediator of the electromagnetic force; (e) the gluon, the quantum of strong interaction; and (f) W or Z, mediators of the weak force. At the bottom, on the left is an example of a Feynman diagram that describes the electromagnetic annihilation of an electron and a positron into a virtual photon that then converts into a pair of positive and negative real muons. On the right we show a so-called box-diagram, where two real photons interact with a virtual fermion loop—constituted by any electrically charged fermion—giving rise to another pair of real photons.

respective calculations, is what happens at the vertices, where the fermions come into contact with the mediating bosons. Formally, the interaction takes place at that point, whether it is weak or strong or electromagnetic. With the vertices, we introduce a multiplicative constant proportional to the intensity with which the mediator couples with the fermion (called, precisely, the *coupling constant*). Its value is higher for stronger interactions (e.g., electromagnetic versus weak) and the probability for the process to occur increases, as well. These constants are specific to the physical reaction in question. Consider, for example, the case of electromagnetism. By now, we know that two charged fermions (electron and positron in figure 10.3g) are subject to a reciprocal force by virtue of the fact that they have an electric charge equal to $+1e$ or $-1e$, where e, as I have had occasion to repeat, has an absolute value equal to 1.6×10^{-19} coulombs, the elementary charge of the electron, of the proton, and of the other particles, except quarks that have a fractional charge of $+2/3e$ or $-1/3e$. The electromagnetic coupling constant, therefore, is proportional to e. From e, we can derive another constant—one of the most important in nature—that plays a key role in myriad physical processes: the fine-structure constant α. This constant is constructed by combining e with other fundamental quantities, the dielectric constant in a vacuum, the Planck constant, and the speed of light. The constant α is dimensionless, and therefore equal to a pure number:

$$\alpha = \frac{e^2}{2\epsilon_0 hc} \cong \frac{1}{137}.$$

The numerical value of α is so important that all physicists remember it. For my part, I use 137 for the combination lock on my suitcase, counting on the fact that any potential thief is unlikely to be a physicist—although after the publication of this book, I should probably come up with a different combination …

Moving back to the diagram in figure 10.3, we add a multiplicative factor $\sqrt{\alpha}$ (hence proportional to e) that takes account of how intense the coupling is. In an entirely similar way, the other interactions also possess their coupling constants. Then we will have that of the strong interaction $\sqrt{\alpha_s}$ and that of the weak interaction g. The reader will recall that the strong force is about 100 times more intense than the electromagnetic one for typical nuclear distances. It is reasonable to expect then that α_s is approximately equal to 1, compared to the 1/137 of its electromagnetic sibling. This is shown concisely in figure 10.4, where we have three examples of Feynman diagrams that describe three specific reactions: electromagnetic, strong, and weak. In the figure, each of the two vertices in a diagram is independent from the other. In other words, process (a) consists of the emission of a photon at the top vertex and the absorption of the same in the lower vertex. The two processes are independent, and each one happens with a probability proportional to $\sqrt{\alpha}$. Hence, the combined probability is proportional to $\sqrt{\alpha} \times \sqrt{\alpha} = \alpha = 1/137$; on this subject, the reader may recall that if the probability of extracting the number 27 or 41 in Lotto is 1/90, the a priori probability of extracting them together, at the same

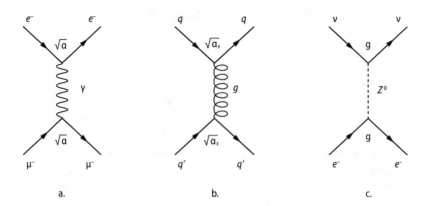

FIGURE 10.4
(a) Electromagnetic interaction between an electron and a muon; the coupling constant is $\sqrt{\alpha}$ for each of the two vertices; (b) strong interaction between two quarks that scatter elastically (in this case, the coupling is defined by $\sqrt{\alpha_s}$); (c) weak reaction relative to the scattering of a neutrino off an electron, through the exchange of a Z^0 and with a coupling constant g. Care should be taken not to confound the *g* that stands for the gluon with the weak coupling constant g!

time, is $1/90 \times 1/90 = 1/8{,}100$. Similarly, the overall probability of the (b) reaction will be proportional to $\sqrt{\alpha_s} \times \sqrt{\alpha_s} = \alpha_s$. For low-energy processes, $\sqrt{\alpha_s}$ is roughly equal to 1, making the probability of the reaction in fact equal to $\cong 1$ again—and therefore extremely high!

Another term included in Feynman's diagrams is the *propagator*, which is constituted by lines that join vertices—virtual particles, either bosons or fermions. To simplify, we can say that the propagator accounts for the probability that, as in the case of figure 10.4a, the photon emitted in one of the two vertices is propagated and reaches the second one. In this case, the propagator, the photon, introduces into the calculations a term equal to $1/q^2$, where q is the momentum transferred by the photon from the first to the second vertex. This implies that the probability of the first reaction in figure 10.4 is globally proportional to $\sqrt{\alpha} \times 1/q^2 \times \sqrt{\alpha}$. It is worth noting that in the process of figure 10.4c, the propagator has a slightly more complex expression due to the large mass of the Z. Therefore, the probability that the physical process in question will take place also depends on the energy at which the interaction (or collision) occurs between the two particles. Finally, it is important to remark that Feynman diagrams represent probability amplitudes. To pass to the total probability, and consequently to the cross section of the specific process at stake, we usually resort to what is called *Fermi's golden rule*, which allows for describing both collisions and decays:

$$W = \frac{4\pi^2}{h} |M_{if}|^2 F. \tag{5}$$

$|M_{if}|$ is the so-called matrix element, which describes to the quantum transition between the initial state of the system i and the final one f, and numerically coincides with the amplitude of the related Feynman diagram. F is a weight that quantifies the availability of final states for the reaction in question and typically takes into account the kinematic constraints of the reaction. The decay of a particle in another two very heavy ones has a weight F that is less than that of the decay into lighter particles for the same type of interaction described by the matrix element. In the second case, in fact, each of the daughter particles has a greater possibility of exploiting the mass of the parent particle to gain momentum and therefore have a more ample spectrum of possible final states. Equation (5) can be combined with the formula $W = \Phi\sigma$ to correlate the cross section of a given reaction with the probability amplitude:

$$\sigma = \frac{W}{\Phi} = \frac{4\pi^2}{\Phi h} |M_{if}|^2 F.$$

The consequence of all this is that for a given process described by a Feynman diagram, the cross section is proportional to the square of the product of the coupling constants multiplied by the propagator. So the final probability (cross section) of the first reaction in figure 10.4 will turn out to be proportional to $(\sqrt{\alpha} \times 1/q^2 \times \sqrt{\alpha})^2 = \alpha^2/q^4$. And this prediction, obtained in a relatively simple manner, is in perfect agreement with the experimental data. Unfortunately, however, the complete calculation of a Feynman diagram is much more complex than what we have attempted to show previously in a qualitative manner, and to which we will do in the following section with other examples. We should not forget that the "fermion lines" are only a graphic illustration that hides the real wave functions of the particles involved in the reactions. Calculating collision cross sections and decay times implies solving complex mathematical integrals and filling pages with calculations. I hope that the reader may at least have grasped the principle, as well as to estimate that the probability of a certain process is proportional to α^2 and not, let's say, to α^3—which already seems pretty gratifying, don't you think? The complete and rigorous calculation of the numerical value of the cross section of a given process, including the aforementioned factor F, will lead to dependence on the same kinematical

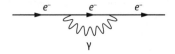

FIGURE 10.5
Example of the self-interaction of the electron with its own electric field. A virtual photon is emitted and rapidly reabsorbed, in accordance with Heisenberg's uncertainty principle.

quantities (such as the mass of the particles involved, the energy of the collision, the values of their spins, etc.).

At this point, we can return to QED to develop together some of its salient aspects. The first is the concept of self-interaction, a strange, purely quantum phenomenon. Our electron in figure 10.2 is not only able to interact with the electric field generated by the nucleus, but also with that produced by itself, from whence comes the term *self-interaction*, as illustrated in Feynman's diagram in figure 10.5. As has already been discussed, in every moment, the electron is able to emit a virtual photon of energy ΔE, which is arbitrarily high, so long as its borrowing is paid in full by the photon in a time Δt that is correspondingly small, in keeping with Heisenberg's uncertainty principle: $\Delta E \Delta t \geq h/4\pi$. The time interval Δt must be so small as not to be detectable: $\Delta t \leq h/4\pi\Delta E$. The poor photon's aspiration to exist is realized only for a single instant, which gets briefer the more energetic its ambition ... But as the saying goes, to lend a finger can lead to losing a whole hand. Because everything is virtual, and everything is possible under Heisenberg's watchful eye, even the reactions in figure 10.6 can quietly occur! Diagram (c) in figure 10.6 is particularly instructive in this respect. The virtual photon here can even transform itself—using jargon, it can convert—into an electron-positron pair (be aware, as ever, that the total electric charge is always to be conserved), which in turn is annihilated, creating a new virtual photon; in the end, this is reabsorbed by the real electron. All of this can happen quietly because of the "Supreme Laws of Feynman diagrams": the electron-positron pair is also virtual, since its particle lines connect two vertices. But then, given that our imagination has no limits, we can look with different eyes at the simple image of figure 10.2 and redraw it as in figure 10.7. It's certainly more complex but formally unexceptionable. Now I am sure that the readers will ask themselves what really happens in reality, in nature: is it the scenario depicted by figure 10.2, or the much more imaginative version offered by figure 10.7, or is it even more complex? The answer is simple. If something *can* happen in nature, then it happens; it is only a question of understanding and calculating with what level of probability.

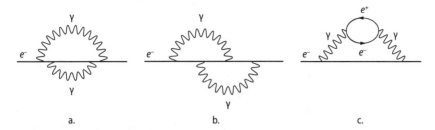

FIGURE 10.6
There is nothing to prevent arbitrarily complex examples of self-interaction occurring around an electron. One or more photons can be emitted simultaneously [processes (a) and (b)]. Before being reabsorbed, a virtual photon can transform into an electron-positron pair (c) that immediately annihilates itself into another virtual photon, which in turn is absorbed by the electron.

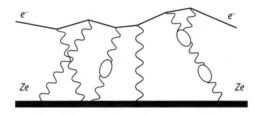

FIGURE 10.7
Coulomb interaction between an electron and an atomic nucleus, interpreted on the assumption that in effect, an infinite series of virtual processes can take place.

Let's try then to estimate the cross section of the process in figure 10.5 and of reaction (c) in figure 10.6. On the basis of what we have learned so far, setting aside the propagator, the probability amplitude of the first graph is proportional to $\sqrt{\alpha} \times \sqrt{\alpha} = \alpha$. There are two vertices, and in each one of them is a factor $\sqrt{\alpha}$. Instead, in the second case, we have four vertices, each with a factor $\sqrt{\alpha}$. It follows that the probability that such a reaction takes place is proportional to $\sqrt{\alpha} \times \sqrt{\alpha} \times \sqrt{\alpha} \times \sqrt{\alpha} = \alpha^2$. If we specify the value of α equal to the inverse of the code for my suitcase, 1/137, the first reaction will have a probability amplitude that is approximately 137 times greater than the second. Therefore, all the possible diagrams happen, but with different degrees of probability. To calculate the real cross section of the electron-nucleus diffusion (figure 10.7), we will have to take into account all of the diagrams that may potentially contribute to a given process: an infinite number. Luckily, as seen in our example, their relative weight diminishes rapidly as the number of virtual particles increases. Hence, as we say in the jargon, we can limit ourselves to calculating the lower-order diagrams—let's say that the ones in which there is a maximum of four vertices or maybe eight. Then we will only have the task of confronting our calculation with the experimental measurement, the numerical value of which is provided directly by nature: the latter, without any difficulty at all, will have in effect instantly calculated all the infinite possible diagrams. The game has clear rules: if we consider many higher-order diagrams, we get progressively closer to the experimental value and can verify how well the QED explains the manifestation of nature.

The first message to be gleaned from this discussion is that to speak of a particle, of an electron, as a well-defined object with absolute properties, is to speak of an illusion. In addition to the devastating effects of quantum mechanics, which prevent us from even predicting exactly the position of the particle before making it the subject of measurement, and the uncertainty principle that underscores how, in the act of a measurement itself, the kinematical variables are irreparably disturbed, there is a cloud of virtual particles that our electron carries with it in its quantum errancy. The simple hydrogen atom is not as simple as we have always imagined it to be. Due to higher-order diagrams, there is a small but measurable probability that instead of an electron orbiting around a proton, we may have two electrons and a positron (figure 10.6c). Let me repeat the point: it is true that virtual particles can have a derisory lifespan, but the diagrams that describe them contribute to the estimate of the total cross section (or, if you prefer, of the overall probability) in a real and quantitative manner.

There are many corrections that we must bring to the basic predictions of quantum mechanics in the calculation of the energy levels of atoms. QED is not just an elegant theory—it is also a powerful one. An illuminating example is provided by the comparison between the experimental value and the QED prediction for the magnetic moment of the electron. We have seen that with all the fermions (and in particular the electrons)—by virtue of their intrinsic spin angular momentum—can be imagined (albeit simplistically) revolving around a quantization axis. The latter is defined in the act of the measurement when we force the quantum state to collapse into one of the two eigenstates (for a spin 1/2 particle) with spin up or down, depending on the two possible directions along the axis. The electric charge of the electron "in rapid (quantized) rotation on itself"—this is inexact, but it gives the right idea—creates a magnetic dipole moment, a little elementary magnet associated with the electron. The value of such a magnetic moment, the Bohr magneton, is a fundamental constant of physics, and as often happens, it can be derived by combining some of nature's other constants:

$$\mu_B = \frac{eh}{4\pi m_e}.$$

To tell the truth, we have here almost all the fundamental constants: the elementary charge, the Planck constant, and the electron mass. Taking advantage of the arguments used to discuss the self-interaction of the electron with its electric field, we can see how it also interacts with its own magnetic one. In figure 10.8a, this process is described by its fundamental diagram, while the graph in figure 10.8b is of the next order. Regarding the first, there are two supplementary $\sqrt{\alpha}$ factors that bring a correction to the probability amplitude equal to $\sqrt{\alpha} \times \sqrt{\alpha} = \alpha$. If we were to add an order of diagrams (figure 10.8c) we would arrive at an additional correction of the order of $\sqrt{\alpha} \times \sqrt{\alpha} \times \sqrt{\alpha} \times \sqrt{\alpha} = \alpha^2$. It is implicit in this discussion, but it should be reasonably evident, that the number of diagrams to be considered (and to be computed) increases significantly with the order that we are considering ... poor unfortunate physics students! Once again, due to the smallness of α, the corrections contributing to the base process are always small—a fraction of them adding, the other subtracting. The surprising thing is that the experimental value of the electron magnetic moment is measured today with extreme precision and accuracy, and departs from that which corresponds to the fundamental diagram (equivalent to the

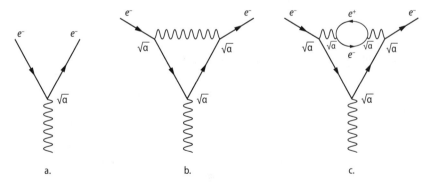

FIGURE 10.8

(a) Feynman diagram describing the basic value of the magnetic moment of the electron; (b) diagram corresponding to the principal correction; (c) an example of higher-order correction, of little probability, given the large number of (smaller than 1) coupling constants.

Bohr magneton) by a factor of around 1 over 1,000. Even more incredibly, such a factor is estimated by calculating a very large number of high-order Feynman diagrams (10th order graphs), and it coincides with the measured one within one part per billion! This impressive accuracy fully demonstrates the predictive power of QED, which for me merits the title of the most beautiful theory in the world. I hope that Einstein, Heisenberg, Dirac, and all of the others will forgive my choice. It was an arbitrary one—though when properly considered, perhaps, not so arbitrary after all ...

But the surprises that lay in wait for us with quantum electrodynamics do not end there. Imagine the Feynman diagram describing an electron that is stationary in space and that propagates only in time. The electron can be represented by a horizontal line, as seen in figure 10.9a. However, the continuous processes of emission and reabsorption of virtual particles complicate the ideal scheme, and our electron will actually be much more similar to what is shown in figure 10.9b. Now, the infinite virtual electron-positron pairs that appear from nothing and return to nothing have nonetheless, at least on average, enough time to polarize themselves.

Let me explain. The positrons will be attracted by the negative charge of the real electron, while the virtual electrons will tend to distance themselves, as illustrated in figure 10.9c. This implies in turn that if I observe (measure) the electron from a certain distance—remember: the shorter the distance, the higher the energy of the probe—it will appear to be screened by the cloud of virtual negative electrons. The larger the distance from which I examine the electron, the stronger the screening effect, given that the number of electron-positrons dipoles that I observe grows with the distance. At the limit, the screen practically ceases to grow, and our experiment with low energy (large distance) will measure values of the electric charge and of α that are equal, respectively, to the canonical 1.6×10^{-19} coulombs and to $1/137$.

Things change, and not a little, if we pass from the laboratories of atomic physics, or in any case from low energy measurements, to experiments with high-energy elementary particles. Let's imagine that we are causing our electron to collide with another one that has been previously accelerated at high energy. The latter will come very close to the former, in keeping with the relation $\lambda_{elect.} = h/p_{elect.}$. At this point, the screening effect diminishes because in the vicinity of the target electron, there are relatively fewer virtual pairs (dipoles) present, and the electric charge that is "perceived," and with it α, appreciably increases, actually diverging for very close inspections. To provide some parameters, α grows from the asymptotic value at a low energy of $1/137$ to around $1/128$ for an electronic projectile of

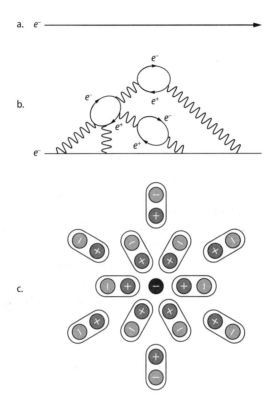

FIGURE 10.9
(a) Feynman diagram of a stationary "ideal" electron; (b) diagram of a real electron surrounded by its cloud of virtual particles; (c) polarization effect of the virtual electron-positron pairs and of the corresponding screening of the bare electron charge.

approximately 100 GeV. Needless to say, all this is predicted perfectly by QED—another proof of its power, if one were needed ...

In an equivalent manner, virtual photons can transform themselves in particle-antiparticle pairs of any type of charged fermions: muons, *up* quarks or *down* quarks etc., so long as Heisenberg's uncertainty principle that correlates time and energy is respected. As a result, the contribution of muon-antimuon pairs ($\mu\bar{\mu}$) becomes influential only at high energy due to their mass, which is larger than that of the electron-positron pairs. This means that the effects of higher order graphs depend on the energy or, equivalently, on the distance at which we probe the original electron, creating in the end the so-called running of the coupling constant. These considerations lead us to formulate a question of principle: What is the "real" value of the electron charge? Is it that which is measured in the classical laboratory experiments, of low energy ($\alpha=1/137$), where in effect we take into account all possible diagrams of every order that nature calculates for us? Or is it the charge that we measure, ever growing and completely divergent, in high-energy experiments? The question is far from trivial—and so, too, is the answer. Every beautiful theory such as QED hates the "infinites," detests the denominators that have become extremely small and the numerators that balloon out of proportion. The issue is complex. Suffice to say that, thanks to the physical-mathematical mechanism of renormalization, the nonmeasurable "real" charge is replaced in the calculations by the one corresponding to the full screening effect that is obtained in low-energy measurements. And as if by miracle, there are no more infinites and the theory provides accurate and reliable predictions. Renormalizability is a very important property of both QED and of the so-called gauge theories. Besides, as in classical mechanics and electrostatics, the potential may always be renormalized by manually attributing to it an arbitrary value. What always matters, in the end, are the potential differences. Later on, we will come back to talking about constants that are not constant, just like α and the charge of the electron, when describing the theory of strong interaction, another gauge theory (with a similar name to QED that I will not unveil here so as not to spoil the surprise).

But now is the time to move on to some typical processes in quantum electrodynamics for predicting the corresponding cross sections—at least considering the fundamental Feynman diagrams, those of lowest order—that is, with fewest vertices. We have verified that the cross section of Coulomb diffusion between two electrons is proportional to $(\sqrt{\alpha} \times 1/q^2 \times \sqrt{\alpha})^2$ $=\alpha^2/q^4$, if we limit ourselves to the lowest-order diagram. We will use this process as a reference and will compare it to the other results. In passing, it is worth noting here that the $1/q^4$ dependence indicates a cross section that diminishes very rapidly with the increase of the transferred momentum, or (if you prefer) of the energy of the interaction. It is rather as if, with the increase in energy, our electrons could hardly make head-on collisions. Reprising the topics explored in chapter 8, the emission of a real photon by a free electron is impossible because in the reaction, momentum and energy would not be simultaneously conserved. But the process does occur if the electron is not free; that is, if it is subject to a force and hence accelerated, as in the case of synchrotron radiation. A similar case of a radiation due to the emission of photons is commonly known among physicists as

Bremsstrahlung (i.e., braking or deceleration) radiation. Braking? No problem. By *acceleration*, we mean a change in the velocity with time. This implies that electrons and positrons can radiate photons both if they are energized in a particle accelerator or slowed down by passing through a piece of metal. On the other hand, we talk of acceleration even if only the direction of the velocity vector changes, and not its magnitude, as in the case of a particle orbiting inside a circular accelerator, with the modulus of the velocity being constant but with a trajectory continuously changed by the action of magnetic fields.

The Feynman diagram for *Bremsstrahlung* is shown in figure 10.10. Put simply, let's imagine that our free electron, in passing through a piece of material (e.g., a metal plate), "decides" regardless to emit a photon. We observe that both particles (electron and photon) are real, given that the lines that describe them are free to extend themselves in space and time at one of the extremities. As soon as the photon is emitted, however, the electron becomes virtual. Of course, we know that virtual particles can do anything, or almost anything, so long as their eccentricity does not exceed the limits set by Heisenberg. No sooner said than done—the nucleus of an atom of the medium characterized by an electric charge Ze and a consequent coupling $Z\sqrt{\alpha}$ supplies to the virtual electron a photon, also virtual. *Voilà*: the electron returns to being real and continues calmly on its way. At the end, the momentum of the (real) emitted photon will be balanced by the loss of momentum of the incoming (real) electron. In this reaction, the atomic nucleus is called the *spectator*, perhaps to indicate its role in registering the momentary virtual journey of the incident electron. In reality, all the events described here occur virtually at the same time; ours is only a stratagem for describing the features of the process. In the end, the only entities that we observe experimentally are the electron, before and after the emission, and the (real) radiated photon. At this point, therefore, once the relevant diagram of the reaction—of the first order—has been identified, we calculate the cross section behavior. Setting aside the propagator term introduced by the momentarily virtual electron, it is proportional to $(\sqrt{\alpha} \times \sqrt{\alpha} \times Z\sqrt{\alpha})^2 = Z^2\alpha^3$. From this, we learn two things: first, the process is intrinsically rarer than the electron-proton diffusion that we have used as a measure (α^3 is much smaller than α^2); and second, photon radiation becomes more probable as the atomic number Z of the medium in which the electron is traveling increases (even with the square of Z); that is, the total charge of the spectator nucleus. For this reason, to facilitate experimentally the production of *Bremsstrahlung* photons, for example, it is customary to make a high-energy electron beam pass through an appropriate thickness of lead characterized by $Z=82$.

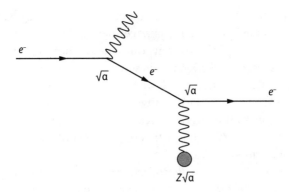

FIGURE 10.10
Feynman diagram of the *Bremsstrahlung* process, namely the radiation of a photon on the part of an electron that interacts with the electric field generated by an atomic nucleus.

Another interesting process explained by QED is the production of an electron-positron pair by a real photon. The reaction is illustrated in figure 10.10, observed in two inertial reference frames. The first (figure 10.10a) is the one in which the photon arrives at full speed from the left in the laboratory; after which it has no choice but to travel at speed c ... At a certain point, the photon transforms (converts) itself into the electron-positron pair. The two fermions also move in the same x direction to conserve the momentum of the photon along that axis. Momentum is also conserved along the y-axis: it is equal to zero before the reaction, and equal to zero after it, because the electron goes upward and the positron goes downward (or vice versa) with momentum components that are equal and opposite: $p_y(e^-) = -p_y(e^+)$. Good—the reaction is apparently correct. The total initial electric charge (the photon) is zero, and it is thus afterward as well (plus+minus=zero). The photon spin, equal to 1, is also converted into the two 1/2 spins of the pair elements. In this way, the conservation of energy *seems* to be happening for the best: the photon momentum is transformed into the mass of the electron and the positron, as well as into their individual momenta. This implies that the minimum momentum required for the photon to produce the pair must be equal to nearly 1 MeV/c, at least corresponding to the sum of the electron and positron masses (~ 0.5 MeV/$c^2 \times 2$) with the two particles being at rest, in practice without momentum.

There is a problem, however, and once again, it is quite a serious one. Due to Lorentz invariance, the results of the reaction must be the same in the second reference frame shown in figure 10.10b, in which the two electrons are

The Most Beautiful Theory in the World 181

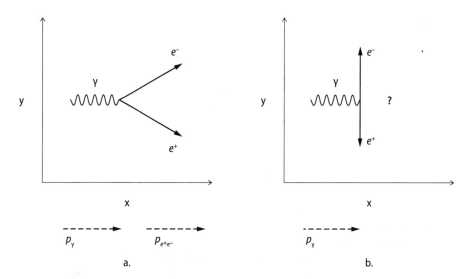

FIGURE 10.11
(a) Scheme of the conversion of a photon into an electron-positron pair; (b) the same reaction seen in the reference frame in which the electron and the positron have no momentum along the x-axis. It is evident that the total momentum of the reaction is not conserved because whatever the reference frame, the photon always travels at the speed of light.

back-to-back along the x-direction, with null total momentum. Because the photon always goes at the speed of light in any reference frame, it will still possess a nonvanishing momentum. This is the opposite of the case where a massive particle—hence traveling at a speed lower than c—would decay into an electron-positron pair. In this reference frame, therefore, one cannot conserve simultaneously energy and momentum ... What is missing is a particle to balance the photon momentum: another photon, perhaps, emitted by an accommodating spectator?

This is an instructive example because it shows that if one of the two great conservation principles is not satisfied—specifically that pertaining to the conservation of the momentum—even a reaction that would satisfy the other required conditions cannot take place. In this and in other similar cases, nature, having an infinite number of choices at its disposal, allows the process to happen, but technically only if we can have another particle involved, which allows the conservation principles—which would otherwise be violated—to be maintained. Paris is worth a mass! Remember the spectator in the *Bremsstrahlung* reaction? That's what we need now. A corollary, a photon that propagates freely in empty (or low-density) space will

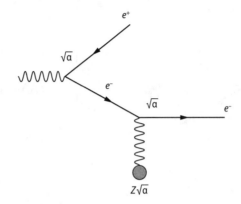

FIGURE 10.12
Feynman diagram of the production of an electron-positron pair from the conversion of a photon. The process is possible thanks to the presence of a spectator atomic nucleus that supplies a virtual photon, which in turn makes it possible to conserve the momentum in the reaction.

never be able to achieve the conversion, just as a similarly free (nonaccelerated) electron is not permitted to radiate a photon. The Feynman diagram that explains the process of pair production is shown in figure 10.12. The arguments are similar to those used to discuss photon radiation. A real photon emits a real positron and a virtual electron propagates between two vertices. The electron becomes real when an extra momentum is provided by the spectator and the total momentum is thereby conserved, as is required.

There is an alternative but wholly equivalent way of reading this diagram: a real electron coming from the future—equivalent to our positron that goes forward into the future—emits a real photon; in the interaction, it becomes virtual and begins to go forward in time; in the end, thanks to the contribution of the spectator that recoils in the opposite direction to that of the virtual photon, it becomes real again and continues on its way. Not bad—but also, as the arrows show, everything appears due to the propagation of a single electron ... back and forth in spacetime, partly real, partly virtual ... To be precise, according to this interpretation, the diagram is analogous to the one in figure 10.10, except for some rotation of the spacetime plane. Consequently, it is justified to expect the cross section of this process to have the same dependence of the *Bremsstrahlung;* that is, to be proportional to $(\sqrt{\alpha} \times \sqrt{\alpha} \times Z\sqrt{\alpha})^2 = Z^2 \alpha^3$. This is the power of Feynman's diagrams, and of QED.

The third process that I would like to describe unquestionably has had a major impact on and been of great interest to the media, if only in the context of science fiction movies and novels, not least *Angels and Demons,* as mentioned earlier in this book. In this film, one gram of antimatter is stolen from CERN and is made to explode above the Vatican. It's a rather special explosion. When matter and antimatter are brought into contact with each other, there is a certain (high) probability that the respective constituent particles will undergo a process described by quantum electrodynamics, known as *electromagnetic annihilation.* Unfortunately—or fortunately—there is quite a distance between actual science and science fiction. At the start of our account, we calculated that to completely convert 1 g of matter into energy through $E=mc^2$ would produce a total energy of around 10^{14} joules, corresponding in fact to that of an atomic bomb. With CERN's current capabilities, to create such a mass of antimatter with the potential of being annihilated by as much matter—hence being able to generate 2×10^{14} joules—, we would need to run the accelerators constantly for a period of about ... a few billion years—not a very practical prospect. We are, therefore, quite a way off from using both the peaceful applications (alas) and the bellicose ones (fortunately). While we await further developments, let's confront the problem as physicists by studying the elementary process of electromagnetic annihilation through the formalism of QED.

The basic annihilation reaction involves the fusion of an electron and a positron into two photons. As is evident from the diagram in figure 10.13a, the requirement of two photons rather than one is once again dictated by the principle of conservation of momentum. If there were only one exiting photon, there would be an imbalance along both the collision axis and the transverse one. This would imply the nonconservation of the total momentum, which was null to start with and must remain so after the annihilation. In figure 10.13b, the corresponding lowest-order Feynman diagram is also shown. Above all, we can assume that electrons and positrons are virtually at rest, or very nearly so. We will then see that as the energy of the collision increases and each of the members of the pair also has very high momentum, other processes will occur that are more probable than the simple electromagnetic annihilation. The mass of each of the colliding particles is equal to 0.511 MeV/c^2: this means that each of the two photons will gather a momentum of 0.511 MeV/c. The diagram in figure 10.13b should be read in the following way. The electron emits a real photon and consequently becomes virtual; the positron does the same. The central line, which corresponds to a virtual fermion, can therefore belong indifferently

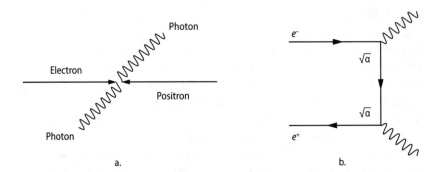

FIGURE 10.13
(a) scheme of the electromagnetic annihilation between an electron and a positron that produces two photons; (b) the first-order Feynman graph of the reaction.

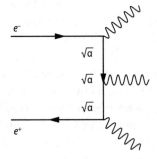

FIGURE 10.14
Rare electron-positron annihilation into three photons.

(simultaneously) to the electron and to the positron, which reciprocally transfer their whole mass there in order to pay the debt incurred by the momentum of the photons. The cross section of the annihilation process is proportional to $(\sqrt{\alpha} \times \sqrt{\alpha})^2 = \alpha^2$. In the previous discussion, as always, we have considered the fundamental diagram of the lowest order that is the most probable one and that contributes most to the total cross section. But it is clear that the illustration in figure 10.14 is certainly possible, so long as a further factor $(\sqrt{\alpha})^2$ is introduced when calculating the cross section. This means that the reaction with three photons should occur in no more than 1 percent (approximately equal to 1/137) of the cases.

Now let's consider the collision of high-energy electrons and positrons. Let's imagine that our particles have been energized in an accumulating ring, such as up to a few gigaelectronvolts (GeV). This is a very different

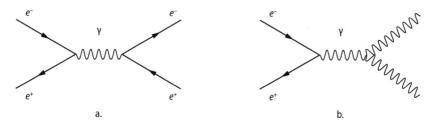

FIGURE 10.15
(a) high-energy electron-positron annihilation into another electron-positron pair, through the intermediate production of a virtual photon; (b) the process shown here cannot happen: the photons couple only to particles that are electrically charged, while here, the three-photon vertex only connects neutral particles.

condition from the one described before, in which the annihilation takes place between particles that are substantially at rest or nearly so. In that case, electrons and positrons sometimes manage to constitute a very special atom, the *positronium*, formed by the two particles orbiting around each other, before colliding and transforming into energy. The lifetime of positronium can be quite long on the particle scale: even up to a hundred nanoseconds. If the collision is of high energy, therefore, the two particles have a considerable momentum in relation to their mass. In these conditions, the process described by the Feynman diagram in figure 10.15a may occur. In this diagram, we have something that could not happen in the low-energy process described in figure 10.13; that is, the production of only one photon rather than two. The single photon on this occasion is virtual, and propagates the momentum of the colliding particles until it is reconverted into another electron-positron pair. The violation of the momentum conservation discussed previously does indeed take place, but usually for the fleeting life span of the virtual photon, while everything is settled after the conversion of the latter into the new pair. Note that the final state can be an electron-positron or a muon-antimuon pair, and so on, so long as we have enough energy from the incoming particles to transfer to those exiting and create (at least) their mass. Obviously, the virtual photon is not able to convert into a pair of photons, as shown in figure 10.15b. The reason for this is that only charged particles can couple with photons with an intensity of $\sqrt{\alpha}$, while in figure 10.15b, the virtual photon is coupling with two photons that are real, but obviously neutral, and the electromagnetic interaction cannot take place.

The diagram in figure 10.15a, therefore, describes the annihilation of two particles (fermions) and the consequent creation of another two. The mass and the momentum of the latter are created from the energy balance

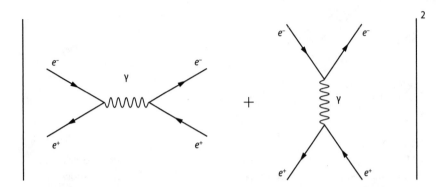

FIGURE 10.16
The two lower-order Feynman diagrams that describe electron-positron annihilation at high energy. The graphs represent the two amplitudes that must be added quadratically to obtain the cross section. This is a rule that must be applied every time the initial and final states are the same for different diagrams.

of the first two; that is, from their mass opportunely added to the momentum. We observe that the same process can happen through another diagram of the same order, and should therefore be considered together with the one in figure 10.15a, because it is equally possible. Both are shown in figure 10.16. Dealing with two diagrams that generate the same result and are of the same order, to arrive at a numerical value of the cross section of the reaction $e^+ + e^- \to e^+ + e^-$, we have to add the two amplitudes and square it. From this, a term emerges that describes the quantum interference between two processes that are equally possible: the *double product*. This should not surprise us, given that our particles are described by complex (i.e., including imaginary numbers) wave functions. It is precisely like the propagation of light waves, or those that ripple the surface of water, that overlap and combine, occasionally canceling or reinforcing each other; the probabilities of various processes exhibit this property, too. But be careful—this special sum of diagrams is justified only in that they have exactly the same initial and final state, and consequently, we will never be able to know which of the two takes place in each of the collisions. Given that they are indistinguishable, they must both be taken into account in the mathematical computation of the cross section. Needless to say, the result of the calculation perfectly agrees with the experimental data.

Before concluding this rather complex chapter—medicine is sometimes bitter—we consider that the electromagnetic electron-positron annihilation can produce various final states, but in exchange for the total available

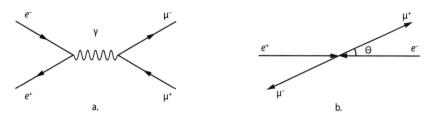

FIGURE 10.17
(a) The Feynman diagram representing an electron–positron annihilation into a muon–antimuon pair; (b) the collision as schematically observed in the laboratory frame.

energy in the collision being sufficient to at least create the masses of the two final particles, and that the process is permitted by the various other conservation principles. The diagram in figure 10.17 shows the electromagnetic production of two muons, one positive and one negative. Given that the mass of each of the two particles is equal to 106 MeV/c^2, the energy of the two colliding beams must be at least a little more than 106 MeV, including the mass of about 0.5 MeV/c^2 of the electron and of the positron. In the figure, we have the Feynman diagram on the left, and the interaction scheme represented in the laboratory reference frame on the right. After the head-on collision, the two muons are emitted one against the other, at an angle θ with respect to the direction from which the electron and positron originated. The value of the angle is obviously not always the same, nor is it predictable for each collision, which responds to the aleatory laws of quantum mechanics. Nevertheless, the statistical distribution of the values of the θ angle is perfectly predicted by QED. As we shall soon see, in the collision between electrons and positrons, quark-antiquark pairs may be electromagnetically produced (e.g., instead of a pair of muons). In this case, we must take more care in making calculations, although the principal characteristics of the process are once again perfectly interpreted by quantum electrodynamics. In conclusion, what can you say: it really is a beautiful theory!

11 THE COLOR OF ELEMENTARY PARTICLES

Every morning, I go out into the streets, looking for colors.

—Cesare Pavese

At this point, let's take a leap back in time to 1964, when Murray Gell-Mann and George Zweig were postulating the existence of quarks and aces, pointlike entities of fractional electric charge—a mathematical tool more than anything else, capable of explaining the complex variety of hadrons that were gradually being discovered. A few years later, the experiments taking place in California in the Stanford Linear Accelerator Center (SLAC) would provide the quark with the required passport to enter with full official identity and rights into the world of elementary particles. But not everything was running so smoothly. *Fractional* charge? No one had ever observed it. And besides, *why did the quarks not reveal themselves in the experiments as "free" particles?* In effect, only the hadrons, definitely composed of quarks, had manifested in nature. And there was something even more disturbing ... Remember the multiplets that were quickly constructed to catalog the hadrons on the basis of their quark content and other quantum numbers? Let's take the decuplet shown in figure 7.4 in chapter 7, and in particular, one of the particles that appear at the vertices of the triangle, the Δ^{++}. Being made of three identical u quarks, each with an electric charge $+2/3$, this has a total charge equal to $+2$. Given that its total spin is equal to $3/2$, the individual spins of the three quarks will also be identical—that is, pointing in the same direction, ↑↑↑ or ↓↓↓. In the end, given that the orbital angular momentum is null (the Δ^{++} only possesses that of spin), the particle finds itself at the fundamental energy level, as a result of which the three quarks will occupy with identical probability the same space, even the same specific point. In chapter 4, we illustrated the salient aspects of quantum mechanics. The reader will recall the so-called Fermi-Dirac statistics which half-integer spin fermions obey, and the Bose-Einstein statistics that describe bosons of integer spin. An important implication is provided by the Pauli exclusion principle, which demands that the total wave function of a statistical system of identical fermions must be antisymmetric for the exchange of any pair of such fermions. Let's try to understand the consequences of this principle. The Pauli principle requires that the overall quantum state that describes a set of fermions—let's take the particular case of our Δ^{++}, made up of three u quarks—must be represented by a total wave function that is antisymmetric after the exchange of each one of the pairs of three quarks. In other words, if the first u quark is described by the function Ψ_1, the second by Ψ_2, and the third by Ψ_3, then the total state of the particles (i.e., that which corresponds to the whole Δ^{++}), will be constituted by $\Psi = \Psi_1 \Psi_2 \Psi_3$. By imposing the condition of antisymmetricity for the exchange, we have ($\Psi_1 \Psi_2 \Psi_3$), which *must be* equal to $-(\Psi_2 \Psi_1 \Psi_3)$, and so on, for every permutation of quarks within the Δ^{++}. In the light of what has been said thus far, however, this requirement is not satisfied in our case. The three quarks are completely indistinguishable, and exchanging a pair must necessarily return us to the same total wave function: $\Psi_{before} = \Psi_{after}$, with a symmetric wave function, not an antisymmetric one. If you prefer, we can rewrite the Ψ in an equivalent way, in terms of its three components: space, flavor, and spin:

$$\Psi(\Delta^{++}) = \Psi(\Delta^{++})_{space} \times \Psi(\Delta^{++})_{flavor} \times \Psi(\Delta^{++})_{spin}.$$

But then, what can we do to distinguish in some way between the three *identical* quarks, to render the total Ψ function antisymmetric for any of the arbitrary quark permutations and save the exclusion principle for a fermion such as our Δ^{++}? As you will have gathered by now, the main method is to add a new quantum number, a new quantity that has a different value for each of the three quarks, and thereby to resolve the conundrum. Think of a family consisting of a mother, a father, and their completely identical triplets. How can we manage to distinguish between them? The answer is given by Huey, Dewey, and Louie: giving them a red, a blue, and a green cap, respectively, elegantly solves the problem. The three boys are now distinguishable, and their exchange of places at the dinner table no longer goes undetected by their parents, consequently changing their overall wave function and making it antisymmetric.

The need was already identified in 1964—by Oscar Greenberg—to add in some way a new quantum number. These were truly special times for the physics of elementary particles. Working independently of each other, various theoretical physicists began to improve the original quark model, with the aim of fixing the evident problems, and developing a real dynamica theory capable not only of explaining their geometric combination, but also their interactions. It was the dawn of Quantum Chromo Dynamics (QCD), the strong sibling of Quantum Electrodynamics (QED), another gauge theory based on transformation groups and symmetries (see chapter 10). I did not use the example of the colored caps by accident. Toward the end of the 1960s, Gell-Mann called the new quantum number "color." This has nothing to do with the colors we are familiar with, but no matter: there is no connection either between our taste buds and the "flavor" of quarks. And yet in reality, there is a kind of link to conventional color ... The solution to the problem of the Δ^{++} is the following:

$$\Psi(\Delta^{++}) = \Psi(\Delta^{++})_{space} \times \Psi(\Delta^{++})_{flavor} \times \Psi(\Delta^{++})_{spin} \times \Psi(\Delta^{++})_{color}.$$

The antisymmetric nature of the fourth term, which is guaranteed by the existence of three different colors, also assures compliance with the Pauli principle. The three quarks will be formally distinguishable: one will have the red *color charge*, another the blue, and the third the green. So many fundamental contributions to the construction of the theoretical framework of QCD came thick and fast—from David Gross, Frank Wilczek, David Politzer, Steven Weinberg, Moo-Young Han, Yoichiro Nambu, William Bardeen, Gerard 't Hooft, Harald Fritzsch, and my colleague in Bern, Heinrich Leutwyler, to mention only the major contributors, with many Nobels among them. The scenario rapidly became clearer, and the previously mentioned Rutherford-type experiments of the early 1970s contributed decisively to enhancing the credibility of the nascent theory.

We have already observed in chapter 8 that the intensity of the strong force is due to the high value of its coupling constant. In the case of the decay of unstable particles, their lifetime is significantly different—for strong and electromagnetic interactions: it is in the order of 10^{-23} s for the first, and 10^{-19} s for the second. As you will remember, the lifetime of a particle is inversely proportional to the decay amplitude Γ, which may be measured experimentally by studying its decay products. Now, remembering Fermi's golden rule, we have the following:

$$\Gamma = \frac{h}{2\pi\tau} = \frac{hW}{2\pi} = 2\pi |M_{if}|^2 F.$$

Given that the expression M_{if} contains the coupling constants of the specific interaction that connects the initial state (i) with the final one (f), the expected lifetime τ of an electromagnetic or strong process will be inversely proportional to the electromagnetic coupling constant $\sqrt{\alpha}$ and to the strong $\sqrt{\alpha_s}$, squared, respectively. To make explicit the relation between the two constants, we derive the following formula:

$$\frac{\alpha_s}{\alpha} \sim \sqrt{\frac{10^{-19}}{10^{-23}}} = 100 \to \alpha_s \sim 1.$$

This is a result that we had anticipated, together with the fact that the mediator of the strong interaction is the gluon (see figure 10.4b in chapter 10). QCD gives precise prescriptions as to how the strong interaction is to be mediated by the gluons. In the first place, we have introduced the color quantum number to safeguard the Pauli exclusion principle. While the single electric charge responsible for the electromagnetic interaction may assume values of $+1e$ and $-1e$ (and $-1/3e$ and $+2/3e$ for the quarks), there are three color charges that are the generators of the strong force, dubbed the red, blue, and green. We could easily have called them *a, b,* and *c,* but as I have already hinted, the decision to use three fundamental colors makes sense in physics: each of the three color charges may assume two values that we shall call, continuing the conceit, color and anticolor (red-antired, blue-antiblue, green-antigreen). To be more precise, we should have defined the charges as "plus red" and "minus red," "plus blue" and "minus blue," and "plus green" and "minus green," in the same way that we speak of positive and negative electric charges. But using colors and anticolors is certainly illustrative.

At this point, each quark inside a hadron can carry a given red, blue, or green color, while the antiquarks exhibit an anticolor: antired, antiblue, and antigreen. Given that, as we shall shortly see, the color charge is in some way confined to the interior of the hadrons, the hadrons themselves seem devoid of it; that is, to continue the analogy with the classical theory of colors, they turn out to be white. Let's take a proton. Each of the three quarks inside it will have one of the three colors and these, combining, will produce a completely white proton (i.e., with a null strong color charge). We could ask ourselves: "If the proton and the neutron are white and have no net color charge, how do they manage to keep together to form an atomic nucleus?" The atom, too, is electrically neutral, however, and yet the atoms attach themselves to each other electromagnetically to form molecules. The apparent enigma can be solved, in this latter case at least, with the so-called interatomic forces between different atoms, and the intermolecular ones between different molecules (these also being neutral). The former make it possible to construct molecules composed of atoms, while the latter enable molecules to unify among themselves in a more or less weak way to form liquids or solids. As shown schematically in figure 11.1, when two atoms

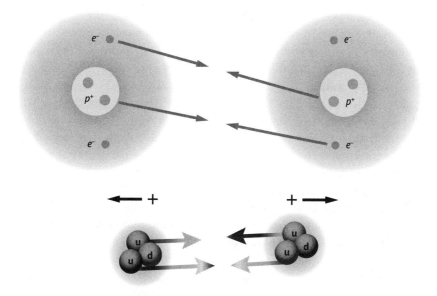

FIGURE 11.1
The interatomic electromagnetic (above) and strong (below) forces that bond electrically neutral atoms to form molecules, and "white" protons, without net color charge, respectively, to constitute the atomic nucleus. The overall strong force between quarks of different protons can turn out to be attractive and more intense than the repulsive counterpart due to their identical electric charge.

are sufficiently close, the protons of one attract the electrons of the other, and vice versa. In a similar way, the respective electrons repel each other, as do the protons of the two nuclei. Under certain specific conditions determined by the distribution of the electric charges of the atoms involved, the resulting force can be attractive, even though the atoms may have a null electric charge. This happens if the energy level corresponding to the bond is lower than that relating to the two separate atoms: complex systems in nature always tend toward a configuration of minimum energy. There is nothing strange, then, about the fact that such a mechanism can occur for two white protons that are neutral for the strong force (figure 11.1). In this case too, the strong interaction that acts between the quarks of two protons that are close together can generate a bond by means of an attraction—once again corresponding to a minimum of the energy of the system—that is more intense than the repulsive force due to their electric charge. The white color may also be obtained by combining a color with its respective anticolor or, if you prefer, combining a positive color with the corresponding negative color. In this way, any meson composed of a quark and an antiquark will also be white because its quark will be a particular color (let's say blue), and its antiquark necessarily would be antiblue.

A peculiar feature of QCD is that, in contrast to QED, its mediating quantum—the electrically neutral gluon—itself has a color charge. The photon has no electric charge, while each gluon simultaneously carries a color and an anticolor. In a similar way to the electric field generated by the electric charge, in QCD we have the color field generated by the color charges. This is illustrated graphically in figure 11.2, which shows the corresponding Feynman diagrams for the strong interaction between one blue quark and one red. The diagram on the left describes the interaction generically: a blue quark and a red one exchange a gluon that must carry the two red and antiblue colors, or blue and antired, as shown in the two diagrams on the right. As it is in QED, so it is in QCD: the charge must be conserved at every vertex of the Feynman diagrams. The result of the interaction is that the blue quark becomes red and the red one becomes blue. Naturally, this process can take place for the quarks that constitute the hadrons. Through this constant exchange of gluons and the corresponding change of suit on the part of the quarks, the system turns out to be bound and stable, at least compared to other phenomena that could take place.

As I've said, QCD is constructed starting from very general symmetry considerations, based on the mathematical group theory. The symmetry that generates QCD is said to be exact, in the sense that the strong force is independent of the specific color charge of the quarks. The topic, once again, is too complex for our present purposes: nevertheless, one can intuit that, having at our disposal three charges (and three anticharges) in order to obtain the independent gluons that can connect colored quarks, we need to

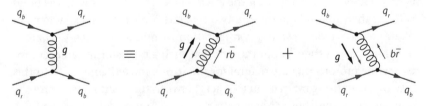

FIGURE 11.2
Feynman diagrams describing the strong interaction between a red and a blue quark. The mediating gluon can equally carry the two red-antiblue or blue-antired color charges. In the second graph, the gluon is emitted by the red quark; and in the third one, it is emitted by the blue one.

take into account the possible permutations of the three fundamental colors. In the end, it can be shown that eight independent gluons are required to guarantee all the possible interactions (with letters referring to their colors):

$$r\bar{g} \quad r\bar{b} \quad g\bar{r} \quad g\bar{b} \quad b\bar{r} \quad b\bar{g} \quad \frac{1}{\sqrt{2}}(r\bar{r} - g\bar{g}) \quad \frac{1}{\sqrt{6}}(r\bar{r} + g\bar{g} - 2b\bar{b}).$$

The first six gluons have a color and an anticolor with a different charge. The seventh and the eighth instead have a combination of colors and anticolors of the same charge, with opportune weights described by the respective numerical coefficients. A ninth gluon, which is possible from the point of view of permutations of the three fundamental color charges, is discarded because it turns out to be white and therefore is not useful for transmitting the color charge between quarks. In short, it isn't active in nature. In calculating the Feynman diagrams relating to a specific strong interaction between quarks, it will be necessary to always remember the presence of $\sqrt{\alpha_s}$ at the vertices, as well as the aforementioned coefficients of the gluon (or the gluons) responsible for the particular interaction. Such coefficients, therefore, are equal to 1 for the first six gluons, equal to $1/\sqrt{2}$ or $-1/\sqrt{2}$ for the seventh, and $1/\sqrt{6}$, $1/\sqrt{6}$, or $-2/\sqrt{6}$, respectively, for the three terms of the eighth. For antiquarks, the numerical factors have opposite signs. Thanks to the specific values of the coefficients, the resulting interaction may be attractive or repulsive based on the sign of the final probability amplitude. If the resulting interaction between quarks is negative (minus sign) the system (the hadron) can actually exist as a quark bound state; otherwise, in the case of the interaction with a plus sign, the combination of quarks is not bound together (stable) and therefore simply does not exist in nature. The great success of QCD lies in its capacity to show quantitatively that three quarks of different colors produce bound systems, the baryons, and at the same time that a quark of one color and an antiquark of the corresponding anticolor generate the mesons actually observed in nature. In the same way, we verify that it is not possible to build up, for example, mesons with two quarks or baryons with two quarks and one antiquark. To be more precise, a baryon can be formed indifferently with the sequence red, blue, green—but also with blue, green, red, and so on. All the appropriately normalized permutations occur spontaneously inside the particle in exchange for keeping the total color function globally antisymmetric. These considerations are shown graphically in figure 11.3. The unexpected aspect of QCD is that it also predicts the existence of exotic states made up, for instance, of two quarks and two antiquarks or four quarks and one antiquark. Even particles formed by two or more gluons are

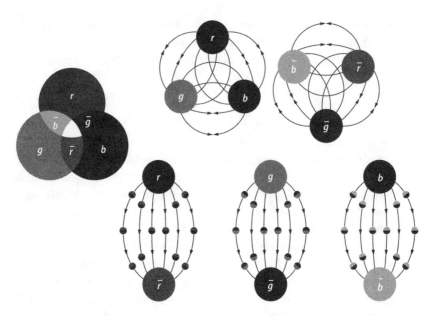

FIGURE 11.3
Illustration of the meaning of colors and anticolors in QCD, and of the related color field. Shown on the far left is the way in which, analogous with the classical color theory, the color charges are combined with each other. Above left, we have the structure of a generic baryon: each of the three quarks possesses one of the three elementary colors (strong charges) and is connected with the other quarks by gluons that are also colored. Above right, we have an antibaryon with quarks endowed with anticolors. Below, there are three possible mesons: the gluons that connect them have the same color and anticolor content as the two interacting quarks.

possible, the so-called glueballs, reconfirming the predictive power of the theory. Eventually, the first hadrons made up of four or five quarks have been identified in experiments, showing that quarks on their inside behave a little like the atomic electrons, getting excited and performing transitions between different energy levels. With the discovery of this new zoology of particles, we have opened another window onto the unknown ...

The fact that the gluon itself carries two color charges generates a very interesting phenomenon. In discussing the α constant, we have seen that due to the effect of screening by the virtual electron-positron pairs continuously created around a real electron, the electric charge perceived and hence α notably increases as the energy grows—or in an equivalent manner, as distance diminishes. In this way, α passes from a low energy value

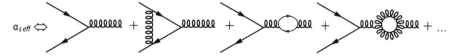

FIGURE 11.4
Feynman diagrams that contribute to the effective value of α_s. Virtual gluons are created and exchanged between quarks; other gluons can be converted into quark-antiquark pairs and into gluons, by virtue of their color charge.

of 1/137 to about 1/128, for a high-energy probe that penetrates the virtual cloud in depth. In QCD, something comparable happens, as shown in figure 11.4. To the theoretical value of α_s we compare the effective value $\alpha_{s\,eff}$ that takes account of the higher-order diagrams associated with virtual particles. Virtual gluons are exchanged between quarks, and other gluons may transform themselves into quark-antiquark pairs in diagrams of up to an arbitrarily high order. The special characteristic of QCD is that, due to the color charge carried by gluons, they can couple with other gluons in diagrams such as the one shown at the far right in figure 11.4. These diagrams are conceptually similar to the one in figure 10.15b in chapter 10; the latter cannot occur, however, because the photon does not have an electric charge and cannot interact with its peers. As Frank Wilczek points out, this is the reason why the light sabers in the *Star Wars* movies could not possibly work in reality ... Actually, wanting to be picky, the light sabers could actually work thanks to a higher order process, as depicted in the last graph of figure 10.3, where the two initial photons interact indirectly through a "box diagram," although with relatively lower probability.

In the case of QCD, the quark-antiquark pairs screen the initial charge (let's say red) of the original quark, precisely as happens in QED for virtual electron-positron pairs. In both cases, this effect produces a weakening of the effective coupling constant for distant or low-energy collisions. Something different happens, however, due to the aforementioned coupling between gluons, which is specific to QCD. The diagram in figure 11.5 illustrates an example of the interaction between three gluons. We start from a red-antiblue gluon ($r\bar{b}$), emitted by a hypothetical red quark, which converts into a pair of two $r\bar{g}$ and $\bar{b}g$ gluons. The initial red charge of the quark is conserved in the process and is available for the subsequent interactions. This effect indicates that, starting from an initially red quark that freely emits virtual gluons that carry its original (red) charge, such as $r\bar{b}$ or $r\bar{g}$, there comes to be formed a cloud of virtual gluons around the quark, with

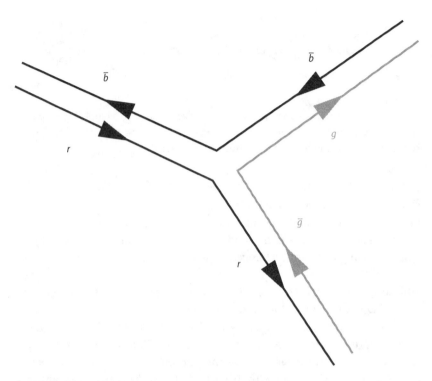

FIGURE 11.5
Diagram describing a vertex with three gluons. The initial red and antiblue colors propagate themselves, while the green charge is globally null.

a predominant presence of a red charge (figure 11.6), amplified by subsequent conversions. At this point, if we closely approach the initial red quark with a probe (high-energy interaction), we perceive a lesser volume of the red cloud; on the other hand, the more we distance ourselves (low-energy interaction), the greater the red charge that we observe. Having more color charge implies a higher intensity of the strong force; having less implies a weakening of the force. The explanation of the physical reason why this occurs is quite complex. The contributions to the effective value of α_s due to gluons, which are bosons, are opposed to those of the quark-antiquark pairs, made up of fermions. The effect of the gluons, therefore, goes in the opposite direction to the screening previously described, also present in QED, and is even dominant, determining an antiscreening effect. Consequently, as in the case for α, α_s is also not constant as a function of the collision energy, but its behavior is opposite to that of α, as clearly illustrated in figure 11.7.

We begin to speak then of the asymptotic freedom of high-energy quarks and of their confinement at low energy. For small distances, or equivalent

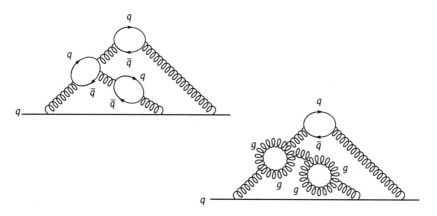

FIGURE 11.6
A quark with a red color charge propagates itself over time. Around it, the usual virtual quark-antiquark pairs are discernible, generated by gluon conversion. The gluons, however, can also generate other gluon pairs: because of this mechanism, the initial color charge of the red quark is propagated and distributed around it. The production of gluons predominates over that of quark-antiquark pairs and has the opposite effect. The result is an "antiscreening" that provokes the so-called asymptotic freedom of QCD.

high-energy interactions, quarks are almost free given that α_s decreases (small interaction probability). At low-energy, α_s increases until it almost becomes equal to 1 when the quarks are at distances precisely comparable to the dimensions of the proton or neutron. To have a coupling constant equal to 1 in practice renders every interaction 100 percent probable, regardless of the order of the diagrams that describe it. In this regime, when α_s is at a maximum, infinite diagrams of every order take place with the same probability. Inside the comfortable volume of the hadron, quarks are confined and create a stable quantum system, constituting nuclear matter in fact. This mechanism, which is typical of QCD, was discovered in 1973 by David Gross, David Politzer, and Frank Wilczek, earning them the Nobel Prize in 2004.

In a more quantitative manner, we can compare the behavior of the electromagnetic and strong forces as a function of the distance between the particles. The first exhibits the well-known dependence on $1/r^2$. As the distance r between two electric charges grows, the intensity of the force decreases rapidly, becoming virtually null at large distances. In the case of the strong interaction described by QCD, the total force is made up of two components: the first, similar to the electromagnetic one, decreases with the square of the distance and is dominant at small distances; the second represents a force that remains constant as the distance between the quarks

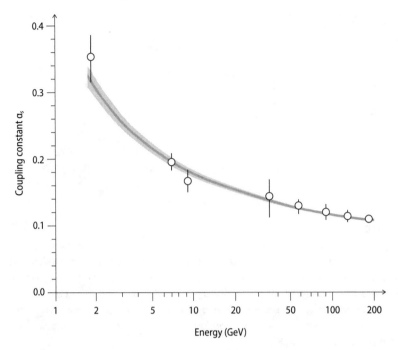

FIGURE 11.7
Behavior of α_s as a function of the energy. The curve shows the theoretical prediction of QCD (with its band due to the theoretical uncertainties), and the open circles indicate the experimental data. The correspondence between measurements and theory is excellent. The strong coupling constant tends toward the limit value of 1 for low energies, corresponding to a distance between quarks that is nearly equal to the diameter of the proton. At high-energy levels, α_s diminishes appreciably, making in practice the interaction between quarks most unlikely, and the quarks "almost free" (asymptotic freedom).

increases and is only relevant for the typical dimensions of the hadronic particles. To this constant force, therefore, corresponds an energy that grows with distance, being the potential energy proportional to the product of the force for the distance between the particles. This particular process is illustrated in figure 11.8, which envisages distancing a quark from an antiquark in order to make them leave their hadronic "container"—a meson in this particular case.

In the illustration, the quark and the antiquark exchange a gluon within the interior of the meson. To try and distance one from the other, we must apply energy via the collision of a projectile with one of the two particles. As the distance between the quarks increases, the strong force described

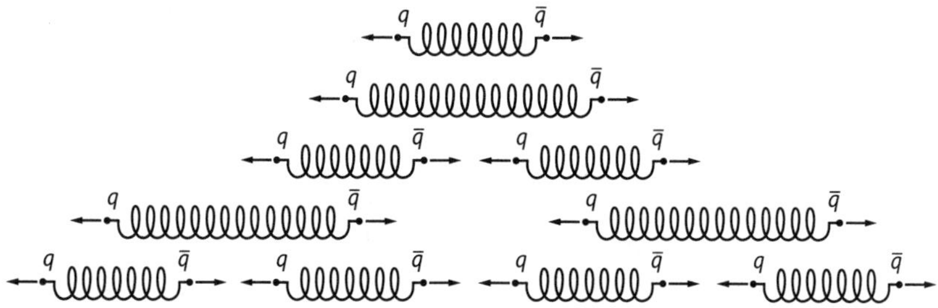

```
        q  ⁀⁀⁀⁀⁀⁀  q̄
     q  ⁀⁀⁀⁀⁀⁀⁀⁀⁀⁀  q̄
   q ⁀⁀⁀⁀⁀ q̄   q ⁀⁀⁀⁀⁀ q̄
  q ⁀⁀⁀⁀⁀⁀⁀⁀⁀ q̄       q ⁀⁀⁀⁀⁀⁀⁀⁀⁀ q̄
q⁀⁀⁀q̄  q⁀⁀⁀q̄   q⁀⁀⁀q̄   q⁀⁀⁀q̄
```

FIGURE 11.8
Trying to separate two quarks bound to each other (in this case, a quark and an antiquark) causes the potential energy that increases with distance to accumulate in the quark-gluon-quark system. At a certain level of stored energy (such as for the case of a spring tension), it becomes convenient for such energy to be converted into the mass of a new quark pair, or more than one, making it in effect impossible to isolate a single quark.

by QCD that's constant in intensity produces an increase in the energy accumulated in the tension of the gluon "spring," which mediates the action. This energy accumulation in the quark-spring-quark system makes it economical, from the point of view of the energy, that this should be converted into the mass of a new quark-antiquark pair, and into a new (additional) meson in fact (the third line in the illustration). If the energy initially accumulated in the collision is sufficient, the process may continue for a new generation, or even more than one, each time creating new particles from energy-mass conversion. The conclusion is surprising: we will never be able to individually isolate free quarks, no matter how intense our energy effort. What we manage is the fragmentation (the technical term) of the energy injected into the initial system into new hadrons. This result supports the previous consideration of the confinement of the quarks into the hadronic bag and illustrates the special nature of the strong force and of the interpretation of it provided by QCD. As in the case of atoms and nuclei, the size of the hadrons is therefore determined by the condition of equilibrium between the forces in question, corresponding to a minimum energy of the system. The confinement of quarks extends naturally to gluons. This explains why the particles that we observe in laboratories are all "white"; that is, composed of suitable combinations of quarks that are colored but globally have no net color charge. The mechanism of fragmentation deserves particular attention, through which the quark and gluon yield the accumulated energy to the color strings to create quark-antiquark pairs, which in turn form hadrons that reveal themselves in the experiments.

The production of quark pairs can happen thanks to both the electromagnetic and the strong force, as described by QED and QCD diagrams, respectively. To a significantly lesser degree, this can also take place by weak interaction, a subject that we will return to later in this book. The principal diagrams are shown in figure 11.9. In the case of high-energy electron-positron annihilation (figure 11.9a) a virtual photon is created that can transform itself into a quark-antiquark pair. The first vertex has a coupling constant equal to $\sqrt{\alpha}$, while for the second, we have $Z\sqrt{\alpha}$. The factor Z is often omitted from diagrams because it is worth 1 for the production of an electron-positron or muon-antimuon pair, for which the particles have a unit charge. In the case of the creation of a quark and an antiquark, however, the charge is fractional; hence, Z assumes the value corresponding to the specific quark pair that is produced. In the case of *up-antiup*, $Z=+2/3$, while for *down-antidown* and *strange-antistrange*, $Z=-1/3$. The diagram in figure 11.9b describes the production of quark-antiquark pairs mediated by strong interaction. The initial pair that can result from the collision of two hadrons fuses into a virtual gluon with a coupling constant of $\sqrt{\alpha_s}$. It is important to remember that for a high-energy interaction, α_s has an appreciably lower value than a low-energy one. From figure 11.7, we can see that around 10 GeV, α_s is worth approximately 0.2: greater than α, but it remains lower than its value at low energy, $\alpha_s = 1$.

For both graphs in figure 11.9, the quarks in the final state, once created, tend to move away from each other. At this point, the mechanism described in figure 11.8 is activated. The energy transferred to the pair by the gluon transforms into the mass of new quarks that combine with the original ones to make up hadrons without color charge. This is precisely the process of fragmentation, of which two examples are shown in figure 11.10. In the diagram on the left, a *down* quark and a *down* antiquark annihilate each other into a gluon, which produces an *up* quark-antiquark

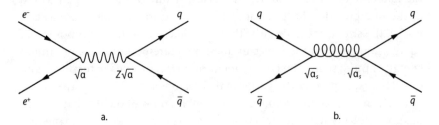

FIGURE 11.9
Diagrams of the electromagnetic (a) and strong (b) production of quark-antiquark pairs.

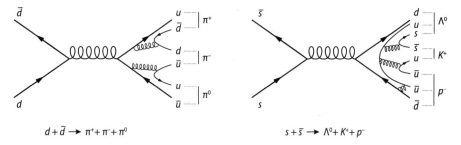

FIGURE 11.10
Feynman diagrams that describe the process of fragmentation that follows the production of quark-antiquark pairs.

pair. The energy accumulated in the color string converts into two *down-antidown* and *up-antiup* pairs, respectively. At this point the quarks reassemble, forming hadrons without net color charge and respecting the various conservation principles to which strong interaction is subject. It should be noted that the process of reassembly into the hadrons happens during relatively long intervals of time, compared to those very fleeting ones in which the initial quark-antiquark pair is created. In the specific example given here, three pions are produced with zero, positive, and negative electric charges. It is worth pointing out that each quark-antiquark pair that materializes is connected through secondary gluons in order to transfer the color charge correctly. The example on the right of the illustration is more complex. In this case, the quarks recombine to create two baryons (one Λ^0 and one antiproton) and a K^+ meson.

In the two examples that have just been discussed, we have assumed that the interaction occurs between quarks belonging to two colliding hadrons. More generally, the other quarks not directly involved in the interaction nevertheless also play a role in the final reorganization, as illustrated in figure 11.11, in this case not describing a collision, but rather a decay mediated by strong interaction. The Δ^{++} baryon decays into a proton and a pion. In the process, an *up* quark emits a virtual gluon that converts into a *down-antidown* pair. The other spectator quarks are appropriately reassigned to the hadrons produced in the final state.

The process of fragmentation therefore hides the experimental existence of quarks. Due to the specific nature of QCD, in our experiments we only observe hadrons, which in the wake of fragmentation materialize as detectable particles. Nevertheless, especially at high energies, these hadrons show quite clearly that they "remember" the production direction

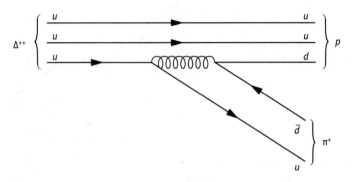

FIGURE 11.11
Graphic illustration of the strong decay of the Δ⁺⁺ baryon.

of the quark-antiquark pair originally produced at the second vertex, as illustrated in figure 11.10. In figure 11.12, let us hypothetically imagine that we find ourselves at the point where two protons beams collide. Two quarks of the protons interact, exchanging a gluon, while the other quarks act as spectators, eventually fragmenting at a small angle with respect to the beams' directions. The scattered quarks produce secondary hadrons with relatively large angles. Due to momentum conservation, these propagate from the collision point in opposite directions. The two groups of hadrons coming from the interacting quarks appear as two "jets" of particles that are well collimated in space. In this example, the collision occurs between two quarks and not between a quark and an antiquark, as illustrated in figure 10.4b in chapter 10.

Figure 11.13 shows a computer reconstruction of one event with two jets obtained with the ATLAS experiment running at the LHC at CERN. The event shows the response of various types of detectors placed around the collision point of the two extremely high-energy proton beams. The histograms along the jet tracks indicate the energy released (and measured) by the various particles present in the jet. The other tracks, nearly collinear with the direction of the protons, originate from other almost simultaneous interactions, which have not produced jets with sufficiently large angles. Events with jets can also provide other important information. In 1979, thanks to the Positron-Electron Tandem Ring Accelerator (PETRA) in Germany, the gluon was experimentally discovered. The argument is conceptually simple. If one of the outgoing quarks decides to radiate a gluon (figure 11.14a), the latter will also fragment and the particles of its jet will remember the direction from which it was originally emitted. An example is illustrated in figure 11.14b. Two of the three jets originate from quarks and antiquarks,

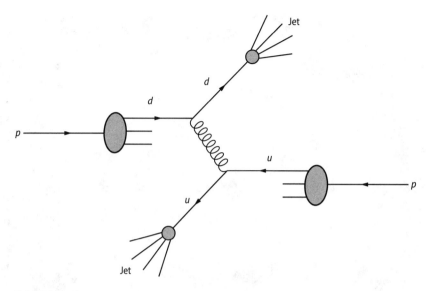

FIGURE 11.12
Illustration of the jets produced from the fragmentation of quark pairs.

respectively, but the other is likely produced by a radiated gluon. The relation between the cross section of the production of a gluon with respect to the basic graph, in which only two quarks are created, is obviously proportional to $(\sqrt{\alpha_s})^2$ (i.e., to α_s). With the energies in play with the PETRA accelerator, $(\sqrt{\alpha_s})^2$ is approximately equal to 0.1. In perfect agreement with this prediction, the events with three jets appeared in only about 10 percent of the total number of events with two jets.

Hence, it became clear that, with the formulation of QCD, strong interactions would come to be described by a complete and predictive theory. This is true especially at high energy, given that in this regime the coupling constant is sufficiently small to predominantly involve diagrams of lower order. This makes the higher order diagrams only a relatively small correction. Decays and collisions at low energy are a different case altogether, where α_s (and therefore $\sqrt{\alpha_s}$) is in practice equal to 1. This makes equiprobable the diagrams of all orders, with an arbitrary number of vertices. In these conditions, carrying out calculations becomes very complex, and today, we still need to use complicated mathematical devices and numerical calculation algorithms to arrive at quantitative conclusions to compare with the experimental data.

Now, we could ask ourselves, despite the aesthetic beauty of QCD and the previously mentioned necessity of a new quantum number (color), whether we really have solid experimental proof that it is truly *the* theory of strong

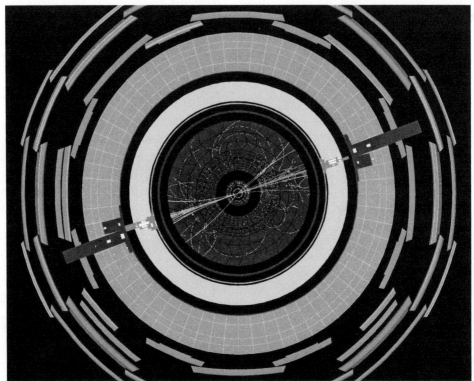
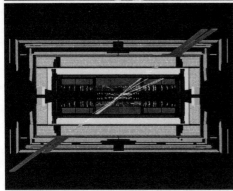
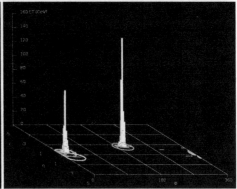

FIGURE 11.13

Computer reconstruction of a proton-proton collision event at extremely high energy, recorded by the ATLAS experiment at the CERN's Large Hadron Collider (LHC). The event shows the production of two hadron jets, each made up of various particles. Above there is the view along the beam direction; below on the left a side view; on the right the quantity of energy released as a function of the angle of the two jets. Courtesy of CERN.

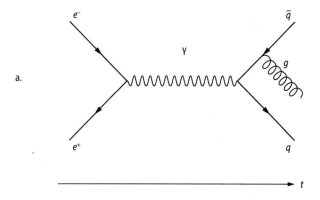

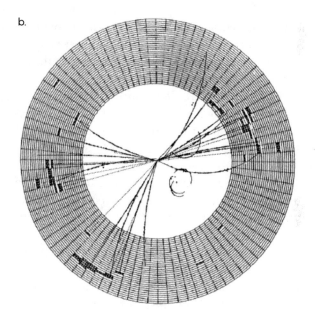

FIGURE 11.14
(a) Feynman diagram of the production (radiation) of a gluon from one of the quarks in the final state. The probability amplitude of the process has an additional factor $\sqrt{\alpha_s}$, corresponding to $(\sqrt{\alpha_s})^2$ in the cross section. (b) One of the first examples of these events obtained toward the end of the 1970s with the PETRA accelerator; the event is characterized by the presence of three jets, one of which is likely attributed to the creation and subsequent fragmentation of the gluon. The large detector is represented in a view with the colliding beams of electrons and positrons perpendicular to the plane of the image.

interaction. In fact, a monumental amount of experimental evidence supports the theory. Clear proof of the existence of the color charge arrived as early as the beginning of the 1970s, also thanks to the contributions of important experiments and theoretical studies conducted at the Frascati Laboratory of the National Institute of Nuclear Physics (INFN) in Italy. At this time, a large number of brilliant young physicists gravitated around the center of Frascati, picking up the virtual baton from the fathers of the Via Panisperna group, headed by Enrico Fermi. Many of them provided notable theoretical and experimental contributions, including Ferdinando Amman, Nicola Cabibbo, Raoul Gatto, Giorgio Salvini, and Bruno Touschek. Since the postwar period, the INFN has been a renowned international center known for its excellence in the study of the physics of elementary particles. Several generations of valued researchers have contributed to fundamental discoveries and important theoretical works in the development of pioneering accelerators and detectors, as well as to a significant number of applications that have been of benefit to society. In the early 1970s, one of the processes that was being studied in Frascati with ADONE, the electron-positron accelerator devised by Bruno Touschek, was the production of hadrons in electromagnetic interactions. This reaction was shown in figure 11.9a. Electrons and positrons that are accelerated to energies of the order of a few gigaelectronvolts were made to collide head on at specific points in the circular machine, surrounded by electronic detectors capable of identifying the particles produced by the energy-mass conversion, and of measuring their characteristics: electric charge, mass, momentum, etc. Besides the production of quark-antiquark pairs, ADONE studied processes such as the production of electron-positron and muon-antimuon ones by means of electromagnetic interactions. These reactions are conveniently reported in figure 11.15. It is evident that the initial states are the same, and thus they have an identical coupling constant. But the final ones are different: it is possible to produce both leptons (electrons and muons) or hadrons (quarks of different types). From the three diagrams shown in the bottom half of figure 11.15, we note that the coupling constant of the final state presents the coefficient Z equal to 1 for particles of integer charge, and to a fractional number (+2/3 or −1/3) for quarks. Let's now define the R ratio between the cross section for the production of quarks and that for the creation of muons in electron-positron collisions:

$$R = \frac{\sigma\,(e^+ + e^- \to q + \bar{q})}{\sigma(e^+ + e^- \to \mu^+ + \mu^-)}.$$

The reaction at the denominator is described by the central diagram in figure 11.15; the process at the numerator instead proceeds through the three diagrams shown at the bottom. For this reason, in calculating the R ratio,

The Color of Elementary Particles

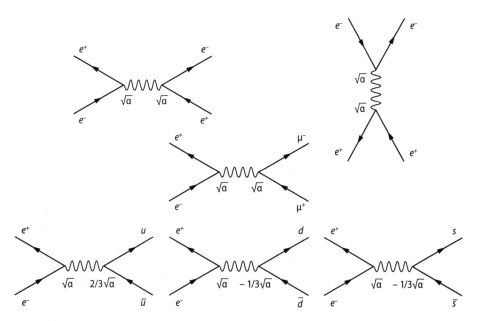

FIGURE 11.15
Feynman diagrams relating to high-energy electron-positron collisions. At the top, the two fundamental diagrams for the production of electron-positron pairs are shown (the one on the right does not show the annihilation but the electron-positron diffusion—or scattering; in the middle, the creation of muon-antimuon pairs is illustrated; and below, there is the production of quark-antiquark pairs. It is worth noting that in the latter three diagrams, the coupling constant takes into account the fractional value of the charge of the quarks.

the only difference between numerator and denominator will be given by the different final state, which, in the case of quarks, contains the fractional charges that multiply $\sqrt{\alpha}$. In the end, we will have:

$$R = \frac{\alpha^2 \times \left[\left(\frac{2}{3}\right)^2 + \left(-\frac{1}{3}\right)^2 + \left(-\frac{1}{3}\right)^2\right]}{\alpha^2} = \frac{2}{3}.$$

This is a reasonable result because while there is only one way of producing the muon pair, there are three for creating quarks, each of which has a different multiplicative factor of $\sqrt{\alpha}$, according to the value of the fractional charge of the specific quarks involved. At this stage, if we represent in a diagram the experimental value obtained measuring R as a function of the total energy available in the collision (the sum of the energy of the colliding beams), we get the result shown in figure 11.16. The experimental points (open circles) are in total disagreement with the prediction $R=2/3$, giving it as around 2 instead. However, we have not yet factored in the color issue. Every quark-antiquark pair can be created with all of the three combinations of strong charge: red-antired, blue-antiblue, and green-antigreen. This means that there are not three states to add up, but nine; and that R goes from 2/3 to $3\times 2/3 = 2$, just as measurements indicated. This is very compelling proof, in the end, of the whole color hypothesis.

It should be noted, however, that the experimental points in figure 11.16 lie a little above the predicted value of 2. In fact, this apparent inconsistency reinforces the QCD even more: in our discussion we have not taken into account the correction due to the production of gluons through graphs of a higher order, such as that in figure 11.14. Needless to say, taking this consideration into account, the agreement between data and predictions improves, to the glory of QCD. Furthermore, this agreement is in effect also proof that the attribution of fractional charges to the quarks is correct. If this had not been the case, we would certainly not have obtained the value $R=2$. But what is it that guarantees that it's precisely the *up* quark that has

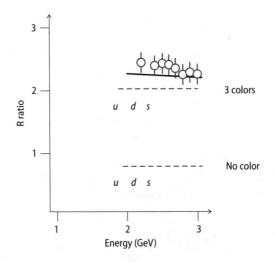

FIGURE 11.16
Value of the R ratio as a function of the energy available in the electron-positron collision. In the standard quark model, one expects that $R=2/3$ with the three quarks u, d, and s. Assuming the existence of color charge, the ratio rises to 2, as indicated by the experimental data (open circles). The small difference between the measured points and the value $R=2$ (dashed line) is understood by taking into account higher-order diagrams involving gluons (continuous line).

the charge of +2/3, and not the *down* or *strange* one? This property of quarks has also been verified with a series of experiments that are to some extent similar to that of the R ratio—and I myself participated in one of these as a young researcher, in a previous century ... These experiments involved the collision of π mesons (of both positive and negative electric charge) with a metal target made up, therefore, of protons and neutrons. Among the various processes that can occur, those are selected in which an *up* quark of a nucleus of the target is electromagnetically annihilated with an *up* antiquark composing the π^+, and producing a pair of opposite sign muons. For the historical record, this underlying theoretical model has been named after the two physicists who first proposed it in 1970: Sidney Drell and Tung-Mow Yan. In the light of the arguments discussed before, the cross section will be proportional to the square of the *up* quark-antiquark charge; that is, $(2/3)^2 = 4/9$. If we then collide negative pions that possess a *down* antiquark with a *down* quark of the target, and if all goes as predicted by the theory, the resulting cross section should be proportional to $(1/3)^2 = 1/9$. Therefore, we find that the probability of the first case should be four times larger than the second. And this turned out to be precisely the result obtained experimentally.

The more solid part of our journey is concluded thus. Battles and revolutions lie ahead: fire and flames ... of high energy.

12 THE NOVEMBER REVOLUTION

> The revolution is the harmony of form and color and of everything that exists, and it moves under a single law: life.
>
> —Frida Kahlo

I remember November 1974. I was beginning my studies at the Federico II University of Naples and starting to follow Ettore Pancini's courses in general physics. My baptism of fire coincided with one of the major discoveries in the history of particle physics. The excitement at the Institute of Physics on via Tari was palpable. Everyone was talking about it, and I could not understand why: my attempts to do so were hampered by the limitations of my competence at the time, and I hardly got beyond sharing something like that sense of national celebration after the country's soccer team has won the World Cup: we all celebrate, though relatively few understand the technical aspects of the victory, and only eleven or so people actually played in the final match. The "final" in question was played between two groups of American physicists, the leaders of which were jointly awarded the Nobel Prize. The Italians of Frascati took third place—narrowly missing out on a higher one due to some major bad luck. But what did this victory, this "November revolution," consist of exactly?

Four years earlier, in a scientific paper published in *Physical Review D*, Sheldon Glashow, John Iliopoulos, and Luciano Maiani had solved a complicated puzzle that was disturbing the sleep of many physicists at the time. The then-prevailing theoretical models predicted a certain kind of process between particles (which we will touch upon in the next chapter) called *flavor-changing neutral currents* (FCNC)—but alas, no one had ever observed it experimentally. Therefore, the theory had to have something wrong or incomplete about it, and so it needed to be modified in order to take account of the nonobservation of the process. This was a serious problem that the three solved elegantly with the theoretical mechanism that became known as GIM, from the initials of their name. The new hypothesis canceled the undesirable FCNC, as nature was demanding, but it had a small collateral effect, a tiny insignificant detail ... Nothing less, in fact, than the need to introduce a fourth quark in addition to *up*, *down*, and *strange*. The new fundamental particle was the *charm*, originally postulated a few years before by Glashow and James Bjorken. As in the case of its three siblings, the *charm* quark had been regarded as carrying an additional quantum number C, conserved as the strangeness S by the strong and electromagnetic forces and violated in weak interactions. This was an intriguing proposal—let's go so far as to say a revolutionary one—but at the same time it was completely unsupported by experimental data. And no matter how much the GIM model led to the solution of the problem of the FCNC, the demand for a new quark seemed a little extreme.

In these years, Samuel C. C. Ting, an American physicist of Chinese origin, was a professor at the Massachusetts Institute of Technology (MIT); he was carrying out an experiment known as E598 at the Brookhaven National Laboratory on Long Island (figure 12.1). Ting wanted to explore a relatively large interval of energy searching for resonances in the invariant mass spectrum of electron or muon pairs of opposite sign produced in the interaction of 30-GeV protons with a target made of beryllium. As discussed in chapter 8, the search for resonances makes it possible to identify particles with very short lifetimes; their width in terms of energy—that is, the width of the peak of the resonance—is inversely proportional to the mean lifetime of the particle that is eventually produced. Very soon, Ting and his collaborators observed something anomalous around a mass of approximately 3 GeV/c^2. Yes, a resonance was being registered—but something did not quite add up. The resonance appeared to be very narrow, and this implied a long lifetime, not compatible with those of the many broad resonances corresponding to particles—especially hadrons—decaying via strong interaction in very short time intervals on the order of 10^{-23}–10^{-24} seconds. In parallel to the Ting experiment, at the Stanford Linear Accelerator Center (SLAC) laboratory in Stanford, on the opposite coast of the United States, Burt Richter (figure 12.2) and his collaborators were studying head-on collisions between electrons and positrons at energies of up to 4 GeV with the SPEAR accelerator. In this case too, the previous measurements indicated an anomalous excess of events in that energy region, compared to what should have happened if there had only been "known" physics.

Events succeeded each other rapidly. Having clarified that it was not reasonable to envision instrumental errors or significant systematic uncertainties,

FIGURE 12.1
Sam Ting and some collaborators in the laboratory of Brookhaven, in the days of the "November revolution."

on November 11, 1974, Ting met with Richter in California, and together they decided to announce the discovery: both experiments independently observed the production of a huge resonance in the invariant mass of electron-positron and muon-antimuon pairs, and of hadrons, centered on a value of about 3 GeV/c^2. What was unexpected was the relatively high cross section of the resonance, and above all, its extremely narrow width, indicating a very long mean lifetime for the new particle and one apparently incompatible with its own hadronic decay, which nevertheless occurred

FIGURE 12.2
Burt Richter.

copiously along with that in pairs of leptons with opposite charge. The new particle was dubbed *J* by Ting. The story goes that the choice was dictated by the fact that the original name of the physicist is written in Chinese with a character that resembles the letter *J*. It seems that Richter wanted to call it *SP* instead, after the first initials of the SPEAR accelerator. In the end, he was persuaded to invert the two letters to *PS*, and then to use onomatopoetically the Greek letter Ψ (*psi*). Given that the discovery was in effect made by both groups simultaneously, the physics community decided to exercise the judgment of Solomon by calling the revolutionary particle *J*/Ψ. Sam Ting sent the article that described the discovery to *Physical Review Letters* on November 12; Richter and his colleagues submitted theirs to the same journal the very next day.

Let's now try to understand the significance of this discovery, the reason for its exceptional status. On the left of figure 12.3, the result of Sam Ting's experiment is shown, on the right that of Burt Richter, just as they were presented publicly in the two original articles. In the graph on the left, the invariant mass distribution of electron and positron pairs is shown in the interval between 2.5 and 3.5 GeV/c^2. It is evident that the number of events between 3.0 and 3.2 GeV/c^2 is much larger than in the adjacent bins of the histogram. The resonance is nevertheless very narrow, which is typical of an

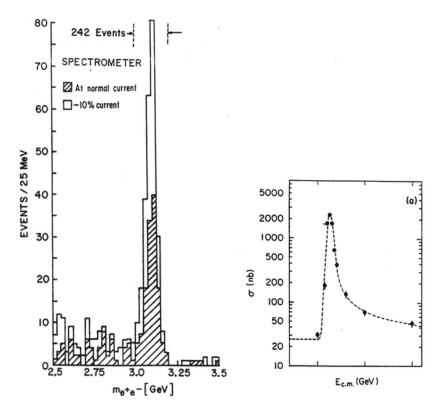

FIGURE 12.3
On the left is the distribution of the invariant mass of electron-positron pairs measured in the experiment of Sam Ting at Brookhaven. A large resonance is observed at 3.1 GeV/c^2: the new particle, named J by Ting. The same resonance for hadronic decays was observed by Richter in the SLAC laboratory and called Ψ (right).
The hadronic width is also unexpectedly small, an indication of the fact that the decays into hadrons are strongly suppressed.

electromagnetic decay. The same characteristics are evidenced in Richter's result, presented in the histogram on the right of figure 12.3 for hadronic decays. In this case also, the J/Ψ is very narrow, and this indicates that the decays into hadrons are in some sense disfavored. Having many possible ways of decaying makes the resonance wide; having a few causes a slow process with little probability, and thus with small probability amplitude, just as had been observed experimentally. The width, in effect, corresponds to a mean lifetime of 10^{-21} s, 100–1,000 times larger than in a normal hadronic decay generated by the three existing quarks …

The measured mass of the particle turned out to be just below 3.1 GeV/c^2. We are talking about the particle which at the time had the highest mass ever observed. It was really difficult to construct a mass more than three times that of the proton with the three quarks that were known at this time. Soon it was proved that the J/Ψ was a hadron—more specifically, a meson. The idea that it was made up of the fourth quark, *charm*, and of its antiquark ($c\bar{c}$), as suggested by the GIM trio, seemed to be increasingly plausible. At SLAC, other J/Ψ states of higher mass were immediately found that corresponded to the excited energy levels of the *charm-anticharm* quantum system. The paradigm was coherent, and many other proofs pointed unequivocally to the discovery of a new quark. Now, however, we must return to November 11—the date of the announcement of the discovery.

ADONE had already been in operation for several years in Frascati, and it had produced important results (figure 12.4). The Italian physicists had even searched for resonances in the interval of energy accessible to their accelerator. As ADONE was an electron-positron accumulating ring with colliding beams each of 1.5 GeV, the total available energy was of 3.0 GeV—enough to produce particles of masses up to 3.0 GeV/c^2—just about 0.1 GeV/c^2 less

FIGURE 12.4
Photograph of ADONE, the electron–positron collider of the INFN laboratory in Frascati. © INFN Frascati National Laboratories.

than the mass of the J/Ψ! My colleague from Frascati, Mario Greco, tells me that he found himself at SLAC at the moment of the announcement of the discovery. He immediately phoned Giorgio Bellettini, the director of the Frascati laboratory, to inform him of the value of the mass of the new particle. Together with Claudio Villi, the president of the already-mentioned National Institute of Nuclear Physics (INFN), they decided to "wring the neck" of ADONE, so to speak. Between November 11 and 13, the Italian physicists managed to earn the mere 0.1 GeV necessary to reach the mass of the new particle. Two days spent gathering data were enough. An enormous signal appeared before the Frascati researchers. On November 18, the scientific article was dictated by phone, word by word, to the office of *Physical Review Letters*, and was published immediately after Richter's own. So what's the moral of the story? If you have to build a new accelerator, increase the maximum design energy by a small amount, perhaps with the odd minor technological tweak, since you never know ... it could lead to a Nobel.

The reason for the suppression of the hadronic decays of the J/Ψ and the consequent narrowness of the resonance is quite complex. Basically, the more natural (and in principle dominant) particle decays are precisely those that are forbidden by the huge mass of the *charm* quark, in practice equal to approximately 1.5 GeV/c^2. The J/Ψ should decay into a pair of mesons—each comprised of a charm quark (antiquark)—whose total mass is larger than its own ... impossible. The other alternative decays are, if not forbidden, at least severely suppressed, this time due to the small value of the coupling constant $\sqrt{\alpha_s}$, which is small because, as a result of the large mass of the J/Ψ, the energy in play during the decay is correspondingly high (see figure 11.7 in chapter 11). In short, we have a complex and coherent picture, which opened a Pandora's box containing the existence of new quarks. The whole mechanism for assembling hadrons based on the magic number three—three quarks, each with three color charges—was well and truly destroyed, together with the underlying mathematical symmetry. The new particle demanded the existence of a new quark (and of its related antiquark), shifting the satisfying triangular symmetry toward something much more complex in kind. In effect, during this period, four fundamental leptons were known: the electron, its electron neutrino (v_e), the muon, and its muon neutrino (v_μ)—we will encounter later how it was discovered that for every charged lepton there is a corresponding neutrino partner. Four leptons, then, and—why not—four quarks! George Zweig himself, by calling aces the entities that constituted the hadrons (the quarks), had left open the possibility that such particles were four in number ... hearts, diamonds, clubs, and spades. The magic number chosen by nature to describe

FIGURE 12.5
Kiyoshi Niu, the first to detect a particle including a *charm* quark.

the microcosm seemed now to be four rather than three: numerology, mere accidental numerology, as quickly became clear. For the moment, however, it is necessary to take another proverbial backward step, in both time and space, to 1971 and to Nagoya, Japan.

In this year, Kiyoshi Niu (figure 12.5) conducted an experiment with truly momentous results. Thanks to a detector consisting of nuclear emulsions in flight on a cargo plane, he collected events induced by primary cosmic rays at high altitudes. The experiment aimed to study the production of pions and protons in the aftermath of collisions of cosmic rays of extremely high energies, but as often happens in physics, the results showed something unexpected. Careful analysis of the emulsions revealed evidence for a particle that did not fit into the existing scheme of those that were already known. Niu named it X, perhaps in accordance with its mysterious character. He measured its fundamental characteristics, including its lifetime that was much longer than that of any known particle at the time (e.g., one of the many resonances produced by strong interaction). Today, we know that it involved the first evidence of a particle containing a *charm* quark, which was still unknown at the time, apart from in the theoretical hypothesis of the GIM trio. The news of the probable discovery did not circulate widely beyond Japan, perhaps because of a lack of trust in the old-fashioned technique of nuclear emulsions, or perhaps because the community of cosmic ray scientists was becoming sidelined by the nascent and rather more assertive one of those using particle accelerators and complex electronic detectors in Europe, the Soviet Union, and the United States. Three years later, Ting and Richter would go on to discover the J/Ψ, for which they would be awarded the Nobel Prize in 1976, while only a few years later it began to be acknowledged that perhaps the discovery of the *charm* quark should be attributed to the brilliant and self-effacing professor from Nagoya. One of my Japanese colleagues tells me that among the reasons for Niu being overlooked for a Nobel was the opposition of eminent Japanese physicists belonging to the accelerator camp and the science that was emerging from the use of accelerators.

The newly born *charm* quark was immediately associated with the three that had already been identified. The symmetry with four components, as foreseen by the GIM model, involves an organization into two doublets: the *up* quark, with electric charge +2/3, and the *down* one, with a charge of −1/3; the *charm* quark, with a +2/3 charge and the *strange* one, with −1/3. Quite soon, various hadrons were detected containing combinations of all four quarks. One example is the D meson, composed of a *charm* quark (or antiquark) together with an *up* or *down* antiquark (quark). Predictably, the discovery of the new quark inaugurated a hunt for other potential siblings. In 1977, in an experiment conducted at Fermilab in the United States, Leon Lederman and his colleagues detected the Y (upsilon), a high-mass resonance (around 9.5 GeV/c^2), while studying the production of muon-antimuon pairs in the interaction of high-energy protons (400 GeV). The story of the J/Ψ particle was being repeated exactly. Once again, the unexpected narrow resonance and large cross section pointed to the existence of a new quark (figure 12.6). It was named *bottom* by the Israeli physicist Haim Harari. Nowadays the *bottom* quark, with its relative quantum number, is also called *beauty*.

The notion of an extended family of quarks had already been posited in an embryonic way in 1973 by the two Japanese theorists Makoto Kobayashi and Toshihide Maskawa, and Lederman's result seemed once again to be the chronicle of a discovery foretold. The *bottom* quark, with an electric charge of −1/3, destroyed the newly established symmetry of four hadrons and four fermions. The existence of five quarks indicated that the real choice of nature had to be even more complex, thanks to its predilection for the principles of mathematical symmetry. In strong support of this hypothesis, between 1975 and 1977, Martin Pearl and collaborators obtained—once again with the SPEAR accelerator—evidence of the existence of a new lepton, a brother with higher mass than the electron and the muon: the tau or tauon (τ), not to be confused with Powell's τ, later renamed the *kaon*, a member of the meson family. The large mass of the τ, around 1.8 GeV/c^2, accorded well with the large estimated mass of the *bottom* quark, equal to about 4.5 GeV/c^2. Incidentally, it is worth remembering here that the experimental search for a third (heavy) lepton had already been conducted, without success, in 1967

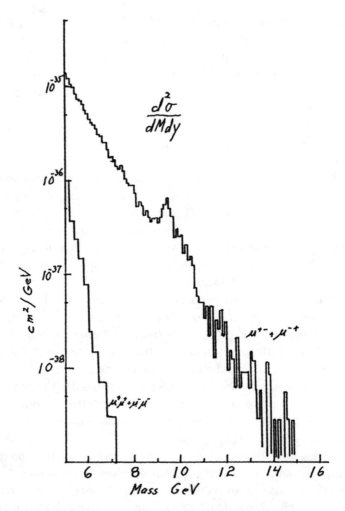

FIGURE 12.6
Distribution of the invariant mass of muon-antimuon pairs obtained in 1977 by Leon Lederman's experiment. Shown here is the original histogram produced by Lederman, from which the existence of a resonance for mass values close to 9 GeV/c^2 is shown: namely, the particle Y. Similar to the results of the experiments by Ting and Richter, this proved the existence of a new quark, the *bottom* (*b*).

by Antonino Zichichi, using the ADONE accelerator in Frascati. As was the case in 1974 with the failure to discover the J/Ψ, the energy of the machine unfortunately had not been sufficient to reveal the new particle. In fact, a pair consisting of τ^+ and τ^- can be produced in electron-positron collisions if the total energy of the two colliding beams is at least double the mass of the tau: $1.8 + 1.8 = 3.6$ GeV (!).

Needless to say, the fact that there were five quarks immediately appeared unsatisfactory from an aesthetic point of view. Three quarks with a charge of $-1/3$, and two with $+2/3$? And in parallel to this, three charged leptons (electron, muon, and tau) and only two accompanying neutrinos? If a symmetry existed, it could not produce truncated doublets of particles that were really elementary. It was obvious to everyone that at least two particles were missing from the record: the tau neutrino and a new quark with a $+2/3$ charge that could act as an accompaniment to the *bottom*. And what else could it be, other than the *top* quark, as Harari had prophetically called it? Once again, the requirement to respect deep-seated physical-mathematical symmetries imposed the existence of new particles, anticipating in some way the choice that nature had effectively made.

The *top* quark was discovered in 1995 at Fermilab, using collisions between protons and antiprotons of the same energy (1 TeV, namely 1,000 GeV), energized in the Tevatron accelerator. Two large and complex experiments, CDF and D0, recorded data over a period of years, in what amounted to a complicated scientific undertaking. Consider that the sixth quark has an incredibly large mass—about 173 GeV/c^2, roughly that of a gold nucleus. Such a heavy particle has an extremely short lifetime that in practice does not give it time to form a bound state, like a meson—and yet it is pointlike! The *top* quark decays immediately, as soon as it is produced, in a cascade of lighter particles. Its identification, therefore, is possible only through the Carthusian work of physicists, who are capable of reconstructing the vase by piecing fragments together, as is made clear by the diagram in figure 12.7. The third neutrino v_τ, partner of the τ lepton, was also discovered at Fermilab in 2000 by Kimio Niwa, and his collaborators, the successor of Niu at Nagoya, thus bringing the matter (temporarily?) to a conclusion. Strangely, perhaps, the discovery was made by utilizing once again the well-known, old-fashioned emulsion technique.

And so with this, we have reached the present day. Figure 12.8 shows a summary of the properties of quarks and their specific quantum numbers. Naturally, the latter have opposite values as their corresponding antiquarks. The current scenario of elementary particles appears to be very symmetrical, with six quarks and six leptons with their respective antiparticles. The

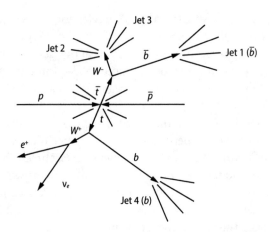

FIGURE 12.7
One of the possible production and decay schemes of a *top* quark-antiquark pair at the Fermilab Tevatron. The two heavy quarks are created by the strong interaction of two quarks of the colliding proton and antiproton, respectively. The two *top* quarks can each decay differently. In the example shown here, the antitop decays into a W^- and an antibottom. The latter produces three hadronic jets. The *top* quark decays into a W^+ pair and into a *bottom* quark, from the decay of which one obtains a jet, an electron neutrino, and a positron. The complexity of the final state makes clear how difficult it is to attempt a reconstruction of such events.

	I	I_z	S	C	B	T	Q/e
up	1/2	1/2	0	0	0	0	+2/3
down	1/2	−1/2	0	0	0	0	−1/3
strange	0	0	−1	0	0	0	−1/3
charm	0	0	0	1	0	0	+2/3
bottom	0	0	0	0	−1	0	−1/3
top	0	0	0	0	0	1	+2/3

FIGURE 12.8
A table summarizing the six quarks and their respective quantum numbers.

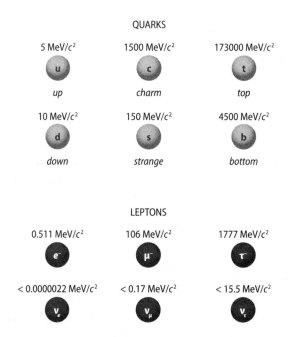

FIGURE 12.9
Table of the elementary particles (fermions) according to our current knowledge. The particles are grouped into three families (in the columns), each one consisting of four elementary fermions. The electrically charged particles (excluding the neutrinos with null charge) have a mass that increases with the order of the family. For the neutrinos, we only have upper limits to their masses. For each particle, there is an antiparticle (antimatter). Ordinary matter (atoms) is essentially composed from particles of the first family.

elementary and pointlike fermions are organized in doublets and constitute three generations of replicas of increasing mass both for leptons and for quarks, as summarized in figure 12.9. In particular, the "uplike" quarks (i.e., the *up*, the *charm*, and the *top*) have an electric charge of +2/3 (in units of *e*); and the "downlike" ones (i.e., *down*, *strange*, and *bottom*) have a charge of −1/3. This classification is in keeping with experimental measurements, as demonstrated in the experiments carried out on the Drell and Yan model described in the previous chapter.

It is really phenomenal that after approximately a century of research on the structure of matter, we find ourselves in possession of such a simple and comprehensible scheme. Finally, some light at the end of the tunnel: it seems almost as if we have returned to fire, earth, water, and air. Of course, we have learned by now that we should never take our conclusions too seriously: perhaps a new experiment will discover another quark, or an additional lepton, thus spoiling the beautiful symmetry of our twelve fermions. But perhaps it will not be like that: maybe there really are just twelve fundamental fermions ... A strong indication of this, if not quite proof of it, will emerge from chapter 16—but we will need some more patience before we get there.

We will speak more extensively about leptons in the next chapter, but there is a good deal still to be said about quarks, in particular about their masses. In the first place, because quarks are not directly and independently observable in experiments, the values of their masses are calculated on the basis of the properties of hadrons that are made up of them. In figure 12.9, note that the *up*, *down*, and *strange* quarks have masses that are much smaller than that of the hadrons that they build up. The *charm*, *bottom*, and *top* quarks, called *heavy flavors*, instead have a large mass that is for the most part equal to the one of hadrons to which they contribute to the construction of. This leads to a surprising and unexpected conclusion: if we consider the ordinary matter that constitutes our world and everything that we observe around us—stars, planets, and galaxies—we can reasonably assert that this mass essentially resides in the nuclei of the atoms, composed in turn of protons and neutrons. If we are justifiably ignoring the extremely small mass of the electrons for these, and thus we add the mass of the *up* and *down* quarks, we obtain only around 2 percent of the atomic and in practice nuclear mass. So what about the vast majority that remains, the 98 percent? Such mass is generated by the color field that keeps the quarks together in the nucleon bags, the interaction energy between quarks and gluons, including the quark-antiquark pairs that are continuously created and absorbed, as illustrated in figure 11.6 in chapter 11. This fact renders the internal structure of the proton and the neutron much more complex than could simplistically be imagined by just combining the three principal quarks, called *valence* quarks. In addition, there is a cloud of virtual gluons and a sea of similarly virtual quarks (of all flavors), which are continually created and absorbed and which nonetheless play a very concrete and primary role in the creation and definition of the mass of the hadrons. Figure 12.10 presents this situation graphically, and in doing so allows us to make another observation. Up until now we have frequently referred to "the inside of the atomic nucleus," or "the interior of the proton," as if there existed some kind of nuclear container, or protonic one, giving a misleading impression that the graphic and symbolic representations of the particles conspire to reinforce. There is obviously no surface that delimits the space occupied by the constituent particles. The volume of the proton,

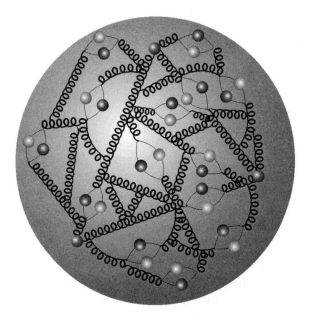

FIGURE 12.10
Representation of the inside of a proton, of a neutron or of any other hadron. In the case of the proton, the three valence quarks (*up, up,* and *down*) are accompanied by a "boiling cloud" of virtual gluons and quarks continuously created and reabsorbed—the so-called sea. These latter (or, to be more precise, the energy associated with their interaction) dynamically constitute approximately 98 percent of the energy-mass of the proton.

for instance, is defined by the average dimension of the quantum trajectories of quarks and gluons, which in turn are determined by the strong forces between them.

In figure 12.9, we also see that the masses of the neutrinos are unknown and we can only set limits; that is, for example, that the mass of the v_e is less than around 2 eV/c^2—and that of the v_μ is less than 0.17 MeV/c^2, as indicated by the experiments. We will see later on that in reality, it makes little sense to talk of the mass of this or that type of neutrino. But the important thing on which to reflect is that, nevertheless, the definition of a limit in an experimental measurement is not necessarily a determinant of the quantity in question. Let me explain. To say that the mass of the v_τ is less than 15.5 MeV/c^2 and that of the v_μ is smaller than 0.17 MeV/c^2 must not lead us mistakenly to believe that the tau neutrino is heavier than that of the muon. These values only reflect the different sensitivities of the experiments. To declare that a mosquito weighs less than 1 ton and that I am lighter than 100 kilograms certainly does not imply that the mosquito weighs more than me!

From Figures 12.8 and 12.9 we learn other things. As has already been observed, each quark carries with it a specific quantum number that characterizes it—its "flavor." The *up* and *down* quarks possess isospin which, like the spin, has a value of 1/2 and two components, +1/2 and –1/2. At the quark level, this property is transferred directly to the proton and to the neutron; that is, two particles that are equivalent from the point of view of the strong force, that is, two different states of the nucleon, one with an isospin value of +1/2, the other with –1/2. The flavor quantum number of the *s* quark is the strangeness *S*; that of the *c* quark is the *charm C*; of the *b* quark it is the *bottomness* (or *beauty*) *B*; and finally, of the *t* quark the *topness T*. Our interest in the flavor quantum numbers lies in the fact that, like other quantum numbers that characterize the elementary particles, they are useful instruments wherever they correspond to conservation principles. Specifically, the isospin is conserved in the strong interaction but violated in electromagnetic and weak ones. The other flavor numbers are conserved in strong and electromagnetic interactions but violated in the weak—do you remember the production and decay of the strange particles? These principles guide and help us to study the processes of decays and collisions between particles.

The quantum numbers may be algebraically combined to obtain others. Let's start with a very important quantity associated with the baryons, universally conserved, the baryon number *BA*, which is equal to 1 for the baryons and to –1 for the respective antiparticles, the antibaryons. This means that the principle of matter stability holds: to say that the matter that constitutes the universe is stable (that is to say, does not decay) is altogether equivalent to assuming the conservation of *BA*. Let me explain this: the proton and the neutron have a baryon number equal to 1, as do the heavier baryons (Λ, Ξ, etc.) that we have encountered in previous chapters. In every reaction, the baryon number must be the same both before and after it. If a neutron decays into a proton, because the former is slightly heavier that the latter, then the baryon number is conserved without any difficulty (1 before and 1 afterward). Because the proton is the lightest baryon, it *cannot* decay into other baryons of lower mass since they do not exist, and it cannot decay even into mesons, leptons, or bosons because in that case, the baryon number would not be conserved (for them, $BA=0$). The fact that the proton does not decay renders ordinary matter stable. And it should be remembered that the free neutron decays into a proton in roughly 900 s of average existence—but if it is tied to an atomic nucleus, it too becomes stable, at least for nonradioactive elements.

However, there are theories that foresee that the proton may also decay within time scales enormously longer than the age of the universe, which is approximately 10^{10} years. We are talking about truly unimaginable amounts of time—longer even than 10^{36} years! And the thing that is even more incredible is that in our experiments, we managed to verify that the average lifetime of the proton, if it really experiences an ending, must be at least longer than 10^{34} years. As we have seen, to say that a particle has a certain average life expectancy does not mean that all particles of a given group decay at the same moment. There is a distribution of decay times with a well-defined average value. The experimental trick consequently entails taking an enormous number of protons and waiting a reasonable amount of time—such as a few years—to see if any of them "die." From the observation of some premature decays, it will be possible to extrapolate the value of the average lifetime (remember figure 8.7 from chapter 8). These experiments are based on gigantic apparatuses weighing even tens of thousands of tons. Let's take as an example the large Japanese Super-Kamiokande neutrino detector, which has a mass of approximately 50,000 tons of water. In it, there are more than 10^{34} protons, sufficient to realize measurements that are very sensitive, even assuming small detection efficiencies for possible proton decay events.

Particle	Lepton number L	Electron lepton number L_e	Muon lepton number L_μ	Tau lepton number L_τ
e^-	+1	+1	0	0
ν_e	+1	+1	0	0
e^+	−1	−1	0	0
$\bar{\nu}_e$	−1	−1	0	0
μ^-	+1	0	+1	0
ν_μ	+1	0	+1	0
μ^+	−1	0	−1	0
$\bar{\nu}_\mu$	−1	0	−1	0
τ^-	+1	0	0	+1
ν_τ	+1	0	0	+1
τ^+	−1	0	0	−1
$\bar{\nu}_\tau$	−1	0	0	−1

FIGURE 12.11
Table of the values of lepton numbers.

The baryon number is extended to the quarks that make up the various baryons. Given that there are three of them in every baryon or antibaryon, each quark possesses a value of BA of $+1/3$, and every antiquark of $-1/3$. At this point, with an approach that may seem a little esoteric, we construct the following equivalence—a useful one for understanding which reactions can take place and which cannot—that links the charge of a composite or elementary hadron to the other quantum numbers: $Q/e = I_z + (BA + S + B + T)/2$. We also define a special quantum number for the leptons, also conserved in the reactions they are involved in, similar to what we have observed for the baryon number. The lepton number L is defined as 1 for leptons and as -1 for their antiparticles. The electron will thus have $L = +1$ as its neutrino, while the positive muon has $L = -1$, the same as all the antineutrinos. Hadrons and bosons have $L = 0$. In contrast to the baryon number, the lepton one may be defined separately for each of the three lepton families shown in figure 12.9. Hence, we will have the electron lepton number L_e, the muon one L_μ, and the tau L_τ. Experimentally, not only is L conserved in all reactions, but L_e, L_μ, and L_τ are conserved as well and separately; consequently, $L = L_e + L_\mu + L_\tau$. The list of the values of the lepton numbers is summarized in figure 12.11. Once again, we have a conserved quantum number, which guides us to discriminate between reactions that may or may not occur, at least until we discover a particular process that violates the specific one pertaining to the conservation principle. Never say never … and, once again, patience is a virtue.

13 WEAK, BUT VERY INFLUENTIAL

Our strength grows out of our weakness.

—Ralph Waldo Emerson

We should be thankful for weak interaction because it makes the Sun work, the source of almost everything that exists on Earth—of vegetable and animal life, including ourselves. In the very dense and hot matter of the solar core, hydrogen gas nuclei (protons) are compacted against each other, while the electromagnetic repulsive force tends to pull them apart. But thanks to the *tunnel effect*, two protons sufficiently close can overcome the Coulomb potential barrier and merge together to form a bound state. However, this pair is very unstable, and generally it dissociates again. But it can happen, albeit rarely, that one of the two protons of the pair can undergo a process of weak positive beta decay, with the transformation of the proton into a neutron and the creation of a (stable) deuterium nucleus, a positron (immediately annihilated) and an electron neutrino. Starting from deuterium, cascade reactions are triggered which initiate the so-called nucleosynthesis, with the creation of increasingly heavy atomic nuclei. The most relevant for we human beings is the production of helium, which starts with the fusion of deuterium and an additional proton. The helium production chain in the end has a favorable energy balance and generates photons (heat) and neutrinos both hitting the Earth; the former are useful for photosynthesis, heating, tanning, and many other applications, while the latter are fundamental on both the cosmological and microscopic scales and are very important for physicists (the author among them) who have been studying this very special particle since 1930.

The influence of weak interaction is extremely remarkable for the physics of the elementary particles, including the eccentricities and peculiarities that distinguish it from the other fundamental forces. As anticipated in chapter 5, it was Enrico Fermi who first proposed a theory of (weak) β

decay, thereby involving Pauli's neutrino, which had just been proposed. The weakness of the force is caused by its very short range, which in turn is determined by the large mass of the mediating bosons W^+, W^-, and Z^0: these, in general, are virtually exchanged between the particles that interact weakly and impose very short lifetimes in order to be compatible with Heisenberg's uncertainty principle.

We have also observed that the particles are very often subject to more than one of the four interactions in nature. All fundamental fermions (i.e., all the known particles that constitute matter), along with their sibling replicas of higher mass, are sensitive to weak interaction. Naturally, this occurs with a much-reduced probability than for the other more intense forces (notably the electromagnetic and strong ones), at least for the typical distances between elementary particles in our experiments—let's say of the order of the dimensions of an atomic nucleus. A negative pion, for example, will collide with another hadron, such as a proton, predominantly via the strong force; only for a small fraction of the interactions will the two particles exchange a photon, the quantum of the electromagnetic force, and only in a very small number of cases will the weak interaction be activated through the exchange of a W or Z boson. Obviously, we ignore here the force of gravity, which is too weak to be relevant to our purposes. In the case of a neutral pion (π^0), the strong interaction would be the dominant one in a collision, but followed by the weak one, given that the electromagnetic force would not be able to take place due to the null charge of the particle. (The neutral pion, incidentally, decays electromagnetically into two photons.) In an analogous way, the electron-proton interaction may proceed through the electromagnetic or weak force—but not, given that the electron is a lepton, through the strong interaction. And if we caused a neutrino to collide with a proton? No strong interaction would occur (the neutrino is also a lepton), nor any electromagnetic force (the neutrino is electrically neutral); and there would be a derisory gravitational interaction for the equally derisory mass of our *Fantomas*. There is no choice: the only interactions relevant when neutrinos are involved are weak ones. And this makes neutrinos the particles that are unquestionably best suited for studying this rather special force of nature.

Another characteristic of the weak force, in a negative sense, is that due to its weakness and short range of influence, it is the only interaction that is incapable of creating bound states of fermions. The gravitational force manages to keep the Earth in orbit around the Sun, and both the celestial bodies are composed of protons, neutrons, and electrons, all of them being fermions; the electromagnetic force binds electrons and protons to form the hydrogen atom, just to give one of the very many possible examples;

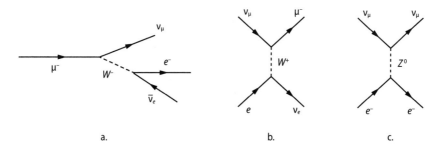

FIGURE 13.1
Feynman diagrams relating to three typical weak reactions: (a) muon decay; (b) neutrino-electron interaction; (c) another neutrino-electron interaction, mediated this time by a neutral Z^0 boson.

the strong color force, in the end, ties the quarks together, thus assembling all the hadrons. No aspect of any of this happens due to the weak force. We have never observed a bound system made up of two neutrinos, or of an electron and an antineutrino. The force is too weak to oblige the particles to spend part of their existence together under its influence.

A really special aspect of weak interaction lies in the fact that unlike the other forces, it can transform the fundamental fermions into other, different fermions (but this doesn't always happen). Expressed in simple terms, while a u quark that exchanges a gluon with a d quark always remains u after the interaction, and an electron remains the same after having exchanged a photon with a proton, a muon neutrino that interacts with a quark emits a W boson and is then transformed into a muon. Alternatively, a weak interaction can proceed through the exchange of a Z boson. In this case, as for the other forces, the particles involved do not change their nature. This is all shown graphically in the Feynman diagrams in figure 13.1. Let's start with figure 13.1a. A muon emits a virtual W^- and generates a muon neutrino. The W^- decays in turn into an electron antineutrino—note the arrow that indicates the exiting antiparticle—and into an electron, both of them real. We have just described the process of muon decay, of which we have spoken on numerous occasions in this book. Given that what we are dealing with is a weak process, the mean lifetime of the muon is rather long (approximately 2.2 μs). The reason for this is that the probability that the muon will emit a virtual particle with a mass that is as large as the W (a good 80 GeV/c^2) is very small. And it is obvious that the W must be virtual, given that the mass of the muon of only 106 MeV/c^2 is not sufficient to produce a real mediator of a hugely larger mass. Figure 13.1b illustrates the process described in the earlier example. In the interaction with an electron, the muon neutrino is converted into a negative muon. In a symmetrical way, at the point of the absorption of the W^+ mediator, the electron also changes, this time into an electron neutrino. One should note that figure 13.1b could proceed equivalently through the exchange of a positive or negative W, respectively, if it is assumed to be emitted by the neutrino or by the electron. Finally, in figure 13.1c, we have a similar reaction to those produced by the other interactions: the neutrino remains a neutrino, and the same applies to the electron because what has been exchanged in this case is a Z^0 boson.

But for the moment, let's concentrate on the weak processes that the W is involved with. The organization of the "Magnificent Twelve" elementary fermions shown in figure 12.9 in chapter 12 is not accidental: the six doublets are in effect connected to the properties of the weak interaction mediated by W bosons. We define the latter as *weak charged currents*, or simply as *charged currents*. By analogy, the reactions involving a Z^0 take the name of neutral currents. Hence, we can say that the action of a charged current is to transform a member of one of the six doublets into the other, thereby changing its electric charge. And the same goes for quarks. An example? The well-known β decay. This process happens for radioactive nuclei for which a neutron transforms itself into a proton with the simultaneous emission of an electron and an antineutrino. This is the only possibility, given that the small difference in mass (energy) between neutron and proton—939 MeV/c^2 compared to 938 MeV/c^2—is barely sufficient to produce an electron and antineutrino pair. In reality, the decay takes place at the quark level, as illustrated by the diagram on the left of figure 13.2. One of the two d quarks of the initial neutron emits a (virtual) W^- and is transformed into a u quark. As in the case of muon decay, the W can decay into a (real) electron electron-antineutrino pair. In the reassembling of the quarks into the final state, the newly appeared u quark joins the two noninteracting or *spectator* quarks

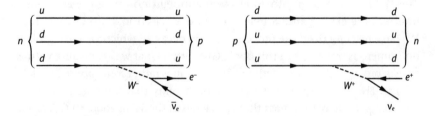

FIGURE 13.2
Two possible β decays: on the left is the one termed negative, and on the right is the positive.

that are required to form a proton. The same process can occur in inverse fashion, as shown in the illustration on the right. A proton is transformed into a neutron, with the emission of a positron and an electron neutrino. This reaction may occur only if we supply additional energy to the proton, as its mass is smaller than that of the neutron; we have in this case a positive β decay, with the emission of a positron. This last process happens for radioactive nuclei where the transformation of a proton into a neutron is energetically allowed by the fact that the latter goes to occupy a quantum state characterized by a lower energy level. In this way, the energy difference between the mass of the neutron and that of the proton is provided by the nuclear binding energy.

Weak decays are very important in particle physics. Let's take, for example, those of the π^+ and π^- charged pions. We have already mentioned that the charged pion decays into a muon and a neutrino (see the results of the Powell experiment shown in figure 5.15 of chapter 5). Now we can interpret this process in terms of the weak interaction. The mean lifetime of a pion is long (almost 30 ns), which is typical of the weak force. Let's take a π^+. Its composition in terms of quarks is $u\bar{d}$. The quark-antiquark pair forms a bound state, which is quite stable thanks to the strong force mediated by gluons. The two quarks continuously exchange gluons and wander chaotically in the interior of the volume of the pion. Now it can happen that in their quantum errancy, the quarks pass close to each other to render possible the activation of the weak force (i.e., the interaction between the quark and the antiquark mediated by a W boson). At this point, the pion decays as shown in figure 13.3, generating a positive muon and a muon neutrino through an intermediate state, constituted by a virtual W^+. Naturally, the decay of a negative pion would have produced a negative muon and a muon antineutrino.

Weak processes are important reactions and also relatively intelligible: in fact, it is customary to say that the physics of weak interaction is "clean,"

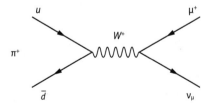

FIGURE 13.3
Diagram describing the decay of a positive pion into a muon, that is also positively charged, and a muon neutrino.

in contrast to that of the strong force, which is "dirty" because of the gluons, the fragmentation of the great number of quarks and particles usually involved—and also, above all, from the experimental point of view. It is one thing to detect leptons, and quite another, more complex undertaking to identify and measure hadrons. Thanks to Feynman's diagrams, we can compute amplitudes and cross sections of weak processes, collisions, and decays, almost exactly as we have done for electromagnetic and strong interactions. The principal difference consist of the fact that in this case, the propagator is a boson endowed with a large mass, and this renders the interaction weak in practice, independently of the coupling constant g that appears at the vertices of the corresponding diagrams.

But the surprises that the weak force has for us do not end there. We have already dealt with one of them when we talked about parity violation. The fact that neutrinos have only negative helicity—that is, when the direction of the momentum is opposite to that of the spin, along the propagation axis—and the antineutrinos positive helicity—that is, having the same direction for the two vectors—induces parity violation for all weak reactions, where it is conserved for the other three fundamental interactions. The concept of helicity, very clear for neutrinos, becomes more complex for massive fermions. In chapter 9 I explained that for particles with mass the helicity can be changed by "overtaking them" with a suitable Lorentz transformation. We therefore define a new quantity called chirality, intimately linked to the quantum description of particles with spin, which represents a sort of intrinsic helicity. Helicity and chirality coincide for massless (or almost ...) particles, such as the neutrino, but differ for particles with mass. The thing may seem, and perhaps it is, somewhat artificial, but it is necessary to understand the characteristics of weak charged current reactions. In these reactions, the W bosons only couple to fermion wave functions of negative chirality. These in turn describe physical particles with both a negative and a positive helicity component. The second is vanishing if the mass of the fermion is zero—essentially almost nothing for neutrinos, given their very small mass. The antiparticles are instead always generated by negative chirality functions, but, if they have a small mass, they will mostly have positive helicity. For example, this is the reason why in the decay of the W^- in an electron and in an antineutrino (as happens in beta decay), the electron will normally have negative helicity and the antineutrino a positive one. These considerations do not apply to neutral currents, which involve particles of both helicities.

Furthermore, from the point of view of cataloging weak reactions, the processes described so far may be classified as purely leptonic (without

FIGURE 13.4
Nicola Cabibbo.

hadrons), for example the muon decay; or as semi-leptonic, as in the case of β decay. There nevertheless exists another category of charged currents, those that are purely hadronic, that relate only to transitions between members of the quark doublets. For these reactions, as was to be expected, a further complication of the weak force comes into play. Nicola Cabibbo (figure 13.4), a theoretical physicist of great importance who emerged from the previously mentioned Frascati-Roman school, proposed in 1963 that with the weak interaction there was a kind of quantum mechanism of rotation between the various quarks—that is, between the wave functions that describe them, in their abstract mathematical space—for which, in the transitions of charged current that take place between the members of the doublets *ud*, *cs*, and *tb*, we must also take into global account the quarks of other doublets, according to appropriate numerical coefficients. Obviously, since the energy conservation principle holds, this prevents the transformation of light quarks into heavier ones. To be more precise, reprising the arguments that we have discussed when referring to quantum mechanics, there exist two groups of eigenstates: those that have the mass of quarks such as eigenvalues (in effect, the mass eigenstates) and the flavor eigenstates, corresponding to specific quark flavors relevant for the weak interaction. The quantum rules allow us to express every eigenstate of a group as a mathematical combination of the eigenstates of the other. For example, the d quark, which interacts weakly, transforming into a u, consists in quantum terms of a linear combination that includes the mass eigenstate of all three quarks that are placed at the bottom of the doublets (d, s, b). We are dealing here with the quantum mechanism of *mixing*. This idea was introduced by Cabibbo to remedy some experimental evidence that indicated an apparent difference in the intensity of the weak force depending on the flavor of the quarks in question, as opposed to what happens with leptons for which, independent of their family, the intensity of the weak force is the same, in accordance with the principle known as *the universality of the weak interaction*.

In Cabibbo's time, only the three u, d, and s quarks were known, and as a result, his model envisaged transitions between an appropriate combination of d and s toward the u quark. Years later, Makoto Kobayashi and Toshihide Maskawa extended Cabibbo's theory to a system of six quarks. The coefficients of the relative weights relative to the transitions constitute a 3×3 numerical mixing matrix, known as CKM from the initial letters of the surnames of the three physicists. The elements of the matrix must be determined in experiments by measuring the relative probability for the various reactions observed in nature. Such coefficients appear at the vertices of the relevant Feynman diagrams and multiply the weak coupling constant g. I cannot resist observing that when in 2008, the Nobel Prize was awarded to Kobayashi and Maskawa, and not to Cabibbo, it left, as the euphemism has it, a bitter taste in the mouth ...

In figure 13.5, to be as complete as possible, I am including some diagrams of purely hadronic charged currents. In figure 13.5a, we have the weak decay of the Λ into a proton and a negative pion. Notice that the s quark makes a transition into the u quark, given that the c quark (*charm*), a member of the same doublet, has a higher mass than that of the *strange* from the outset. The W^-, with an electric charge of -1, decays into a quark-antiquark pair ($d\bar{u}$) that also has a total charge of -1. Figures 13.5b and 13.5c illustrate two possible decays of the B^- meson constituted by a b quark (bottom) and by an $\bar{u}$ antiquark. In the two vertices associated with each of the transitions between quarks mediated by the W boson, the respective V coefficients of the CKM matrix appear explicitly. The possibility of cascade-type transitions between quarks of a given flavor into different quarks, albeit of smaller mass, $b \rightarrow c \rightarrow s \rightarrow u$, explains why the quantum flavor number is not conserved in the weak interaction. You will recall, for example, what was said on the subject of strangeness at the end of chapter 5. The characteristic of the hadron charged currents changing the flavor of the interacting particle is not echoed in the reactions induced by neutral currents. It is said that the former change flavor, but the latter do not. In chapter 12, we observed that the hypothesis that the hadronic neutral currents can also

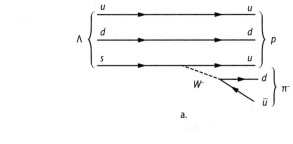

FIGURE 13.5
Examples of purely hadronic decays, mediated by the charged current weak interaction.

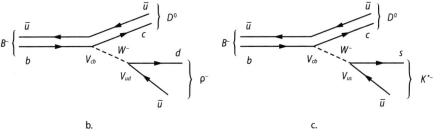

FIGURE 13.6
Experimentally determined values of the coefficients of the V_{CKM} Cabibbo-Kobayashi-Maskawa matrix (CKM). It is noticeable that the diagonal terms are much larger than those off the diagonal. This makes the weak transitions between *tb*, *cs*, and *ud* quarks more favored (more probable) compared to the others.

change quark flavors—flavor-changing neutral currents (FCNC)—allowed in principle in a theoretical scheme with only three quarks, was rightly rejected from the previously mentioned GIM mechanism, leading to the introduction of the fourth quark, the *charm*.

Our current knowledge of the CKM matrix is summarized in figure 13.6, which derives from a great number of experiments, that over the years have increasingly refined our understanding of the weak interaction between hadrons. In particular, the matrix also includes a term that is responsible for the aforementioned CP violation, occurring for weak transitions involving quarks. This is a rather complex argument for which we invite the interested readers to consult specialized textbooks. It is, finally, particularly important to note that the diagonal elements of the matrix are appreciably greater than those outside the diagonal, making the transitions between the corresponding quarks more probable than the others. This characteristic clashes with what we shall see occurring in an analogous matrix of mixing that involves precisely the usual suspects: the neutrinos ...

14 THE CHAMELEON

The moment of change is the only poem.

—Adrienne Rich

The time has now come to discuss in more detail the sovereign particle of the weak interaction, the spinning top made of nothing, the smallest imaginable entity of matter, the ghost particle, chameleon, mutant, the cross that physicists bear, and yet also their delight. I am referring to the neutrino, which shares the responsibility for the eccentricities of the weak force—and which is itself the artificer of surprising behaviors. The story of the neutrino is from the outset intertwined with the development of the weak interaction, and the two paths have intersected on many occasions, producing scientific results of enormous value and thereby contributing to our knowledge of elementary particles. Twenty-five years elapsed between the Pauli hypothesis in 1930 and the actual discovery of the neutrino, but during all that time physicists never once doubted its existence, even if the hope of one day being able to detect it seemed, because of its negligible cross section with matter, to be minimal. Pauli himself, at the time of his "desperate" hypothesis, said that: "I have done a terrible thing, something that no theoretical physicist should do: I have proposed the existence of a particle that will never be detected!" To give an idea of the scale of the problem, a neutrino of 1 MeV energy could travel through an imaginary tube made of lead one light year long without interacting with it in any way... Unbelievable!

During World War II, the Manhattan Project at Los Alamos represented an enormous organizational and financial effort for the development of the first atomic bomb: if undertaken today, the total cost of the project would be upward of $20 billion. However, the project was also a moment of encounter between the best minds of the period. Among them was Enrico Fermi, who was very much interested in those aspects of fundamental physics that the project entailed. The first bomb exploded at the Alamogordo polygon on July 16, 1945. The result showed all the scientists who were present the enormous energy hidden within the atomic nucleus, and anticipated the horror that its subsequent military application would hold for humanity. But somebody managed to see its scientific potential: it was understood that such an explosion was that great producer of neutrinos with which we could overcome the limitation of their extremely low interaction probability that had prevented their detection. The American physicist Frederick Reines discussed with Fermi the possibility of installing a neutrino detector close to an atomic explosion to identify this particle for the first time (!). Such a fanciful notion did not get very far, although it apparently received a green light from the military, who were perhaps interested in the potential weaponizing applications of the neutrino ... and I am very glad that, however many science fiction and military fantasies have been produced since then, the neutrino still remains a peaceful particle of Olympian and aristocratic harmlessness. In any case, the determination of Reines was rewarded by the fact that, together with Clyde Cowan, he succeeded in detecting the neutrino for the first time, in the mid-1950s, many years after the Pauli hypothesis and the subsequent theoretical systematization by Fermi—proof, if it were needed, of just how elusive our particle is. Following an original idea of Bruno Pontecorvo—the "baby," as he was called by his colleagues of the Fermi group—after the hardly practical idea of using an atomic explosion had been discarded, Reines and Cowan understood that one of the nuclear reactors which at the time were beginning to be used for military applications and for the production of electricity would present an excellent alternative for discovering the neutrino and allow the experiment to be repeated several times. Today, we know that even a modest reactor of one gigawatt produces around 10^{20} (100 billion billion) antineutrinos (and not neutrinos) per second!

On the subject of Pontecorvo, the Pisan physicist who in the 1950s provoked an enormous international controversy in the aftermath of his sudden decision to emigrate secretly to the Soviet Union, it must be recognized that he has been undoubtedly one of the principal protagonists in the history of neutrino physics. I had the great pleasure of meeting him in the summer of 1984, on the occasion of an international conference at the Joint Institute for Nuclear Research (JINR) in Dubna, 100 km from Moscow and situated on the quiet banks of the Volga. It was one of the defining moments of my life as a researcher, giving me strong encouragement to work on the physics of neutrinos. I am still moved today by the recollection of that meeting. There were about ten of us participating in the conference who were non-Soviet physicists, or at least not from the Eastern bloc, and because of this, we were welcomed with open arms by the director of JINR, the Russian academician Alexander Baldin. The director proudly took us to visit the equipment and the labs of the research center. Many of the structures were dilapidated, but in spite of this, our Soviet colleagues did their utmost to work to international standards. A few days later, I was approached by some Russian physicists who announced that Professor Pontecorvo would like to meet me. I could hardly believe it at first—that the great Pontecorvo could be interested in a young Italian researcher. Then I understood that in effect, the professor had a certain appetite for informally meeting and chatting with compatriots, something that was hardly a common occurrence in the Soviet Union at that time. The meeting took place a few days later. I was ushered into an elegant sitting room, and after a few minutes, preceded by some companions, Pontecorvo appeared and welcomed me with great courtesy. We talked about this and that—about physics, but also about the political situation in Italy. The illness with which he was afflicted impeded the fluency of his movements and his speech, but it was a cordial meeting—and a wonderful one for me (figure 14.1). Only when I told him that we were preparing an experiment

FIGURE 14.1
Meeting with Bruno Pontecorvo in Dubna in 1984.

at CERN that would allow us to look for neutrino oscillations did he show a sign of overt pride, saying: "I invented those myself!" But let's return to the idea that had occurred to Reines and Cowan.

Two formidable experimental problems had immediately emerged. The first, which I have already alluded to, concerned the extremely small cross section for (anti)neutrino-matter interactions. The other, which was equally difficult to solve at the time, was that of having an unambiguous way of affirming that a neutrino had really been observed, or better still, of registering the effect of its interaction with a detector. Allowing for the fact that one of the extremely numerous antineutrinos from the Hanford reactor in Washington State, chosen by Reines and Cowan for their fundamental experiment, had finally been able to interact with the matter of the detector, what kind of signal should they have expected? The idea was to develop an experiment that would be receptive to the process of the inverse β decay that proceeds through a reaction of weak charged current: $\bar{v}+p \rightarrow n+e^+$. This describes the interaction of an antineutrino produced by the nuclear reactor with one proton among the many that constitute the detector target. This collision results in a neutron and a positron in the final state. The exchange of the W boson transforms the members of the respective weak doublets: the antineutrino into a positron, and a u quark of the proton into a d, thus creating a neutron, as illustrated in the diagram in figure 14.2. The problem of the neutrino detection becomes then that of identifying the neutron produced in coincidence with the positron. In the absence, or at least nondominance of other processes that are similarly probable and with an analogous signature (i.e., those that physicists designate as background events), we have the signal of the interaction taking

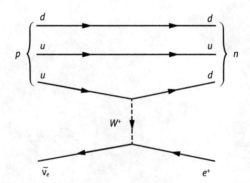

FIGURE 14.2
Diagram of inverse β decay, the process that made possible the discovery of the neutrino—or better, of the antineutrino.

place of a neutrino, and therefore we have detected it. The detailed way in which the identification of a positron and a neutron is carried out is beyond the purposes of this book. Suffice to say that the newly produced positron annihilates with an electron of the detector, hence producing two photons, while the neutron is absorbed by an atomic nucleus, exciting it and obliging it to subsequently deexcite itself by emitting an energetic photon. The observation of the antineutrino is therefore translated into the more accessible detection of three photons with particular temporal, kinematical, and topological features.

The Poltergeist experiment, so called perhaps in order to underline the evanescent nature of the particle, was realized thus—and the first results arrived a few months later in the summer of 1953. Unfortunately, the neutrino again showed a strong reluctance to be observed, and due to an experimental background that was too high, the signal of its interaction was not convincing. After the first failed attempt, Reines and Cowan tried again, this time placing their neutrino detector near the Savannah River reactor in South Carolina. Figure 14.3 shows the two scientists at work in the control

FIGURE 14.3
Reines and Cowan in the control room of their experimental apparatus.

room of their experiment. The experimental apparatus consisted of a large container of liquid scintillator doped with cadmium chloride (an efficient neutron catcher) in which the photons produced in the interaction of the neutrinos were measured by photo-detectors distributed over the internal surface of the vessel. Reines and Cowan managed in 1956 to detect a significant signal of the interaction of antineutrinos. Thus, 13,800 million years after their first appearance, and a quarter of a century after the Pauli hypothesis, humanity finally had proof of the existence of neutrinos. The two physicists sent a telegram to Wolfgang Pauli on June 15, 1956, which reads: "We are happy to inform you that we have definitely detected neutrinos from fission fragments by observing the inverse beta decay of the proton. Observed cross section agrees well with expected six times ten to the minus forty-four square centimeters." Pauli's reply, telegraphed the next day, was lapidary: "Thank you for the message. Everything comes to him who knows how to wait."

It is interesting to note that with another experiment that was proposed by Luis Alvarez and conducted using the same Savannah River reactor by Don Harmer and Ray Davis—the latter being two of the physicists who have contributed most substantially to our understanding of neutrinos—the result was a failure. The two used an alternative detector that was receptive to neutrinos rather than to the antineutrinos generated by the reactor. Luckily for Davis, using the same technique many years later, he would go on to successfully detect the neutrinos (not the antineutrinos!) produced by the Sun: this momentous discovery would garner a well-deserved Nobel Prize, which was awarded to him in 2002 at the venerable age of 88 ... better late than never. One is duty bound to point out that the method employed by Davis had originally been proposed by Pontecorvo, who was developing an original idea by Jules Guéron, and was therefore yet another of the Italian's achievements. The stories of Reines and Cowan, and of Davis, teach a lesson that is in effect one of the leitmotifs of research in general, and of that focused on particle physics in particular: the need to persevere, to not become discouraged, and to proceed without hesitation, backing with a kind of constructive and self-critical arrogance one's own determination to get to the bottom of the matter.

A couple of years after the discovery of parity violation in 1956, physicists confronted another complex problem, and in this case too Bruno Pontecorvo played a key role thanks to his visionary and (to a certain extent) revolutionary ideas. The problem was this: the neutrinos discovered by Reines and Cowan coming from nuclear reactors—in effect, antineutrinos—must be the same emitted in β decay given that, interacting with our detectors, they both give rise to electrons or positrons by way of charged currents between members of weak doublets. But what were we to make, then, of the neutrinos produced in the pion decay ($\pi \to \mu + \nu$) or of the decay of the muon ($\mu \to e + \nu + \nu$)? It's a question that seemed somewhat academic, but it had a solution that would go on to open the way for other important discoveries. Bruno Pontecorvo was the first to speculate that the fact that the muon does not decay simply into an electron and a photon—as we would expect with the hypothesis of the muon as a simple excited state of the electron—concealed something much deeper. He endeavored to suppose that there might be a kind of symmetry between the family constituted by the electron and its neutrino, and that of the muon with another type of neutrino, which today we call a *muon neutrino* but at the time was given the name *neutretto* precisely in order to distinguish it from the neutrino associated with the electron. Although beautiful in aesthetic terms, the proposal seemed almost impossible to verify given the intrinsic difficulty of realizing such complex experiments involving neutrinos—on account of that well-known (but always worth stressing) reluctance to interact with matter.

Once again, Pontecorvo launched a suggestion that would solve the dilemma. He pointed out that the neutrino interaction probability, the cross section, would grow with energy, and that hence it would be appropriate to use the recently built particle accelerators to produce intense and high-energy beams of neutrinos. The idea of a neutrino beam produced by an accelerator had already been proposed in 1957 by the Russian physicist Moisey Markov. Unfortunately, however, the situation of experimental physics in the Soviet Union at the time did not allow the creation of such infrastructure. In 1959, on the other side of the world at Columbia University, the subject was being tackled in lively discussions in the canteen. Tsung-Dao Lee and Mel Schwartz were debating the possibility of creating the artificial neutrino beam that Pontecorvo had conceived. Schwartz proposed using protons, which for the time were of extremely high energy (several gigaelectronvolts). Through collision with a target, pions would be produced—and decaying, they would generate muons and neutrinos in turn. Now, if such interacting neutrinos with an appropriate target-detector had produced muons, this would have been proof of the existence of two types of neutrino: one associated with the electron and the other with the muon.

Lee and his friend Yang began to calculate the cross section of the processes involved, while Mel Schwartz, along with Leon Lederman and Jack

Steinberger (a brilliant ex-student of Fermi's whose doctoral thesis proved for the first time that the muon decays into an electron and two neutrinos—which may justifiably be called a pretty good thesis ...) initiated the project and developed an experiment that Jean-Marc Gaillard would subsequently join. They chose the best available accelerator: the Alternate Gradient Synchrotron, a new proton machine at the Brookhaven laboratory. For the detector, they used a spark chamber, a device recently realized by James Cronin after an original idea by the Japanese Fukui and Miyamoto. This detector was capable of electronically visualizing the tracks of the particles generated in the interactions of the neutrinos and of efficiently discriminating between the long tracks produced by the penetrating muons and the shorter ones generated by electrons. The experimental apparatus was rapidly completed, and in 1962, the data-taking began. Billions of neutrinos were shot toward the detector with the first beam of artificial neutrinos produced by humans. From these billions, scarcely forty interacted visibly with the target. Only six events showed the presence of an electron, a number compatible with the estimation of the experimental background. The remaining thirty-four events provided evidence for a clear muon track. These findings proved that the Pontecorvo hypothesis was correct; the neutrino associated with the muon, v_μ, is different from that which accompanies the electron, v_e. The four particles form two doublets, the elements of which can transform themselves reciprocally, one into another, exchanging a W boson through a process of weak charged current.

In 1988, Lederman, Schwartz, and Steinberger were awarded the Nobel Prize for this discovery (figure 14.4). However, Pontecorvo was not recognized for his contributions to this project. In my opinion, this failure to honor him is all the more bitter if we take into account his other two fundamental contributions to the physics of neutrinos: his already mentioned seminal part in the detection of solar neutrinos, and the hypothesis regarding the transmutation (oscillation) of the particle. Today, the ashes of Bruno Maksimovic Pontecorvo, who died in Dubna in 1993, have been kept half in Dubna and half in the nondenominational cemetery in Rome; figure 14.5 shows the emblematic tombstone that commemorates him and reminds us all of his contributions to science. A final note about Jack Steinberger: in 2008, I invited him to my University in Bern, for a seminar. It proved to be very moving to hear him speak, an 87-year-old youngman who told us about physics, history, and the great scientists of the past. I remember that when one of my students asked him what his current dream or desire was, he replied with a tinge of sadness: "To be your age again!"

FIGURE 14.4
Steinberger, Schwartz, and Lederman, awarded the Nobel Prize in 1988 for the discovery of the muon neutrino.

The discovery of the second type of neutrino, the v_μ, caused notable excitement in the emerging community of particle physicists. A series of experiments directly or indirectly involving neutrinos, conducted in all the major laboratories of the world, began to shed light on the physics of weak interaction, as well as on the internal composition of hadrons. In fact, in parallel with the progressive growth in understanding of the physical mechanisms involving neutrinos, neutrinos themselves began to be used as probes (or projectiles) for the study of the intimate structure of matter. It is intuitively understood that to investigate the interior of the atomic nucleus, or even of a single proton, the neutrino is in many ways more efficient as a projectile than as a proton or an electron; both the composite nature of the projectile proton and its electric charge may produce physical conditions after the collision which are difficult to interpret—remember "dirty" physics? The situation is very different for a beam of neutrinos, which interact by their very nature in a weak but univocal way, penetrating the target deeply and generating events that are relatively simple to understand.

FIGURE 14.5
Gravestone of Bruno Pontecorvo in the nondenominational cemetery in Rome. It commemorates one of the physicist's many scientific achievements: the idea that the muon and electron neutrinos constitute two distinct particles.

At the beginning of the 1970s, there was operating at CERN an intense beam of neutrinos produced by the Proton Synchrotron accelerator, as well as a detector that was innovative at the time—the Gargamelle bubble chamber. An absorbing race conducted with the Americans of Fermilab came to an end in 1973: Gargamelle made it possible to identify a handful of events (one of which is shown in figure 14.6), in which it is evident that the beams of muon neutrinos also produced interactions without muons, which is clear evidence of the existence of weak neutral currents (the corresponding diagram has already been shown in figure 13.1c in chapter 13).

The importance of these neutral currents goes well beyond the discovery of a particular way of applying weak interaction. They had been anticipated by a theory developed toward the end of the 1960s by Sheldon Glashow, Abdus Salam, and Steven Weinberg—an extremely beautiful theory that, above all, is capable of explaining the experimental results. The model envisaged the unification of two of the four fundamental interactions: the electromagnetic force and, needless to say, the weak force, combining in the *electroweak* interaction. I will refrain from adding anything further, so as not to spoil the surprise when we tackle the subject later. Suffice to say that it created a fundamental result for physics, and that the three theorists

The Chameleon

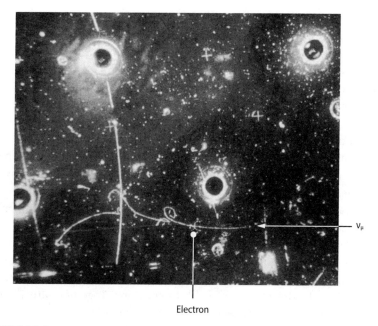

FIGURE 14.6
One of the first groundbreaking neutral current events recorded by Gargamelle. A muon neutrino interacts with an electron of the liquid that fills the bubble chamber, creating an electron and a muon neutrino v_μ (which is not detected). No muon is identified, thereby excluding the possibility of a charged current reaction.

FIGURE 14.7
From left to right: Steven Weinberg, Sheldon Glashow and Abdus Salam, authors of the theoretical unification model between the electromagnetic and weak forces.

obtained the Nobel in 1979 (figure 14.7). The prize was awarded, as it was lucidly put: "for their contributions to the theory of the unification of the electromagnetic and weak interactions, including, inter alia, the prediction of weak neutral currents."

With the discovery of neutral currents, and with their entry with full entitlement into the paradigm of the new electroweak theory, neutrinos were exploited intensively between the 1970s and 1990s as tools for understanding the structure of matter through their interactions. This is a peculiar aspect of research in particle physics: with what had been an unexpected and groundbreaking result just a few years earlier—the discovery and understanding of a physical mechanism—applications are created with great speed and almost immediately become routine. As Richard Feynman used to say: "Yesterday's discovery is today's calibration!" And Valentine (Val) Telegdi, the brilliant Hungarian physicist who had belonged to the Fermi group in Chicago, added: "and it is tomorrow's background." Among other accomplishments, Val Telegdi was the author of very important contributions to the physics of weak interactions. Together with Jerome Friedman and contemporaneously with Madame Wu, he conducted an experiment that evidenced parity violation, using nuclear emulsions as a detector.

As a young man, I had the intense experience of working with Telegdi (figure 14.8), as we were both engaged at CERN on the NA10 experiment, which looked at the production of muon pairs in hadronic interactions for the study of the Drell and Yan model (see chapter 11). A wonderful

FIGURE 14.8
Val Telegdi in 1987 during the seminar at CERN on the occasion of his sixty-fifth birthday. Courtesy of CERN.

relationship developed between us, with his appetite for teaching and my eagerness to learn from him—including from his command of Italian, a language that he loved and spoke without an accent, partly because of his Italian wife, Lia. It was he who instilled in me the dry style that purely scientific works must have, insisting: "Every superfluous word is not only useless; it is damaging." On one occasion—I think it was in 1983, while we were honing one of our articles (he would correct and I would take note of his corrections ...)—I accorded myself a great linguistic satisfaction. Telegdi was praising the deliciousness of "*i gnocchi*" that Lia, who was a great cook, had prepared. When I pointed out that it should be "*gli gnocchi*," together with a pedantic illustration of why this was grammatically correct, I came close to being kicked out of the room. I did not relent, however, having bet the price of a coffee on the outcome of the dispute, and so he resolved to conclude it by phoning Maria Zaini, the legendary secretary of the renowned Italian physicist Antonino Zichichi, asking her to check the dictionary. Maria replied that it was unnecessary, but in response to his (now desperate) insistence, she did so. He hung up the phone, and we resumed working without saying a word. A few minutes later, noticing my satisfied, if respectful smirk, he said: "You see, because you are Neapolitan, you have learned Italian as if it were a foreign language. That's why you know all of these ridiculous rules ...." Joking aside, Val Telegdi gave me another important lesson, which I am glad to pass on to my younger readers. In relation to his precociously assumed scientific vocation, he said to me: "I also decided to become a physicist because then, even as a simple student, I would be able to say to a great scientist such as Fermi: you're wrong and I'm right!"

The time has now come to confront a property of neutrinos that surely identifies them as special particles, and which, studied theoretically and experimentally for years, has made it possible to reach extremely important scientific results to help understand the world of elementary particles. I am referring to neutrino oscillations. The argument is quite complex, certainly from the theoretical point of view, but we will seek to sketch the less formal and most intuitive aspects by focusing our attention on its historical trajectory above all. When we spoke of the Cabibbo, Kobayashi, and Maskawa (CKM) model (see chapter 13), it was said that the quantum states that describe quarks can mix between themselves just as waves, so that from the point of view of the weak force, the transitions between members of the doublets involve all six quarks though the CKM matrix. Something similar happens for neutrinos, in the hypothetical case that they possess mass—which we know today is small enough, but not altogether null. In effect, no matter how many experimental attempts we have made to do so up until now, a mass for the neutrino that is different from zero has still not been measured, despite having indirect proof, as we shall see, that such mass is not null. We have only determined upper and lower limits for it. For the electron neutrino produced in the β decay, we can affirm that its mass is less than a couple of eV/c^2—certainly no more. For the rest, all fermions have a mass, and it would be difficult to swallow the idea that the neutrino uniquely lacked it. Nevertheless, the stringent limits on the neutrino mass present a real cause for head-scratching because they correspond to possible mass values that are so very much smaller than those of the other leptons and quarks. This is, in fact, one of the great problems of particle physics, due to its manifold implications, including of an astrophysical and cosmological kind. In short, then, here's something that it is worth working hard to resolve ... There are, both in the design and running phases, experiments that are potentially receptive to neutrinos with mass as small as less than a tenth of eV/c^2, developed by exploiting the β decay of radioactive sources (figure 13.2 in chapter 13) and conducted through accurate measurements of the distribution of the energy of the emitted electron. We can mention, for example, the large and complex KATRIN experiment at the Karlsruhe laboratory in Germany.

Let us assume then that neutrinos have a mass, and let us apply the same arguments that we used to describe the original Cabibbo model of quark mixing. Obviously, all of the considerations that we shall make apply equally to antineutrinos. Thanks to their flavor, neutrinos interact with the other fermions through the weak force, and consequently, we define the three flavor eigenstates as v_e, v_μ, and v_τ. Each neutrino, by interacting with matter, will produce its charged lepton partner, therefore making it possible to identify its specific flavor. In a wholly analogous way to how it was done before in relation to quarks, in keeping with the hypothesis that the neutrino has mass, we can also define the mass eigenstates v_1, v_2, and v_3, the eigenvalues of which are the three different (hypothetical) quantities m_1 m_2, and m_3. If we take a state of arbitrary flavor, such as v_μ, we can then express it as a combination (or superposition since wave functions are involved) of the mass eigenstates: $v_\mu = av_1 + bv_2 + cv_3$. We thus obtain the result—which at first sight is surprising—that the mass for our muon neutrino (if not null) will not be defined unequivocally, but it will be able to assume each of the three values m_1, m_2, or m_3, corresponding to the three mass eigenstates v_1, v_2, and v_3 through the values of the three coefficients a, b, and c, respectively. These considerations teach us that the question "What is the mass of the muon neutrino?" makes little sense. Every time that a pion decays into a muon and a muon neutrino, the latter can assume, according to the relative weights, each (actually all) of the three eigenvalues m_1, m_2, or m_3. This fact may be difficult to accept, but it is permitted by the bizarre laws of the quantum world. The uncertainty as to which is the specific value of the mass detected in an experiment is dissolved in the interaction is also an effect that is strictly quantum in character.

By extending our arguments to the three flavors of the neutrinos, we will develop a system of three equations:

$$v_e = a_e v_1 + b_e v_2 + c_e v_3$$
$$v_\mu = a_\mu v_1 + b_\mu v_2 + c_\mu v_3 \quad (6)$$
$$v_\tau = a_\tau v_1 + b_\tau v_2 + c_\tau v_3.$$

The various coefficients constitute a matrix that we shall call, unimaginatively, of *mixing*. Such a matrix can also be inverted to obtain another that makes it possible to express the mass eigenstates in terms of flavor eigenstates:

$$v_1 = a_1 v_e + b_1 v_\mu + c_1 v_\tau$$
$$v_2 = a_2 v_e + b_2 v_\mu + c_2 v_\tau \quad (7)$$
$$v_3 = a_3 v_e + b_3 v_\mu + c_3 v_\tau.$$

By now, the reader will have understood that we can equivalently describe each mass eigenstate according to its content of v_e, v_μ, or v_τ. Nature will tell us how likely it is for v_1 to coincide with an electron neutrino or, similarly, what the probability is that a neutrino of mass m_2 appears as a tau neutrino. The mixing matrix has particular mathematical properties, which we will pass over for the purposes of this discussion. I will only say that it corresponds to the operation of rotation in an abstract mathematical vector space that represents the various neutrino eigenstates. For this reason, in an altogether equivalent manner, we speak of both the coefficients of the matrix and the rotation angles θ. It remains understood that the values of the coefficients, such as for the CKM matrix, must be determined experimentally. But how?

A very important phenomenon takes place if we study the propagation of neutrinos in space (and hence in time), keeping in mind their composition as expressed by the mixing matrix. For a rigorous treatment, we would need to fill a page with calculations. The result may be summarized by saying that because the wave function of a given neutrino flavor (e.g., the muon neutrino) includes the three eigenstates v_1, v_2, and v_3, it will be able to assume the three mass values m_1 m_2, or m_3. Given that these latter, by assumption, are different from each other, the three mass eigenstates will propagate in spacetime in an equally different manner. This will result in the instantaneous composition of the muon neutrino in terms of mass eigenstates varying in a wavelike manner during the journey of the neutrino. A result similar to the classic case, in which the superposition of sounds with slightly different frequencies produce the so-called, more or less cacophonous, "beats." The effect of such a propagation is dramatic. After a certain distance L from the point of departure (or of production), a muon neutrino will be able to acquire the composition of mass eigenstates appropriate for its sibling tau neutrino. Hence, an experimenter who places a detector at a distance L will have a good chance of detecting a tau neutrino in place of the initial muon neutrino. Given that such a spacetime dependence has an oscillatory character—from whence the term *neutrino oscillation* comes—after a subsequent length L, a similar detector could again identify a muon neutrino (figure 14.9). The caveat is that not knowing a priori the values selected by nature for the coefficients of the mixing matrix (or the angles) and for the specific mass eigenvalues, if we wanted to collect evidence for such a phenomenon, we would have to proceed by trial and error, performing experiments at varying distances and using neutrinos of different energies. In effect, as well as the distance L, the process depends on the energy E of the neutrinos. To be more precise, it should be said

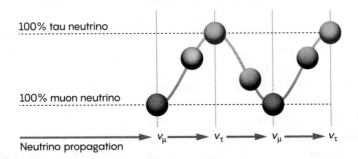

FIGURE 14.9

The mechanism of neutrino oscillations. A muon neutrino during its propagation may be detected, for example, as a tau neutrino, as a function of the distance from the point at which it was produced. The process has an oscillatory behavior (hence the term neutrino oscillation). At the points corresponding to the function's minimum and maximum, the probability of oscillation is lowest and highest, respectively. The scheme shown is an ideal one: in reality, a given oscillation channel (e.g., from v_μ to v_τ, as in the illustration) may not change the flavor of all the starting neutrinos, but only of a fraction of them.

that the oscillation probability is in reality a function of the L/E ratio. But once again, this does not alter the essence of our subject. It is, furthermore, important to note that the detection of neutrino oscillations necessarily implies that they have mass (this is the starting hypothesis), but it does not allow us to derive each mass's value. The oscillations depend in fact on the differences between the squares of the mass eigenvalues: $\Delta m^2_{12} = m_2^2 - m_1^2$ and $\Delta m^2_{23} = m_3^2 - m_2^2$.

From the physical point of view, for the oscillations to take place and be detectable in concrete terms, certain conditions need to be respected (in particular, the relative similarity between the neutrino mass eigenvalues), and these are predefined by nature, along with the level at which mixing occurs. The latter is qualified by the value of the mixing parameters (angles). In fact, we do not know a priori if after an oscillation length, all the starting neutrinos, or just a fraction of them, will have changed their flavor. Only the experiments will be able to tell us if, for the values of L and E selected by us physicists, the oscillations manifest themselves in that specific regime—and then, if so, with what values of Δm^2, and with what intensity. Each oscillation channel that is studied, such as $\nu_\mu \to \nu_\tau$ or $\nu_e \to \nu_\mu$, allows us to determine the relations that relate the values of Δm^2 to the mixing angles θ. It should be fixed in the mind—because *repetita iuvant* in physics—that we talk about angles from the mathematical point of view; the matrix does nothing but rotate the vectors represented by the mass and flavor eigenstates. This concept is illustrated schematically in figure 14.10. The rotation between the two sets of eigenstates of mass and flavor causes it to be the case that each mass eigenstate becomes a combination of flavor eigenstates, and vice versa, just as is illustrated mathematically by equations (6) and (7).

In 1952, when only the electron neutrino was known, Bruno Pontecorvo (yes, him again) was the first to propose the mechanism of oscillation and its application to the system constituted by the neutrino and its antineutrino. Truth be told, it was not even known at the time whether neutrinos existed or not, as Reines and Cowan had not yet carried out their momentous experiment. A few years after the discovery of the ν_μ, in the context of the so-called Nagoya model, Ziro Maki, Masami Nakagawa, and Shoichi Sakata proposed a more general paradigm of mixing between the ν_e and ν_μ flavors, with the two states dubbed "real neutrinos" ν_1 and ν_2. The mixing among neutrinos endowed with mass anticipated by Pontecorvo consequently was assumed in a manner akin to that which happens with quarks, according to the original hypothesis advanced by Cabibbo, Kobayashi, and Maskawa. For the neutrinos, the analogous mixing matrix is known today

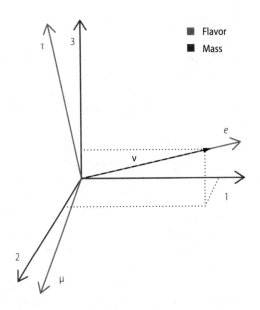

FIGURE 14.10
Graphic representation of the mixing of the mass and flavor eigenstates of the neutrinos. Thanks to the values of the rotation angles (parameters), each flavor eigenstate acquires some components of the mass eigenstates, and vice versa, in accordance with equations (6) and (7).

as (guess what?) *PMNS*—after Pontecorvo, Maki, Nakagawa, and Sakata. It appears that the mixing matrices invariably bear the name of an Italian together with those of two or three Japanese ...

Pontecorvo's original theory remained an academic hypothesis for some time, and in any case, it was largely ignored by the community of physicists—perhaps in part because it was initially published in Russian. But something began to happen toward the end of the 1960s, when Ray Davis undertook a study aimed at the first experimental detection of solar neutrinos: these were believed to be produced by the Sun in its capacity as a gigantic thermonuclear furnace. Davis's experiment, to which we have previously alluded, was one of the longest projects in the history of particle physics, lasting for some thirty years. In all this time, Davis worked determinedly and unstintingly, with the aim of achieving a complete understanding of that which was becoming increasingly highlighted as a terrible problem, a genuine enigma: the "solar neutrino puzzle," as generations of physicists have called it. The problem lies in the fact that with his experiment, Davis collected many fewer neutrinos than he expected according to the theoretical model that described the functioning of the Sun. I've already said that the detection method—a very astute one—was proposed by Pontecorvo in 1946. The idea that forms its basis is that an electron neutrino coming from the Sun can interact with the chlorine-37 contained in a liquid solution consisting (in practice) of a chemical detergent, producing by an inverse β reaction the transformation of a neutron into a proton, accompanied by the creation of an electron: $v_e + {}^{37}Cl \rightarrow e^- + {}^{37}Ar$. In practice, this transforms the atom of chlorine-37 into an atom of argon-37. The latter is radioactive and with an average lifetime of about thirty days it decays into chlorine-37 again, with the emission of a neutrino and of a relatively energetic γ-ray. In principle, extracting the few radioactive atoms of argon-37 by radiochemical methods and identifying them through their decay, we can obtain a measurement of the number of neutrinos that interact with the apparatus, and consequently an estimation of how many are produced by the Sun.

Though apparently complicated, after a few initially unsuccessful attempts, the method seemed to function effectively. The experimental setup consisted of a large metal cylinder filled with about 600 tons of chlorine solution that was installed underground in the decommissioned Homestake gold mine in South Dakota, and of a complex radiochemical system for the extraction of argon-37. Ray Davis worked in close collaboration with another key figure in the saga of solar neutrinos, the American physicist John Bahcall. After pursuing various interests in the field of physics and beyond—he even considered the option of becoming a Reform rabbi—Bahcall decided to devote himself to the theoretical study of the Sun and the mechanism whereby solar thermonuclear reactions produce electron neutrinos of different energy. During the course of many years, he established a solid collaboration and friendship with Davis (figure 14.11) and developed in an equally obstinate and determined way what is known today among physicists, after Bahcall's untimely death in 2005, as the Standard Solar Model, or Bahcall's Model. In the early 2000s I had the pleasure of meeting John at a conference in Naples, and I still remember his kind and affable character, in addition to his outstanding physics skills. Bahcall was convinced of his calculations, which stubbornly failed to match up with the number of neutrinos gathered by Davis. More than half of the neutrinos that were supposed to have left the Sun failed to turn up at registration.

Some time passed before the community of physicists realized that there was indeed a problem—and a very serious one at that. It was quite natural to ascribe to the theoretical estimations, or to the experimental detection, some errors that artificially created a disagreement. But nature often plays

FIGURE 14.11
Ray Davis and John Bahcall in the depths of the Homestake mine, searching for solar neutrinos.

strange tricks: John and Ray were both right. In 1964, they simultaneously published two articles in which they certified the start of the puzzle of solar neutrinos, affirming that to the best of their knowledge, the sums did not add up. Bruno Pontecorvo was the first to suggest a solution to the enigma in 1967, based on theoretical considerations that he had developed years previously—namely, the quantum mechanism of neutrino oscillations. The proposed solution really represented the proverbial egg of Columbus. The Sun, very deep within its interior, produces electron neutrinos, and once they have risen to the surface, they begin a journey of 8 minutes and of 150 million kilometers, which brings them to an interaction with Davis's underground detector. Because of the way it is made, this detector is only sensitive to solar electron neutrinos. You will recall that Davis had attempted with the same kind of apparatus, but without success, to detect the antineutrinos coming from a nuclear reactor. On this subject, it is timely to compare the reaction induced by antineutrinos with the formula: $v_e + {}^{37}Cl \rightarrow e^- + {}^{37}Ar$, relative to the interaction of neutrinos. In the first case, a charged current reaction transforms an electron antineutrino into a positron, and a proton into a neutron. We pass consequently to an element with an atomic number that is smaller by one unit; that is, to the immediately preceding element in the periodic table. In order for the second reaction to take place, a neutrino rather than an antineutrino is required. In this case, the charged current reaction transforms an electron neutrino into an electron, and a neutron into a proton—or, if you prefer, a d quark into a u. Hence, we obtain the next element in the periodic table, argon, which, as it decays because it is radioactive, goes back to being an atom of chlorine. Now if the electron neutrinos oscillate into neutrinos of another flavor after their production (muon or tau neutrinos), they will be inactive from the point of view of the interaction with the Homestake detector, consequently inducing an apparent deficit of solar neutrinos. "Elementary, my dear Watson ..."

The interest provoked by Davis's results inspired the emergence of new proposals for even more sensitive experiments. In particular, it was decided to apply the Pontecorvo reaction not only to chlorine, but also to gallium-71, which transforms into the radioactive germanium-71 in a reaction analogous to that of Davis's experiments, but with a lower activation threshold (233 keV instead of 814 keV): $v_e + {}^{71}Ga \rightarrow e^- + {}^{71}Ge$. In this way, it would become possible to observe neutrinos of much lesser energy that make up over 99 percent of the entire spectrum of solar neutrinos! The problem this time was purely practical: to realize experiments capable of furnishing statistically significant results, we would need to use a quantity of gallium so large that it would require the entire global supply of this element ... The thing that may seem incredible is that in the end, three such experiments were actually carried out, gathering and recycling all the available gallium—around 100 tons—and yielding results that were far more precise than those obtained by Davis. The first two experiments were GALLEX and SAGE (the latter name stands for "Soviet American Gallium Experiment"); SAGE was conducted beginning in 1989 in the underground laboratory of Baksan in the Caucasus mountains and still in operation today. It is worth pointing out that the experiment continues to be called by this name, even after the demise of the Soviet Union and the rebirth of Russia (perhaps because the acronym *RAGE* would not be very appealing). On the subject of SAGE, Victor Matveev, the Russian academician and director of JINR, told me a newsworthy anecdote. At the time of the breakup of the Soviet Union, unscrupulous individuals learned of the uniqueness and commercial value of the gallium stockpiled in the Baksan laboratory and tried to get hold of it. It was only thanks to the courage and cool heads of Victor and his colleague, Vladimir Gavrin, that this (potentially disastrous) plan was not successful. Heroic physicists indeed ...

GALLEX and subsequently GNO (Gallium Neutrino Observatory)—both also successful in proving that the measured flux of solar electron neutrinos is markedly smaller than the expected one—were conducted in the underground laboratory of Gran Sasso in Italy, starting in the 1990s. GALLEX in particular was the first experiment that measured in a statistically significant manner the production of low-energy solar neutrinos, the so-called *p-p*, generated in the basic solar nuclear fusion reaction between two hydrogen nuclei (protons). Its precise measurements already appeared to indicate that it was plausible to attribute the origin of the puzzle to something regarding neutrinos rather than the Sun ... It was a truly significant outcome.

The study of solar neutrinos begun by Davis also attracted the attention of a research group that has played a principal role in modern neutrino physics; that is, the physicists of the Kamiokande experiment that was devised and realized in the Kamioka mine in Japan, under the strong leadership of Masatoshi Koshiba. We will have occasion to say more about this Japanese recipient of the Nobel Prize later; for the moment, however, it should be observed that Kamiokande featured a detection method that was radically different from those described thus far. Radiochemical experiments make it possible to derive information on the flux of neutrinos from the Sun only after a careful procedure of extraction of the radioactive atoms that have been produced, as well as an appropriate calibration of the detector. Kamiokande instead was the first large solar neutrino detector operating in real time; that is, it is capable of providing in a fraction of a second the signal of the observation of a neutrino taking place. The principle on which Kamiokande relies is that of the detection of the light produced by a particle that travels in a medium (such as water, in Kamiokande's case) faster than light would in the same medium, due to its continuous interaction with matter (yes, the constant c designates the speed of light in a vacuum; in any other transparent material this speed can be appreciably lower). It's a bit like when a jet travels through the air at a speed higher than that of sound, producing the so-called supersonic bang. The luminous radiation, electromagnetic counterpart to the jet's bang, was discovered in 1934 by the Russian physicist Pavel Cerenkov. This earned him the Nobel Prize in Physics in 1958, together with the recognition of physicists who named the radiation after him. The emission angle of the Cerenkov light along the direction of travel of a relativistic particle is a function of the refraction index of the material that it moves through. For a volume full of water, this angle is around 40 degrees.

Kamiokande consisted of a large "swimming pool" filled with 3,000 tons of purified water and with its internal surface covered with 1,000 very large photomultipliers. The purpose of the latter was to amplify, and therefore reveal, the faint Cerenkov light produced by the electrons, generated in turn by the interaction of the solar electron neutrinos in the water. The photomultipliers are glass tubes under vacuum, a sort of huge electric "valve" with a metal photocathode exposed to a potential (and weak) luminous source. When a photon hits the photocathode, you will recall that the photoelectric effect (as originally explained by Einstein) may take place. Some electrons are extracted from the metal, and their number is multiplied by a series of electrodes positioned in the tube at high voltage, in an avalanche-like process. In this way, a weak light signal is transformed into a robust electric one, acquired and processed by an appropriate electronic circuit. It is important to note that the electrons produced by solar neutrinos "remember" quite well the incidence direction of the neutrino. This allowed Kamiokande to "aim at the Sun," reinforcing the separation of the signal from the nondirectional background events. The fact that the Cerenkov light signals and the response of the photomultipliers are both very rapid allowed Koshiba's experiment to make the first online detection of solar neutrinos and, what is still more important, the observation of a strong deficit, confirming the result obtained by radiochemical experiments. It is also important to observe that Kamiokande was intrinsically sensitive to the high-energy component of the solar neutrino spectrum, providing complementary information to that of the radiochemical detectors. This is due to the relatively high energy threshold required to "turn on" the Cerenkov effect.

The Kamiokande photomultipliers represented a significant technological development. Koshiba convinced the Japanese Hamamatsu company to develop a new tube of size and performance that had been hitherto unthinkable. The result proved to be a resounding success: what we could call a virtuous collaboration between fundamental science and specialized industry—one of the principal tools for producing real technological innovation. The construction of the apparatus was completed in 1983. In 1987, in addition to the already mentioned observation of solar neutrinos, the experiment achieved one of the major discoveries of neutrino physics—to which we will come shortly. This fact should make us stop to reflect that the reason for the experiment, as conceived and realized by Koshiba, was quite different. The name of the project (its initialism) is a combination of the word *Kamioka*—the underground mine in which the detector was installed—and of the letters *NDE*, for "Nucleon Decay Experiment." The original objective was hence to search for the experimental proof of the instability of matter, which according to the theory of the Grand Unification of fundamental forces, would manifest itself through the decay of the proton, the hypothetical process to which we have already alluded. Here is another proof of the unpredictability of experimental physics research: sometimes great discoveries come from experiments designed with altogether different objectives.

FIGURE 14.12
A spectacular "fisheye" photograph of the underground SNO solar neutrino detector in the Canadian mine of Sudbury. The great ball is a container of around 1,000 tons of heavy water (with deuterium instead of hydrogen) in which the neutrino interactions take place, detected by approximately 10,000 photomultipliers immersed in the liquid. The value of the heavy water, which was loaned to the experiment by the Canadian Atomic Agency, is around $300 million. The cost of the loan was a nominal $1.

The debate about the solar neutrino deficit was to all intents and purposes closed in 2000 when the SNO experiment in Canada (figure 14.12), directed by Art McDonald, simultaneously observed a signal of the disappearance of electron neutrinos and the one that came from the interaction of muon and tau neutrinos, in which the solar neutrinos oscillate in keeping with the predictions of the solar model. This measurement represented the first indication, albeit an indirect one, of the "appearance" of the oscillations. The SNO result would win McDonald a well-deserved Nobel Prize. We should point out that the original idea behind the experiment was developed as long ago as 1984 by Herb Chen, an American of Chinese descent. Chen unfortunately died in 1987, at age 45, before seeing his project realized.

Returning to the deficit of solar neutrinos, the combination of the results from various experiments, each one sensitive to a portion of the entire neutrino energy spectrum, made it possible to understand that things were not going exactly according to the scenario of pure and simple neutrino oscillations. The explanation for this is very technical; it took some time for the physicists to formulate an overall paradigm that could take into account all the various factors in the reduction of the flux of neutrinos measured by the different experiments, and that would allow the determination of the corresponding angles of the PMNS matrix. The solution was found by observing that the neutrinos that propagate from the center of the Sun toward the outside begin to oscillate among themselves when they are still inside the dense (inner) solar matter. It should be noted that in the center of the Sun, where the thermonuclear reactions initiate the production of neutrinos, the density is more than ten times greater than that of the Earth! In 1978, Lincoln Wolfenstein already had calculated that something peculiar happens to neutrinos when they travel through dense matter rather than in a vacuum, or through matter with little density, such as air. Around 1986, Stanislav Mikheyev and Alexei Smirnov developed and extended these arguments, arriving at an all-encompassing theory, an evolution of Pontecorvo's original one, that describes the propagation of oscillating neutrinos through matter. Today, the model describes the so-called MSW effect, from the initials of the three physicists. In short, during the propagation of the neutrinos, a resonance is created due to the presence of electrons in the solar matter. Such a resonance (a mathematical resonance, not one of the short-lived particles we encountered in chapter 8) favors and amplifies the oscillations, with specific values for their parameters, which explain perfectly the different deficits of electron neutrinos measured by the various experiments. This represented a great success for the oscillation theory, as well as for the solar model. In a simplified form, the resonance happens because in the neutrino oscillation from one flavor to another, there exists an asymmetry that is nonexistent in the propagation in vacuum or in an insufficiently dense material. Let me explain. On their journey inside the Sun, the electron, muon, and tau neutrinos can vanish because they interact, producing electrons, muons, and taus, respectively, through the charged current reactions shown in figure 14.13a. All this is fine, so far: in the end, there will be fewer oscillating neutrinos, but the reduction will be roughly the same for the three flavors. Analogously, all three neutrino flavors can be subject to the processes of neutral currents

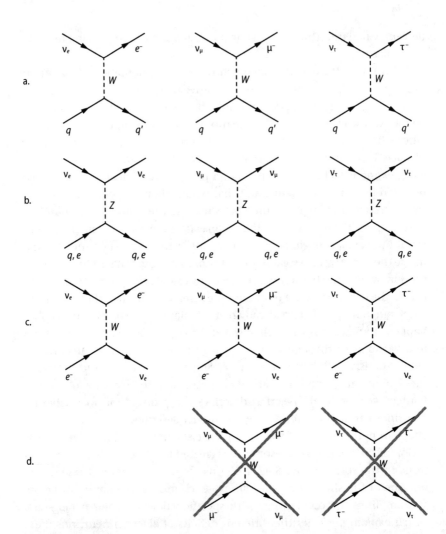

FIGURE 14.13
Neutrino-induced processes relevant for the MSW effect. For reactions (a) and (b), there is a total symmetry between all the types of flavors, but this is not the case for the (c) processes. These, in effect, produce only electron neutrinos, independent of the type of the incident ones. The (d) reactions, which would restore the symmetry, cannot happen because of the absence of muons and taus in ordinary (and solar) matter. The asymmetry generates the resonance at the basis of the MSW effect, privileging the propagation (and hence the oscillation) of electron neutrinos.

with both the quarks of the medium and the atomic electrons, remaining nevertheless active for subsequent oscillations (figure 14.13b). The only difference, which generates the asymmetry between the flavors, and hence the resonant phenomenon, is that all three neutrino flavors can *also* interact through charged currents with the atomic electrons, always producing electron neutrinos when exiting (figure 14.13c), while the muon and tau ones will never be able to be generated (figure 14.13d), given that there are no muons or taus in ordinary matter, let alone in the Sun! The icing on the cake was the Japanese KamLAND experiment, coordinated by Atsuto Suzuki. From 2002, the detector used antineutrinos coming from the Japanese nuclear reactors and verified the oscillation model with artificial neutrinos (i.e., those generated by human activity), proving the exactitude of the results of the solar experiments and their interpretation in terms of the MSW effect.

Now, however, we are obliged to take a step backward because contrary to what might be supposed from our narrative, the first convincing proof (or, better still, discovery) of neutrino oscillations did not occur at the beginning of this millennium, with the definitive solution of the solar neutrino deficit. Rather, it had already happened toward the end of the 1990s, thanks to the study of the so-called atmospheric neutrinos. Such neutrinos (and antineutrinos) exist in all three flavors and have an intermediate energy between that of solar or reactor neutrinos (up to a few megaelectronvolts) and those of the accelerator (from 1–10 GeV upward). They are produced when primary cosmic rays (i.e., protons or light ions of very high energy originating from the depth of the cosmos) interact with our atmosphere—or to be more precise, with the nitrogen and oxygen nuclei of which it is essentially composed. These collisions generate a cascade of secondary particles, typically pions; as they decay, they produce, in turn, electron and muon neutrinos, and their antineutrinos, that reach the surface of the Earth, and in the majority of cases cross it calmly, without any problem. This fact makes it possible to detect those atmospheric neutrinos "that came from below" or are "upward-going," as the jargon says. For example, these may be produced in the strata of the atmosphere above New Zealand, travel the 12,000 km of the Earth's diameter, and with a lot of luck, hit a detector placed in the laboratory at Gran Sasso.

Toward the end of the 1990s, there already was a first indication that something strange was happening with atmospheric neutrinos, and it came once again from the Kamiokande detector. We have just said that atmospheric neutrinos include (mostly) both v_e and v_μ along with their respective antineutrinos. The production chain is as follows (i.e., the pion-muon-electron one that is already well known):

$$\pi^+ \to \mu^+ + \nu_\mu \qquad \pi^- \to \mu^- + \bar{\nu}_\mu$$
$$\mu^+ \to e^+ + \bar{\nu}_\mu + \nu_e \qquad \mu^- \to e^- + \nu_\mu + \bar{\nu}_e. \tag{8}$$

It is evident that, apart from corrections that are in any case easily quantifiable, the ratio between the number of muon and electron neutrinos measurable with a suitable detector placed on the Earth should be equal to 2, as is verified when counting the neutrinos that appear in the equations (8), also adding the antineutrinos. The detectors are usually positioned underground in subterranean physics laboratories. The Homestake mine and the Kamiokande and Gran Sasso laboratories are good examples. The reason for this choice is simple. As well as being traversed and hopefully hit by the elusive neutrinos, the surface of the Earth is hit by those particles that constitute the flux of charged cosmic rays of which we have already spoken, and which have been responsible for many discoveries in elementary particle physics. At sea level, such a flux is essentially made up of muons. I've already said that an approximate estimation foresees that each square centimeter of the Earth's surface is hit by one cosmic muon every minute on average. Incidentally, the energy spectrum of the cosmic rays extends itself well beyond what can be obtained by accelerating protons with the LHC accelerator at CERN. Consider that cosmic rays may reach an energy equal to 10^{11} GeV—roughly 10 million times more than the maximum for the LHC accelerator. Each of these projectiles has the energy of a tennis ball thrown at 100 km/h! That said, these extreme events are really very rare: less than one per century for a square kilometer of the Earth's surface. This, however, helps us to appreciate how powerful the cosmic accelerators are—those cataclysms that take place in the depths of space and are capable of generating incredible quantities of energy in processes that we can't even begin to imagine. It reminds us of the sheer magnitude of the universe ... The flux of cosmic rays, therefore, may undoubtedly interfere with the detection of atmospheric neutrinos—already the proverbial search for a needle in a haystack because they are as elusive as their solar cousins. It is for this reason that detectors are situated underground. In such conditions, as it is poetically expressed, cosmic silence prevails, with muons being effectively absorbed by the overlying rock.

Continuing a line of research that had been carried out unsuccessfully in preceding experiments (such as IMB in the United States, Frejus in France, and NUSEX in a small Italian laboratory in the Mont Blanc highway tunnel),

Kamiokande managed to show a signal of the disappearance of muon neutrinos with respect to expectations. Such a disappearance was not observed for electron neutrinos, the flux of which was found to be very compatible with the expected one. It seemed genuinely to be the story of the deficit of solar neutrinos compared to the Bahcall model repeating itself in this context. The Kamiokande detector, once again used with a different aim than originally intended, was capable of distinguishing the Cerenkov light signal produced by an electron generated by the interaction of an atmospheric electron neutrino from that of a muon originating from a muon neutrino. The latter signal appears much more defined than the former, given that muons, in the act of producing the light of the electromagnetic bang, undergo on average fewer interactions (and hence deflections) than electrons. It's a bit like how the luminous and unstable beam of a bicycle lamp differs from that of a car, which is relatively fixed. This is illustrated in figure 14.14, which shows two events in Super-Kamiokande, the big brother of Kamiokande, due to an electron and a muon neutrino respectively. I'm sure that you will see the difference immediately: our visual organ is particularly well adapted to this kind of image recognition and reconstruction.

The Kamiokande result indicated (albeit in a nonconclusive manner) that the deficit of atmospheric muon neutrinos depends on the point of production in the Earth's atmosphere, which may go from around 20 km for neutrinos produced at the zenith, to around 12,000 for those generated at the antipodes (figure 14.15). Moreover, the detector also made it possible to measure the neutrino energy. As you will recall, the oscillation probability, and consequently the apparent disappearance of the neutrinos, depends on these two quantities: the path traveled by the neutrinos L and their energy E. A rigorous analysis of the data gathered made it possible to conclude that the electron neutrinos did not oscillate, and that their flux turned out to be almost unchanged; muon neutrinos, on the other hand, oscillated with the expected dependence on L and E, probably into tau neutrinos, given that the flux of electron neutrinos certainly did not increase. The follow-up, gigantic Super-Kamiokande detector (figure 14.16) allowed it to take the bull by the horns, as the saying goes. The great amount of collected data, the improved knowledge of the production models of atmospheric neutrinos, and the corresponding possibility of reducing the systematic errors of the measurements made it possible for Super-Kamiokande to arrive in 1998 at the firm conclusion that atmospheric muon neutrinos oscillate into tau neutrinos, with oscillation parameters (the angles of the PMNS matrix) different from those associated with solar neutrinos,

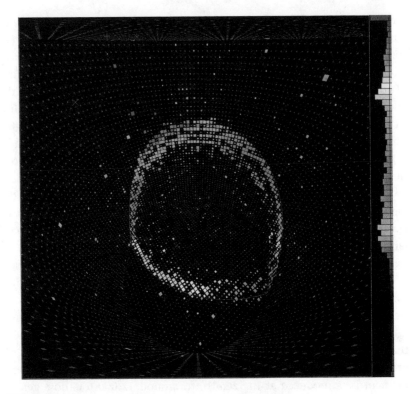

FIGURE 14.14
Two events of the Super-Kamiokande detector. Every luminous dot represents the signal of one of the large photomultipliers immersed in water. The different gray levels indicate the time of arrival of the signal. The rings produced by the Cerenkov light are clearly visible: the one due to the muon is more markedly defined (top), while the one caused by the electron (bottom image) is more blurred. Courtesy of the Super-Kamiokande collaboration.

thus allowing for the determination of some other unknown angles of the matrix. On this subject, we should recognize the great contribution made by Yoji Totsuka, who guided the Super-Kamiokande collaboration for many years after its original conception by Koshiba. Totsuka, who died prematurely in 2008, played a crucial role in leading the scientific collaboration and in making the discovery.

In 1998, at the NEUTRINO98 conference in Takayama, Japan, I had the great pleasure of attending Takaaki Kajita's presentation: having substantially contributed to the analysis of the data on atmospheric neutrinos, he was charged with presenting to the physics community the wonderful results of Super-Kamiokande. It was a historic announcement—the discovery of

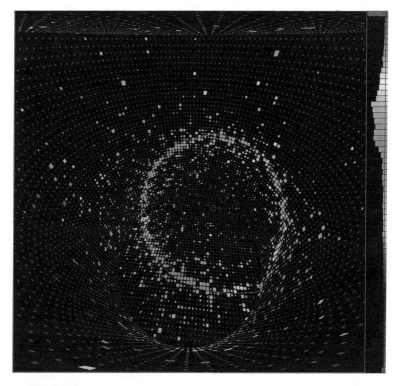

FIGURE 14.14
(continued)

neutrino oscillations, no less. The demonstration that our particle has a mass, and that the lepton numbers L_e, L_μ, and L_τ (as defined in chapter 12) are not conserved separately. This latter consequence derives from the change in flavor of neutrinos during their oscillation. It provides also a proof that the conservation principles do not represent a revealed truth … I remember that Kajita-san's talk received a standing ovation—something that is pretty rare at a physics conference. When only a few years later, the results of SNO and KamLAND concluded in a similar vein the debate about the oscillation of solar neutrinos, it was evident that Pontecorvo's idea was indeed correct, and that particle physicists had the certainty of a new property of neutrinos, capable of rendering these particles even more mysterious. Unfortunately, Bruno Pontecorvo died five years before having the proof that his intuition was correct. As stated previously, in 2015 Kajita and McDonald deservedly received the Nobel Prize for their great contribution to the discovery (figure 14.17). A certain bitterness remains, however,

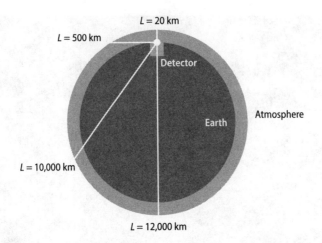

FIGURE 14.15
Diagram of the trajectory of atmospheric neutrinos, from the production point to a hypothetical underground detector. Neutrinos are produced in the approximately 20-km-thick atmosphere and reach the detector after a path ranging between 20 (at the zenith) and, in the case of traversing the full globe, 12,000 km.

with regard to the fact that Totsuka, Bahcall, and Pontecorvo, who would also have merited this recognition, died so prematurely.

The discovery of the oscillations indicated, moreover, the correctness of the mixing model in which all three neutrinos take place, with three mixing angles and two oscillation frequencies, determined by the differences in mass $\Delta m^2_{12} = m_2^2 - m_1^2$ (corresponding to solar neutrinos) and $\Delta m^2_{23} = m_3^2 - m_2^2$ (relative to atmospheric neutrinos); the first of about 0.00008 $(eV/c^2)^2$, the second of 0.0025 $(eV/c^2)^2$. Such values, while these do not provide a measurement of the mass eigenvalues, tell us that the mass eigenvalues are almost "degenerate"; in other words, that they differ very little from each other, given the very small values of Δm^2_{12} and Δm^2_{23}. Combining the two results, we arrive at the statement that the mass of the lightest eigenstates must be larger than around 50 meV/c^2 (but note the lowercase m, so that it's equivalent to saying only 50 millielectronvolt/c^2). This fact, together with the mentioned upper limit of about 2 eV/c^2, limits by a great deal the possible interval for the mass eigenvalues. On the other hand, the two mixing angles θ_{12} (solar) and θ_{23} (atmospheric) are very high, as if to underline that in practice, the mixing, when it occurs, is close to maximal—in contrast to what happens with quarks, as indicated by the CKM matrix. This is the signal of another problem that needs to be

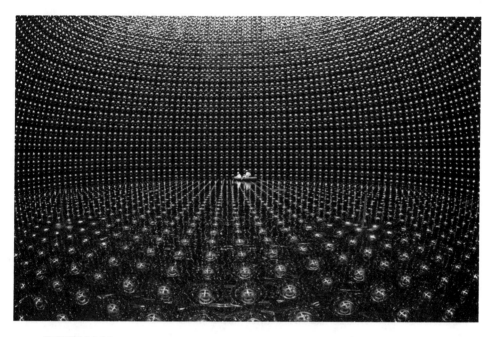

FIGURE 14.16
The interior of the huge pool of 50,000 tons of water that constitutes the Super-Kamiokande detector, photographed during the process of filling it. The large photomultipliers that cover the internal surface of the detector can be seen. In the background are two technicians in a rubber dinghy. Courtesy of Kamioka Observatory, Institute for Cosmic Ray Research (ICRR), the University of Tokyo.

solved by physicists: to understand why nature has decided that the mixing between different quarks and different neutrinos happens with very dissimilar probabilities. But we can't have everything all at once; we must be patient with the mysteries of nature.

What happened next is part of scientific legend—a story rich with puzzles, unexpected results, and discoveries, with physicists, and with their work. The study of the PMNS matrix continued, and proceeds today with strong direction and enthusiasm, thanks to many international research groups. On this subject, I cannot neglect mentioning the OPERA experiment that, together with Kimio Niwa and Paolo Strolin, I proposed long ago in 1997. The experiment situated itself in the grooves carved by projects aimed at verifying the signal coming from natural neutrinos (atmospheric and solar) with particles produced by nuclear reactors and by accelerators, the so-called artificial neutrinos that make it possible to re-create the same

FIGURE 14.17
Takaaki Kajita and Art McDonald, Nobel recipients in 2015 for the discovery of neutrino oscillations.

values of L/E as the natural ones. The aim of the experiment was to identify tau neutrinos originating from the oscillation of muon neutrinos created by an accelerator at CERN. The procedure consists of looking out for the short tracks of the tau leptons (about 1 mm in length) produced by the weak charged currents induced by tau neutrinos. For this task, OPERA utilized the "reborn" emulsion technique on a scale never before achieved, using modern and innovative technologies. A total of 10 million emulsion films for a detector of 1,700 tons were placed in the Gran Sasso underground laboratory (figure 14.18), along the muon neutrino CNGS beam coming from CERN, 730 km away. In 2015, after many years of work, we were rewarded with the discovery of the appearance of tau neutrinos; the same channel as the oscillations of atmospheric neutrinos, and the confirmation of the Super-Kamiokande results: another cherry on the top of the PMNS model. In 2011, as a by-product of the main thrust of research on the oscillations, in the wake of a complex measurement of the velocity of neutrinos sent from CERN to the Gran Sasso laboratory, the experiment seemed to have detected the existence of superluminal neutrinos (i.e., ones

FIGURE 14.18
The OPERA experiment inside Hall C of the National Laboratory of INFN at Gran Sasso. © INFN Gran Sasso National Laboratory/OPERA Collaboration.

that travel faster than the speed of light). Despite the fact that we were not claiming a discovery, but only a very anomalous result, the exceptional nature and impact of the finding immediately unleashed a great outcry in the scientific world—and not just there. The signal was then interpreted by us as being due to an instrumental problem. In that case, the neutrino imparted a very important lesson to everyone: to the scientists, the media, and the public at large. But this too is one of those stories that deserve to be told separately elsewhere.

For now, however, it is worth explaining briefly how a beam of artificial neutrinos created by a particle accelerator actually works. Needless to say, it is impossible to accelerate neutrinos directly—given that they are electrically neutral; they obviously do not feel the electromagnetic force of the accelerating units of cyclotrons and synchrotrons. For the same reason, we cannot bend their trajectories, given their insensitivity to magnetic fields. We are consequently obliged to ask ourselves how we can orientate them so as to send them in the desired direction and, ultimately, from whence they can be sourced in order to obtain a sufficiently intense beam—an essential requirement because of their extremely small cross section. That's a pretty challenging problem, to say the least. In figure 14.19, we can see how a beam of muon neutrinos is produced for experiments such as OPERA. We have seen that muon neutrinos (and antineutrinos) originate from the

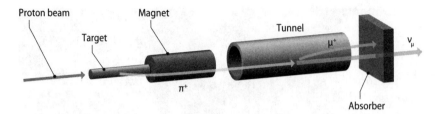

FIGURE 14.19
Schematic structure of a beam of neutrinos produced by a proton accelerator. The length of the infrastructure may vary between several hundred meters and a few kilometers.

decay of pions according to the reactions presented in (8). Let's imagine that we are energizing protons in our accelerator. Once the desired energy is reached, the proton beam is extracted along a tangent of the ring and sent onto an external target; this is typically a cylinder made of a material with small mass number (such as beryllium or carbon), ranging from a few tens of centimeters to a few meters in length, and with a small diameter (less than a centimeter or thereabouts). From the strong interaction of the protons, a little of everything is produced, but predominantly neutral, negative, and positive pions. These conserve the energy of the parent protons, and thanks to the small diameter of the target cylinder, exit undisturbed from within it and proceed in the same direction as the protons in practice. By virtue of the relativistic γ-factor, their lifetime increases significantly. By means of magnets positioned downstream from the target, one has the time to defocus the negatively charged pions and at the same time focus the positive ones—the neutral pions are irrelevant because they decay into two photons and do not produce neutrinos. The beam of (mostly) positive pions is then made to pass through a tunnel, usually between a few tens of meters to a kilometer in length, depending on the pion energy and on the γ-factor, within which vacuum is created. Thanks to this, the pions travel undisturbed, without being subject to interactions with the atoms of the air, and therefore have time to decay into a positive muon and a muon neutrino. And so the die is cast. Thanks to the high pion energy, both the muon and the neutrino are practically collinear with the original collimated pion direction. The muon is stopped by an absorber made of metal and earth, while the neutrino continues calmly on its way toward the experiment that awaits it further downstream. Conceptually, the method reminds us of how a rifle works. But not being able to point the projectile (the neutrino) at the target, we take aim with the butt of the rifle (the decay tunnel) and then shoot. It seems really cumbersome, but as a method, it works—and how. Due to the high intensity of the initial proton beam, the number of neutrinos produced is equally high. Consider that with a modern neutrino beam infrastructure for such as the ones built at CERN, at Fermilab, and in Japan, we succeed in producing more than 10^{20} neutrinos per year. It is timely to remark at this stage that for an experiment over a long distance, such as the beam between CERN and the Gran Sasso, the neutrinos complete their trajectory, crossing the Earth's crust without damage. Their notoriously small cross section means that in the 2.4 ms of the journey from Geneva to central Italy, very few of them are lost along the way by interacting with the rock, and even fewer (unfortunately) manage to interact in the detector placed at a 730-km distance in the cavern of the underground laboratory—just a few dozen per day. The vast majority, billions upon billions, will be lost forever, starting an endless journey through the immensity of the cosmos ...

In conclusion, we can say that the scientific results obtained by neutrino physics in the last twenty years, thanks to the realization of complex international projects, have been genuinely spectacular. They have helped to make giant strides possible in the understanding of the neutrino and its relation to the other elementary particles. Many of the scientific aspects of this advance are too complex to go into here. However, I can tell you that there is a good deal yet to be done, and that as a result, as we will discuss at the end of the book, we already have projects in the pipeline for the next twenty years and beyond. Farsightedness and determination are required ingredients for the realization of future scientific projects, in particular for particle physics research. I would like to end our dossier on neutrinos with a few necessary considerations of their numerical and quantitative relevance—undoubtedly another salient factor pertaining to our particle—and to tell you about an outstandingly important event that took place on a cold February night in 1987, an event that opened the way to new and unexpected fields of research in the physics of neutrinos. But let's not get ahead of ourselves.

We are in no danger of exaggerating when we state that we are literally immersed in a sea of neutrinos—invisible, ungraspable, but enormously numerous. Only a second after the birth of the universe, when it had cooled enough to release the many neutrinos that had been produced up to that point, they began to travel at the speed of light in all directions that space allowed at the time: a space that was rapidly expanding, dragged by the mass of all the superheated matter that filled it, and reducing with extraordinary rapidity the huge density and temperature. In that instant, neutrinos experienced what would happen to photons 380,000 years later. These photons, corresponding in energy terms to visible light, were then finally capable of extricating themselves from the continuous interactions with the surrounding matter. Today, we can observe this original luminous radiation at an extremely low frequency (or high wavelength), typical of radio waves and no longer of visible light, due to a mechanism similar (but not identical) to the so-called Doppler effect (named for Christian Doppler, the Austrian physicist who first described it). Briefly, this effect occurs when a source of waves—of light or sound, it matters little—is moving relative to a receiver or listener. What happens is that if the source is moving away, the measured frequency of the wave is lower, and vice versa. It is like how we hear an ambulance siren: as it gets nearer, the tone is more acute (the frequency increases), while it is less so after it has overtaken us and is moving into the distance. A similar thing happened with the photons produced at the moment of the Big Bang; the rapidly expanding space, in which the photons were propagating, produced the effect of the ambulances. The photons appear today at reduced frequency, and so too does the energy, in keeping with the Planck relation $E = h\nu$. For this interpretation, it is crucial to observe that following the expansion of space, all points distance themselves from all others, as happens in every explosion (such as fireworks). This also occurs to us, who observe all the "ambulances" transporting photons of the Big Bang receding from us, thereby reducing the frequency and energy of the detected radiation. However, as I have already said, the analogy with the Doppler effect is not complete, since from the Big Bang it is the whole of space which expanded, dragging with it the primigenial photons. These photon "survivors" of the Big Bang fill the entire current universe and constitute today the so-called Cosmic Microwave Background (CMB), with an energy which corresponds to a temperature of 2.7 kelvins (around −270 degrees centigrade), as discovered almost accidentally by Arno Penzias and Robert Wilson in 1964. Their 2.7 kelvins is really the temperature of interstellar space, which is minuscule but nevertheless higher than the absolute zero represented by 0 kelvin. As an aside, I would like to point out that inside the LHC of CERN it is possible to reach 1.9 kelvins, a temperature that is lower still than that of cosmic space. The largest and coolest place in the cosmos is in Geneva—that is, if we exclude the potential for accelerator technology of our extra-terrestrial friends …

Something similar to what happened to the CMB also occurred to the fledgling neutrinos of the Big Bang that still fill the universe today. From the original starting value of several megaelectronvolts, these fossil or relic neutrinos became degraded to a kinetic energy, which is minimal—scarcely 0.0005 eV. Their density remained nonetheless large (about 300 units per cubic centimeter), second only to that of photons, and billions of times larger than that of protons and neutrons. These latter appear concentrated and numerous only in the form of stars and galaxies, but they are basically absent from the essentially empty deep space that constitutes the vast majority of the volume of the cosmos. Unfortunately, fossil neutrinos have not yet been detected by physicists. Their discovery, like that of the electromagnetic background radiation by Penzias and Wilson, would represent a milestone in science. It would allow us to shift the study of the archeology of the universe from 380,000 years to only a second after its birth, with the fascinating possibility of "seeing" what our universe was like when it was really young. It is true that with the miniature Big Bangs that we reconstruct in the high-energy interactions at CERN, we attain an energy density equal to those existing fractions of a picosecond after the real Big Bang; but the data that we derive from these events, though extremely relevant, are local and confined to circumscribed, minimal amounts of space and matter. It would be quite a different matter to have a global and cosmological vision of the universe in its infancy, provided by the relic neutrino radiation.

So what is the problem exactly? Why is it so difficult to detect relic neutrinos? Unfortunately, the answer is quite simple, and the problem may be insurmountable. We have had occasion to say repeatedly that in order to be detected, a neutrino must interact with matter and consequently transform itself into its charged lepton partner in the corresponding we doublet through a charged current process; or alternatively, it can supply, by way of a neutral current, sufficient energy to a proton, neutron, or electron so that they can in turn recoil and be detected. But in order for all of this to take place, the neutrinos must possess a minimum energy to produce a measurable effect or simply to generate by means of $E=mc^2$ the corresponding mass of their charged partner. An energy of 0.0005 eV is unfortunately not sufficient to enable any of this. Our fine relic neutrinos, although infinitely numerous all around us, are inexorably elusive … So is there no hope, then? Obviously, many potential ideas have been put forward concerning the detection of Big Bang neutrinos, not a few of them quite fanciful. As far as those that are called *lab based* are concerned, one envisions particular processes for which the decay of radioactive compounds is perturbed—giving rise to something detectable—by the interaction with these neutrinos. A method that is indirect, but probably more realistic, entails accurate physical observations of particular features of cosmic rays. In principle, if we have an extremely low energy particle (our fossil neutrino) colliding with another with extremely high energy—let's say one of the superenergetic protons that wander throughout the cosmos—the energy eventually available after the collision is potentially sufficient to produce particle offspring that could be detected on Earth and perhaps carry with them the signature of interaction with a relic neutrino. Let's think, for example, about the reaction $\bar{v}+p \to n+e^+$. It is clear in any case that we remain far from realizing such measurements, and that we have to continue to dream for a while. To dream, that is, of deriving from fossil neutrino data comparable to that which we have obtained in recent years in the precise study of the cosmic microwave background radiation in important experiments conducted beyond the Earth's atmosphere, such as those of WMAP and PLANCK, placed on space artificial satellites.

Returning to our vast ocean of neutrinos, we have another important producer to take into account. We are referring to supernovae—those huge cataclysmic explosions with which the stars can end their life cycle, dying in spectacular fashion. Such explosions are very frequent in our universe, albeit not just within a single galaxy. The Milky Way, for example, sees only a couple of these events per century. But if we take into account how long our galaxy has existed, we have approximately 100 million supernova explosions to date, which apart from having produced the heavier elements that we have on Earth (and within our bodies), have also generated an awesome quantity of neutrinos of every flavor. Incidentally, the enormous shock wave that follows the explosion of a supernova is also one of the principal engines that accelerate primary cosmic rays. This mechanism was studied for the first time by Enrico Fermi. Without wishing to frighten anyone, it needs to be said that any future explosions that are too close could prove potentially catastrophic for life on Earth. The enormous quantity of radiation produced in the explosion of a particularly close supernova (hundreds of light years away or nearer) would certainly not be very beneficial to the health of the various forms of terrestrial life. Perhaps in the prehistory of the planet, this has already been experienced by some ancient (and now extinct ...) forms of life.

On February 23, 1987, a little starlet at the end of its life, in the Magellanic Cloud—"only" 168,000 light years away—decided to transform itself from an extremely modest, luminous dot in the firmament into the most brilliant object in the night sky apart from the Moon (figure 14.20). In fact, the event had taken place 168,000 years before, and its light had taken this long to cover the abyss that separated us from the star. The estimation of the total energy of the explosion arrives at a truly enormous, impossible-to-imagine figure of more than 10^{46} joule. As a reference, the Sun generates "only" 10^{34} joule in one year of operation. And yet only 0.1 percent of this

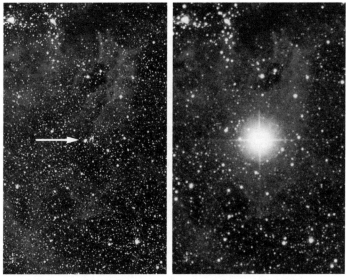

FIGURE 14.20
The explosion of the SN1987A supernova; (left) the star before the explosion, indicated by the arrow; (right) the aftermath. Notwithstanding the remarkable luminosity of the supernova, almost the entire enormous energy of the explosion was transformed into neutrinos.

energy was converted into light. The other 99.9 percent of it was emitted in the form of an extraordinary number of neutrinos. Anyone reading this who was already alive that day would have been doused, despite the abyssal distance, by approximately 10,000 billion of those neutrinos! When Bruno Pontecorvo learned about the event, he said to his nephew Ludovico, now a physicist at CERN: "You know what happened today? 10^{58} neutrinos! All at once!" More significantly, the event marked a fundamental stage in the study of neutrinos in both physics and astrophysics, enabling the birth of the important new discipline of neutrino astronomy. In reality, scientists were not quite ready for an event of this magnitude. Only a few sufficiently massive neutrino detectors were in operation that day—including Kamiokande, the pool of Masatoshi Koshiba. Kamiokande once again did its best in response to an unexpected task. In fact, SN1987A, as the supernova was dubbed, was observed not only by the many optical telescopes thanks to its brilliant light, but also by the photomultipliers immersed within the Japanese detector, detecting the signal produced by a stream of a few but well identified neutrinos through their interactions with the water of the target. The average energy of the SN1987A neutrinos was approximately 5 MeV, as anticipated by the astrophysical models. Needless to say, multiplying the huge explosion energy of 10^{46} joule by the millions of explosions that have taken place in the hundreds of billions of galaxies in the visible universe, we can obtain a measure of the total number, and what is still more astonishing, of the number of neutrinos produced by the supernovas and wandering ceaselessly through space. The observation of those neutrinos earned Koshiba the Nobel Prize in 2002 and opened the way for new and even more ambitious projects, some of which have already been implemented and are operational. If the first Milky Way supernova of this century were to explode, at close (but I hope not too close) range, we will be perfectly prepared to detect its neutrinos and to understand who knows how many further secrets.

The importance of being able to count on neutrino astronomy is self-evident. Observing a cosmic object that is arbitrarily distant places obvious limitations on the principle. For example, the electromagnetic radiation that it can produce is attenuated by its interaction with the few (but not negligible) light atomic nuclei that are found in the intergalactic space. In an analogous way, if we wanted to observe the emission of particles endowed with mass, such as protons, these can interact both with the cosmic background radiation and interstellar gas, and become deflected by galactic magnetic fields, which are weak but very effective over extremely long distances. Neutrinos, by contrast, have an interaction probability that is so negligible that they can reach us undisturbed from the most remote depths of space. Once detected in a sufficiently copious way (and here's the rub), we could reasonably expect to effectively trace back the object that produced it to its origin. Thus there is a need to create neutrino observatories of huge mass, as well as of heightened detection efficiency. We can cite two well-known examples. The first is the huge detector known as IceCube, which has been in operation for several years, and that instead of the pool of Super-Kamiokande, uses millions of tons of Antarctic ice. The Cerenkov light that is thereby generated is read by thousands of large photomultipliers immersed in ice, which turns out, nevertheless, to be quite transparent. The other is the European project KM3NeT, which following the same principle as IceCube, will use seawater from the Mediterranean. The primary aim of such megadetectors is to go beyond Koshiba's discovery, succeeding in observing and studying cosmic sources of high-energy neutrinos. An important result from IceCube has been the observation of a neutrino signal coming from a very distant galaxy, more than 4 billion light years away, with a supermassive black hole in its center. This observation

FIGURE 14.21
With Masatoshi Koshiba in Naples in 2003.

was backed by the coincident electromagnetic signals (visible light, γ-rays) identified by other worldwide experiments.

On the subject of Koshiba, in July 2003, he very graciously responded to my invitation to give a seminar at the Federico II University in Naples (figure 14.21). The lively and engaging Japanese physicist, who was 77 at the time, recounted with passion his life and the results that he had achieved, and began his lecture very appealingly by saying (in Italian): "Good morning, ladies and gentlemen of Naples," before adding, to our surprise, that he knew only one other phrase in Italian: "I have a small stomach." It proved to be an apt one because he availed himself of the phrase several times that evening in the fish restaurant in Borgo Marinari that I had taken him to, while I affectionately offered him *linguine allo scoglio* and *pezzogna all'acqua pazza* ...

Getting back to the subject of neutrino sources, how can we forget the abundant neutrinos produced by the Sun and by the other 200 billion or more stars in the Milky Way? And those generated by the stars of the approximately 50 galaxies in our Local Group? Not to mention the stars of the other maybe 1,000 billion galaxies that constitute the observable universe. These are dizzying numbers, but ones that remind us with scientific objectivity that every star in the cosmos is an efficient producer of electron neutrinos—with the exception of potentially existent far-off galaxies of antimatter, the stars of which would be capable of emitting electron antineutrinos. Keeping our feet on Earth, we must not neglect noting that our very own planet is itself a producer of neutrinos—indeed, of electron antineutrinos generated by the decay of radioactive thorium and uranium trapped in the depths of its crust and mantle. The discovery of these geoneutrinos is recent, having happened almost as a by-product of the study of solar and reactor neutrinos in the Borexino experiments at Gran Sasso and KamLAND in Japan. In this case too, in absolute terms, we find ourselves confronted by enormous numbers. We need only think that the total radiogenic power produced inside the interior of the Earth comprises of between 20,000 and 40,000 GW, of which between 4,000 and 8,000 GW are transformed into neutrinos and the rest into heat. By contrast, you can remember that a standard nuclear fission reactor produces around 1 GW of power. For the rest, it is precisely nuclear reactors spread around the world that provide a small ulterior contribution to the neutrinic balance. If we try to put it all together and place a human being as a point of reference, we can conclude that every second our body is pierced through by approximately 10 million billion neutrino survivors of the Big Bang, by 400,000 billion neutrinos coming from the Sun, 50,000 billion geoneutrinos, 50 billion neutrinos from environmental radioactivity, and between 10 and 100 billion from all the roughly 500 nuclear reactors in the world. These are truly awe-inspiring numbers that would make any kind of life impossible on Earth—or anywhere else, for that matter—if they did not have that extremely small cross section with matter (in this case, living matter). Can we speak, then, of an anthropic principle associated with the neutrino? Incidentally, since we are speaking of the human body, it is worth pointing out that inside each human being, there is a small quantity of potassium-40 present, a radioactive element that emits neutrinos through β decay. On average, each of us therefore produces approximately 300 million neutrinos per day, which leave our bodies at the speed of light and rapidly disappear into the depths of the universe. Consider that the neutrinos produced at birth by an infant who is now four years old have already reached the second closest star to us, Proxima Centauri—a remarkable way, perhaps, of signaling our presence to unknown alien friends, and of continuing with them our study of parity violation begun in chapter 9!

15 SEEING THE INVISIBLE

> Of both visible and invisible things only the gods have certain knowledge; men can only conjecture.
>
> —Alcmaeon of Croton

Over the course of this book, we have frequently encountered particle detectors—and with good reason, because without them, all that has been discussed would remain theoretical, as if suspended in midair. What sense would there be in talking about pions and neutrinos if we had no means of revealing the existence of such more than microscopic entities? In passing, we have crossed paths with some of these instruments: the first cloud chambers, the bubble chambers, the emulsions. It is no coincidence that we are dealing here with detectors that are visualizing and also reassuring: of course, we cannot literally observe the particle, but at least we can see its tracks. Remember the example of the jet plane? Even though it's too distant to be visible, its trail provides unequivocal proof of its trajectory, and ultimately of its very existence.

At the beginning of the 1970s, with the simultaneous progress seen in accelerators and electronics, the old visual detectors were almost retired. Electronic devices began to be developed that were capable of transforming the fundamental interaction processes between particles into electric signals; these, appropriately processed and digitalized by electronic circuits, came to be analyzed by computers that produce digital images of the physical events. This is the current paradigm of modern elementary particle physics, and in figure 11.13 in chapter 11, we have an emblematic example of it, namely, the gigantic and complex LHC detectors. With advances in research, a hunt began for ever-rarer events, for more subtle effects that were consequently also less probable. If the probability of observing a given reaction is low with respect to uninteresting processes, if we are looking for a needle in a haystack and are unfortunately obliged to analyze a considerable number of events that are scarcely interesting, then we must ensure that this happens in realistically short periods of time! The signals of the electronic detectors are intrinsically very fast (microseconds, nanoseconds, or even picoseconds) and the great capacity of modern computers responds to this. Consider an event from a bubble chamber; its analysis could take days, or even weeks. The same event in an electronic detector may be acquired and reconstructed by suitable computer programs in a negligible amount of time, allowing us in just fractions of a second to arrive at a full understanding of the physical processes at stake. As mentioned before, this progress was undoubtedly facilitated by the availability of particle accelerators, which are capable of energizing huge numbers of particles with well-defined characteristics in time intervals that are also extremely short. Particle physics has basically yielded to the tendency of our world and of our society to move ever faster, to not waste a second. I sometimes think about my working days as a young researcher and reflect with a degree of nostalgia that nowadays one of my students can do in an hour what used to take me days of work to achieve, if for no other reason than the exponential increase in the speed and performance of computers.

The electronic detectors may be classified according to the types of particle they detect or, if you prefer, on the basis of the physical quantities they measure: position, momentum, energy, electric charge, etc. Let's try to give a few examples from among the many we have to choose from, dwelling on the physical principles that are at the basis of the detection mechanisms. The argument that we have developed in relation to emulsions and bubble chambers may be applied almost identically to the so-called electric/gaseous detectors. Imagine that you have a metal tube with a wire set at high positive potential at the center, full of an appropriate gas mixture comprised typically of a noble gas (argon, neon, xenon) and a hydrocarbon. The famous Geiger counter, illustrated in figure 15.1, was the earliest example. In the Geiger, the volume originally consisted of only a noble gas. Repeatedly during the course of the previous chapters, we have observed that an energetic and electrically charged particle that crosses through any sort of material produces some ionization, grabbing electrons from the atoms it encounters along the way. These electrons may be identified in the old-fashioned visualizing detectors (emulsions, bubble chambers, etc.). Well then, let's take a step forward. The ionization, as I've explained, is in first approximation much lower, the faster (more energetic) our particle is: the fractional loss of energy per unit length as a function of the particle momentum reaches a minimum, and then in practice remains almost constant for a subsequent momentum increase. In the

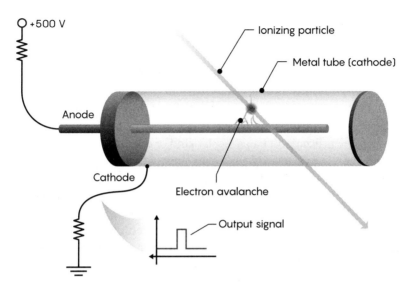

FIGURE 15.1
Operating principle of a gas-wire detector, particularly the Geiger counter.

tube-based detector, the same thing happens: the particle crosses the cylinder and ionizes the atoms of the gas mixture, and the extracted electrons are attracted by the central wire (anode) of high positive potential; given that the filament is very thin, the electric field in its proximity becomes very intense. Therefore, the electrons before being collected are multiplied in a sort of avalanche process: as their energy increases due to the intense electric field that exerts on them a very strong force, they themselves manage to extract other electrons from the atoms they encounter, and so on. In this way, the original few electrons are amplified, eventually producing a robust electric signal—dependent from the total charge of the electrons in the avalanche—collected and analyzed by a suitable electronic circuit. The reason behind the use of noble gases lies in the fact that once hit, their atoms lose energy through ionization rather than through excitations, vibrations, or rotations. The atoms of the hydrocarbon, on the other hand, have the task of moderating the electron avalanche, absorbing the many photons produced in the multiplication, and preventing the generation in the process of too many secondary electrons that would slow down the collection of the electric signal, producing spurious signals. The time it takes to acquire such amplified electrons is extremely short (on the order of a nanosecond). Placing many tubes that are small in diameter next to each other to build a plane, and coupling such a panel with another rotated by

90 degrees, we can create a naval battle type grid that makes it possible to determine the coordinates of the passage of the particle. Then, if the transverse overlapping panels X-Y are of a certain number, we can also follow the particle tracks in the longitudinal direction Z and thereby reconstruct its trajectory in space, as illustrated in the diagram in figure 15.2.

At the time of my undergraduate studies, the Polish-French physicist Georges Charpak was one of my idols. My master's thesis, in fact, concentrated on the realization of a particular kind of detector that he had invented and built approximately two years previously. You can well imagine my excitement when building a similar one, and even getting to see it in operation—after an infinite series of problems, mistakes and torments, of course. Charpak was already famous, because in 1968 at CERN, he had demonstrated that something hitherto considered impossible not only worked, but did so rather well. His idea, which in hindsight appears logical and natural, revolved around putting together very many high-voltage wire detectors in a single gas volume to form what he called a *multiwire proportional chamber (MWPC)*. The chamber is called *proportional* because the generated signal is in fact proportional to the initial release of energy from the particle;

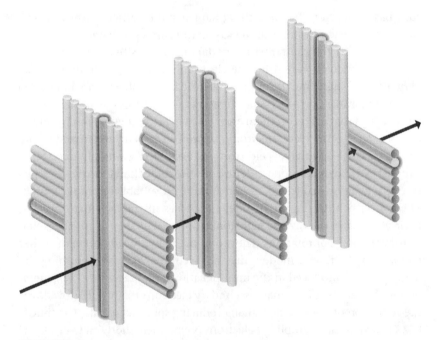

FIGURE 15.2
Consecutive planes of wire tubes make it possible to reconstruct the tracks of ionizing (charged) particles in three dimensions.

and *multiwire* because the structure resembled the strings on a guitar: a series of thin parallel metal wires placed next to each other at a distance of no more than a few millimeters. Following the passage of the particle, we will ideally have a single *hit* wire capable of generating an electric signal useful for detection—two or more in the case of inclined tracks, or of the passage between two wires. Charpak showed to be unfounded fears that the electric signal would diminish across all the wires or would be canceled out by induction phenomena, thereby losing the capacity to determine the actual position of passage of the particle. Figure 15.3 depicts how the MWPC works, and figure 15.4 shows Georges Charpak with one of his detectors.

With the MWPC, particle physics received a very definite boost, because these detectors were immediately used in many cutting-edge experiments, making possible the identification and analysis of a large number of interaction events in extremely small amounts of time. For his invention, and for his many contributions to the physics of particle detectors, Charpak was given the Nobel Prize in 1992, a well-deserved recognition for a great scientist—and for a man who had lived through a large part of the twentieth century, unfortunately one marked not only by scientific progress. Having

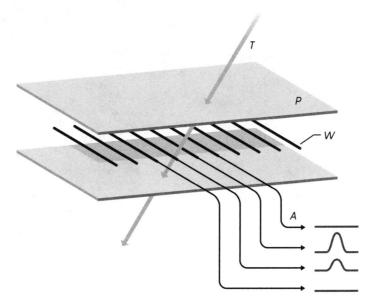

FIGURE 15.3
Operating principle of an MWPC. An ionizing particle track *T* crosses the gas volume of the multiwire chamber. The signal of the electron amplification is registered by some of the wires *W* that constitute the anode *A*. The cathode *P* is formed by the internal surface of the chamber's container.

FIGURE 15.4
Georges Charpak next to one of his wire chambers. Courtesy of CERN.

moved at a very young age with his Jewish family from Poland to France during World War II, Charpak fought in the anti-Nazi resistance. Imprisoned by the collaborationists of the Vichy government, he was deported to Dachau in 1944 and liberated at the end of the war. On this subject, I remember a very significant anecdote. I was having a discussion with him on the margins of a conference in Brazil in 1996. As lunchtime approached, I pointed out to him that everybody was gravitating toward the buffet, and the queue was growing. He politely invited me to go ahead: he would wait until there was no longer a line. Seeing my surprise, he explained that, after his time in the concentration camp, he had resolved never again to "stand in line" in order to eat. Charpak's detectors were immediately followed by others which manifested the creativity of so many physicists dedicated to developing ever more sensitive instruments—ones that were both more efficient and very complex. For instance, they were placed within magnetic fields, making it possible not only to measure the tracks of charged particles, but also to determine their momentum, thanks to the measurement of the curvature due to magnetic force—the larger the radius of the curvature, the greater the momentum of the particle.

To measure the total energy of the particles, we take a different approach. The calorimeters of thermodynamics are in practice akin to thermos flasks like those used to keep coffee warm; they make it possible to determine the quantity of thermal energy exchanged with a body by measuring the temperature variation of the water in which that body is immersed. Stealing its name, in particle physics the calorimeter, as is the case with analogous thermodynamic devices, provides a measurement of the total energy of electrons, photons, and hadrons. The operating principle is based on the physical process of the production of particle showers. The reader is already equipped with the theoretical instruments to understand it: I'd advise you at this point to revisit chapter 10, in particular figures 10.10 and 10.12. Let's imagine that we are starting with an electron of high (and unknown) energy that, in crossing a medium (with many spectator nuclei), radiates a photon. The latter, traveling in the same medium, will be able to convert into an electron-positron pair, and if these have sufficient energy, they too will radiate, producing photons that will be able to convert into other pairs, and so on. It is easy to conceive that this process causes an exponential increase in the number of particles that are sharing among each other the energy of the original electron. This mechanism, called an *electromagnetic shower*, extinguishes when the electrons no longer have sufficient energy to radiate photons (figure 15.5). Obviously, the process of multiplication could very well have begun with a photon instead of an electron. From

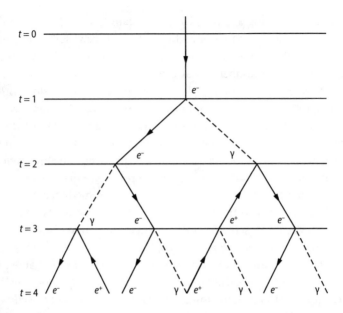

FIGURE 15.5
Conceptual (simplified) development of an electromagnetic shower that takes place when particles propagate in a dense medium. After every fraction of the trajectory traveled ($t = 0, 1, 2 ...$), the number of particles doubles until the shower extinguishes because the energy of the photons becomes too low to produce any more electron-positron pairs.

what has been said, it is easily gathered that the number of particles present in the shower is proportional to the energy of the initial electron or photon; therefore, the total energy released by means of the particle offspring is also proportional to the initial energy. After appropriately calibrating the calorimeter with particles of known energy, it is relatively simple to translate the response given by the detector into the energy value of the incoming particle. From what we just said, it is clear that the resolution in the measurement of the energy improves with increasing the energy, since we have many more offspring particles and hence a reduced statistical uncertainty in their measurement.

In the more general case, calorimeters in particle physics may generate signals that are more disparate: electric, optical, thermal, etc. In a typical implementation of this detection technique, calorimeters are built by alternating layers of a dense material (metal) with detectors of various kinds in order to obtain a signal of the shower secondary particles. We are talking in this case about sampling calorimeters. The example in figure 15.6

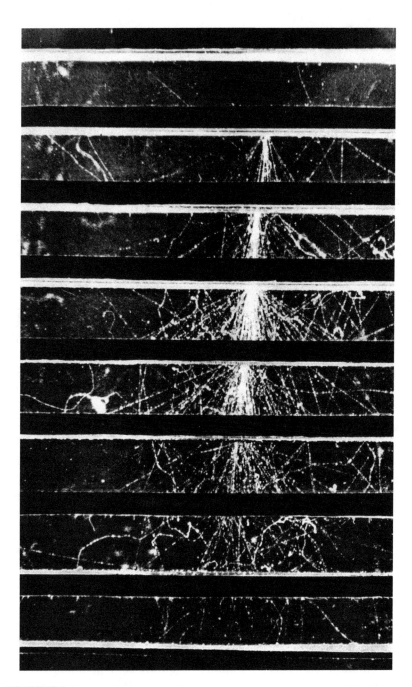

FIGURE 15.6
Electromagnetic shower induced by a primary electron in a sandwich of brass and cloud chambers in alternating layers. After a certain number of stages, the shower reaches a maximum and afterwards starts to become extinguished.

can help us understand. It involves one of the first calorimeters ever developed, constructed by alternating layers of brass and cloud chambers. It is really impressive to have the "visual" proof of what we have predicted, even though the real shower is (rightly) more complex than the approximate model that we described previously. The electron in this illustration, of cosmic origin, starts an electromagnetic shower of which the birth, evolution, and extinction is observed. And all of this occurs in barely a dozen centimeters.

A material that is typically used for the passive layers of an electromagnetic sampling calorimeter is lead. Its high Z value guarantees a large cross section for the processes of photon radiation and pair production, as shown at the end of chapter 10. Furthermore, the high density of lead makes it possible to construct calorimeters that are very compact, and therefore with a thickness of less than 30 centimeters for an adequate interval of energies of the initial particle. Other material used includes copper, iron, and even uranium. As far as the active medium is concerned, physicists have gone to town, utilizing a great variety of materials and devices. I will mention one that is very popular. It entails materials (plastic or crystalline) that are capable of producing so-called scintillation light, from which we derive the term *scintillators*. A scintillating medium has in its composition some molecules that, in the aftermath of interaction with external particles, undergo excitations followed by very rapid deexcitations through the emission of photons, for the majority of the cases within the spectrum of visible light. These weak light signals are generally amplified by photomultiplier tubes about which we have talked in relation to the Kamiokande experiment (see chapter 14). In this case also, the resulting amplification of the light produced by the photomultiplier respects proportionality with the released energy. Figure 15.7 shows a photograph of one of the large calorimeters used at CERN for the LHC experiments. In the top part we can see the mechanical structure, a sandwich of iron plates and plastic scintillator, a material very similar to Plexiglas. The light signal generated in the scintillator material is transported by optical fibers, which are then connected to photomultipliers.

Everything that has been discussed in relation to the measurement of electromagnetic energy (electrons, positrons, and photons) may be extended with the necessary caution to hadronic particles. In the first place, the mechanism for the creation of a hadronic shower, induced by a particle such as a proton or by a high-energy pion, is less predictable and subject to a higher degree of randomness, given that in the interaction of the initial particle, there are many more possible reactions that can occur—typically

FIGURE 15.7
An example of a modern iron-scintillator sampling calorimeter. In the top half of the photo the alternating passive-active material can be seen; below it the scintillation light collection system. Courtesy of CERN.

due to the strong interaction. Second, the dimensions in play are markedly greater. A hadron can travel a long distance before undergoing a secondary interaction. Third, a secondary particle does not always produce a cascade; often it only generates nuclear excitations without a "useful" or detectable production of daughter particles. The final relevant aspect has to do with the fact that in a hadronic shower, electromagnetic particles such as neutral pions or photons can be created that continue the process of energy degradation generating electromagnetic subshowers. In short, this is quite a complex situation, as shown in figure 15.8. The hadronic calorimeters almost always have larger dimensions than their electromagnetic counterparts, and due to the only partial conversion of the initial energy into a detectable signal, their performance is often worse. Nevertheless, at the beginning of the 1990s, in collaboration with a number of international groups, we managed at CERN to obtain a record resolution in hadronic energy (i.e., the most accurate measurement of the incoming hadronic energy). We devised a calorimeter consisting of a dense matrix of lead and plastic scintillating fibers (called the *spaghetti calorimeter* as a result) after an original proposal by my colleague Richard Wigmans. The physical mechanism behind its

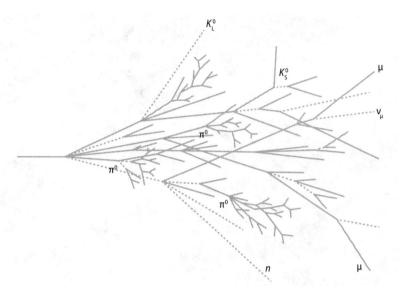

FIGURE 15.8
Scheme of a hadronic shower. It shows the great variety of interactions and the presence of many kinds of particles, in contrast to what occurs in the case of an electromagnetic shower.

enhanced performance is that of compensation. The limited energy resolution in a hadronic calorimeter is substantially due to the various responses to the purely hadronic shower components and to that electromagnetic ones, which are generally more intense. The spaghetti calorimeter, instead, is capable of transforming the majority of the so-called invisible hadronic energy into a detectable signal, thereby compensating for the poorer response to hadrons. In this process, the recovery of the energy transferred to the neutrons produced in the various collisions plays an important role. The latter normally interact sparsely with the active elements of the calorimeter, and we can obviate this by enhancing their secondary interactions with a material rich in hydrogen, such as plastic fibers, for which the neutron cross section is relatively large.

Now let's try to put our information together to attempt to understand what operation principle lies behind one of the most modern and also most gigantic detectors at CERN, and in other laboratories around the world. Before that, however, let's try to answer a question that I think really deserves to be addressed: For what possible reason, in order to detect particles that are even smaller than microscopic, are we obliged to construct such enormous devices, often with dimensions of dozens of meters in length, width, and height and made up of hundreds of thousands (and sometimes millions) of electronics channels for data readout, which weigh 1,000 tons or more and cost a fortune? Well, if we are talking about detecting neutrinos or searching for proton decay, we have already given the answer. In the first case, the cross section of the neutrinos is so extremely small that in order to compensate for this, we are obliged to put together a huge number of target atoms, nuclei, and quarks—and therefore large detector masses. In the second case too, the need for a large number of protons to place "under observation," in anticipation of their decay, implies the need for a very notable mass. But if we collide two proton beams with each other and attempt to show a new particle or a new reaction in the wake of their interaction, what is the justification for gigantism? The reason for this is strictly experimental in kind, as we will see shortly.

One of the main tasks assigned to a large particle physics detector is the determination of the momentum of the charged particles generated in the collision; this is carried out by measuring the curvature of their tracks caused by the action of a magnetic field. The radius of curvature of track trajectories is directly proportional to the momentum, meaning that if the particle is of high energy, even with an intense magnetic field, the curvature of its track may turn out to be very small—the smaller the curvature of the track, the larger the radius of curvature, and vice versa. The radius of curvature is generally measured by sampling the track in given points by position detectors, such as the wire chamber shown in figure 15.9. In the example illustrated in the figure, the bending is measured in three points: x_1, x_2 and x_3. Each of these measurements obviously has an experimental uncertainty that contributes to the error on the final measured momentum. L is the length of the lever arm on which the measurement is made, which in practice is the longitudinal dimension (or thickness) of the detector. One can prove that the determination of the momentum becomes less precise as it increases (something that makes sense ...), that it improves linearly as the value of the magnetic field increases (which is also logical ...), and even—hear, hear!—it goes so far as to improve with the square of L: doubling the length on which the measurement of the x coordinates is performed makes the determination of the momentum four times as accurate! Obviously, all this holds true if the medium through which the particle propagates has little density, a condition that is respected in the case of a gas-based tracking-detector. Otherwise, there appears to be a further factor of degradation in the measurement precision, due to the multiple scattering of the particle with atoms (nuclei) of the medium, but the basic argument is unaltered.

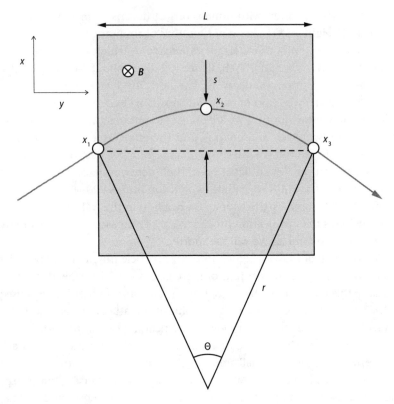

FIGURE 15.9
Measurement of the momentum of a particle. Its trajectory (track) is bent thanks to a magnetic field that is perpendicular to the plane of the figure and exerts a magnetic force on the particle. In the example the track is sampled at three points x_1, x_2, and x_3 belonging to three wire planes of an MWPC. By knowing the value of the magnetic field and measuring the curvature (and the radius r) makes it possible to determine the momentum of the particle.

Another good reason why the scale of the detector is noteworthy is given by the dimensions of the calorimeters. The length of a system consisting of an electromagnetic calorimeter followed by a hadronic one can easily exceed a few meters. Fortunately, this length is just about independent of the energy of the particle that we wish to measure, as it increases only logarithmically with it. Downstream the calorimeter, we must then install detectors for muons, the most penetrating charged particles. These too have a notable thickness along the direction of the emitted particles, if our objective is to measure their momenta accurately, as illustrated above. Hence, it appears

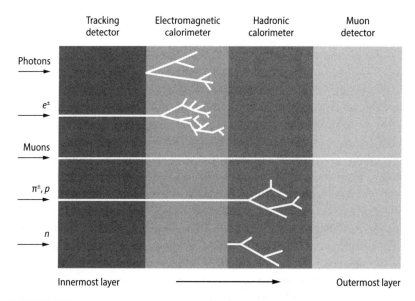

FIGURE 15.10
Diagram of the detection system for various particles produced in an experiment involving a high-energy collider. Each particle provides a specific combination of signals in the various layers of the detector, making it possible for them to be identified almost unambiguously.

clear that for detectors for experiments at high energy, a large bulk is inevitable. To this aspect, we must add a further one that is particularly relevant to experiments at colliders, in which the detectors typically form a cylinder placed around the collision point. Such complex apparatuses are built in the form of an onion, with different detectors as the distance increases from the collision point at its center. This arrangement allows hermetic coverage of the volume around the interaction vertex—nothing must be allowed to escape!—and also the optimization of the detection method of the various particles. I can explain this better by discussing what is shown in figure 15.10.

Imagine that the two particle beams (e.g., of protons) collide at a point along the accelerator ring, and that around it, an onion-shaped detector is installed, with various layers of devices, each with a precise job to accomplish. Immediately around the vacuum tube in which the protons circulate and eventually interact head on, we place tracking detectors (such as wire chambers) shown in dark gray on the far left of figure 15.10. Each wire plane provides a measurement of the position of the tracks generated by electrically charged particles. Electrons, protons, pions, and muons leave behind a wake that's traceable in this detector. Obviously, there is no signal from photons or neutrons. In the next layer, we have an electromagnetic calorimeter (shown in light gray). The protons do not encounter enough matter to generate a hadronic shower in the electromagnetic calorimeter, and as a result, they only leave a track due to ionization energy losses. The neutrons, which are electrically neutral, do not generate any track either in this case. Instead, electrons and photons create showers that eventually extinguish themselves, releasing in practice all their energy in the electromagnetic calorimeter. The next hadronic calorimeter produces hadron showers from interacting protons and neutrons (or even pions). Finally, the wire chambers that are on the outside of the onion structure detect the only charged particles that are sufficiently penetrating to pass through all the upstream detectors—in other words, the muons, which do not shower in the calorimeters because they do not radiate photons (as we have said, due to the mass being relatively high) and, being leptons, are not subject to hadronic interactions. In conclusion, every elementary particle has its distinguishable signature in the detector, according to the combination of signals that it leaves or does not leave at the various detection layers. The presence of a magnetic field would also cause the curvature of the charged particle tracks, making it possible to measure their electric charge (given the direction of the curvature) and momentum (from the radius of curvature).

Clearly, as ever, neutrinos are a special case: they yield no signal in any of the detector layers. For this to happen, we would need to observe a huge number of events and produce an enormous number of neutrinos—requirements that are not commonly met in experiments with colliders, such as the previously mentioned LHC at CERN, where the vast majority of events are produced through strong interactions between colliding protons. In reality, the detection of neutrinos in a collider may be accomplished in an indirect way by measuring the so-called missing energy or momentum: assuming the hermetic nature of the detector and the conservation of the two physical quantities, the observation of an ensuing imbalance in a particular orthogonal direction around the collision point would implicitly indicate the production of an undetected neutrino, or of another unknown particle, which is also noninteracting. The example given in figure 5.15 should be recalled at this stage, in which the decay of the pion and the successive one of the muon—the apparent nonconservation of the momentum—is explained by the emission of invisible neutrinos.

We could fill page after page talking about detectors and their uses in experiments. The ingenuity and inventiveness of physicists have produced

FIGURE 15.11
The CMS experiment at the CERN LHC. The detector is open, and we can see the various sub-detectors of which it is made. Note the cylindrical symmetry around the point of collision of the particles. Courtesy of CERN.

infinite solutions, and refined methods, for the detection and the complete reconstruction of events from particle physics experiments. This is a subject that is too extensive and complex to do justice to in this book, so I once again refer the curious reader to more specialized texts. However, we can show here an example of a large apparatus (figure 15.11): the gigantic Compact Moon Solenoid (CMS) detector at the CERN LHC, which is 15 meters wide, 15 meters high, and 21 meters long, weighs 15,000 tons and has an onionlike structure with tens of millions of independent detectors immersed in a magnetic field of 4 tesla, 100,000 times more intense than that of the Earth. This is a truly complex and efficient machine.

There is an extremely important aspect of experimental research, linked to the use of particle detectors, that needs to be mentioned at this point. We have already had more than one occasion to remark on how a discovery—the evidence of an unknown particle or a new process—manifests itself as something unexpected with regard to known physics: a difference in the expected distribution of a certain kinematical variable, an unforeseen excess of a particular type of events, or a peak in an invariant mass distribution. In other cases, whenever an expected signal is being looked for, or at least something regarded as predictable, it is necessary to know what its manner of manifesting itself in the experimental apparatus could be, in order to be reasonably certain of actually having identified the signal. In all these cases, therefore, it is necessary to know in advance what kind of response we should expect from the detector. Known or unknown physics is described by theories which, through a mathematical representation, make it possible to predict the properties of particles and the distributions of their physical and kinematical quantities as the average energy of the charged particles produced in a reaction, the expected dependence of some background events, or the width of a resonance. But theories only anticipate average behaviors and numerical values, while we have seen that the laws of probability strongly govern the manifestation of nature in the microcosm, at the level of individual events, of the decay of unstable particles or of specific collisions of leptons and hadrons. Moreover, that which we observe with our detectors is almost never what the theory anticipates, but is rather its prediction "filtered" by the senses of the detectors, with their imperfect measurement and particle identification efficiency, which is perhaps better for particles of low energy than for those of high energy; a detector which is, why not, more sensitive to the electromagnetic component of a shower rather than to the hadronic, or maybe such as not to detect particles produced at a certain angle because precisely at that point a wire chamber capable of doing so is missing. In the final analysis even theories need to come up against the hard reality of imprecisions, of technological inadequacies, or simply the omnipresent and unavoidable experimental errors that undermine the measurements.

As we have seen, modern detection apparatuses are very complex pieces of hardware, consisting of various types of detectors, each one with different performance and characteristics, and often endowed with a large number of independent channels for reading the signals, generally of an electronic nature. So how do we take all of this into account? How can we quantify the distortion of the manifestation of nature that occurs in the act of making a measurement? To this end, physicists in particular, and nowadays the scientists of other disciplines as well, have developed software algorithms—computer programs that are generally known as *Monte Carlo method simulations*. The name itself hints at the principle behind it, or rather the randomness of the physical processes and of the response of the detectors, precisely the same uncertainty that occurs when predicting that the ball will land on number 22 on a roulette wheel in a gambling casino … These simulation programs are generally very complex and require both the work of particle physicists and that of information technology (IT) scientists who develop the software that will eventually analyze the data from the detectors, or that carry out calculation of theories and models. Let's try to understand this subject by means of an example. Suppose that we are seeking to determine the angular and energy distributions of the electrons emitted in the decay of a negative muon, just as we would expect it to be from our detector consisting of wire trackers and of an electromagnetic calorimeter. The first are capable of reconstructing the trajectories of the initial muons and of the electrons emitted in the decay, the only identifiable particles in the reaction; the second can infer the energy of the electrons through their electromagnetic showering. It is natural that in order to measure the distributions (regarding angle and energy), it is necessary to have at our disposal many events pertaining to the muon decay reaction. For each of these, for instance, the measurement of the position of the electron will make it possible to determine the emission angle with respect to the propagation direction of the muon. Let's imagine, then, that we are studying experimentally a large number of events, and we ultimately obtain the desired distributions. We will soon be able to realize that these do not in any way correspond to those predicted by the theory of weak interaction! The reasons for this have been explained already: the detectors to a certain extent distort the reality of the physical processes—whatever "reality" means—that is, if "the value" of a given physical variable exists independently of what we do measure, with an experimental error always attached to the result! Anyhow, we can try to simulate the response of our detector to see how the real distributions of the electrons produced in the wake of muon decay may have been modified by the experimental apparatus.

Very generally, a simulation consists of two parts: the generation of physical events and of processed ones. Let me explain what I mean here. The former are the events as hypothetically produced by nature, according to the physics theories that we know. These events, nevertheless, will be markedly different from one another due to the statistical laws of the microcosm (generating fluctuations). For an event, the electron will take, say, 20 percent of the energy of the muon, while the two neutrinos will divide up the remaining 80 percent. Maybe, for a subsequent event, the share will be instead 30 percent and 70 percent. Then, in one event the electron will be emitted at 48 degrees with respect to the direction of the muon, and in a subsequent one at 58 degrees. And so on: we will continue this way, while still respecting for a sufficiently large number of simulated events the average statistical distribution determined by the mathematical predictions of the theory. This is the first task of our Monte Carlo program: to generate a large sample of simulated events, each one with potentially different values—statistically fluctuating—for the physical quantities belonging to the reaction, and therefore in accordance (on the average) with the analytical predictions of the theory. For every event, we will have a combination of variables: the coordinates and the directions of the muon, the emission angles of the electron and its energy, etc.

The second part of the program, the creation of processed events, requires two phases. The first is the simulation of the detector. It is necessary to make a virtual geometrical and functional representation of it. We will need to map the spatial coordinates of the detectors, the possible imperfections, the dead zones, the less performing components, and other elements and translate their responses into an algorithm to be inserted into the program. The second part is more complicated and lies at the heart of the Monte Carlo simulation of our experiment. We start with an event generated in the first phase. Given the coordinates and the production angles of the electron, it will pass in a specific zone of the (simulated!) detector. We will then know whether it crosses a dead region or goes through a zone in which the identification efficiency is larger or lesser than the average design value. Furthermore, by means of our aleatory methods, we always will verify if the energy loss when passing between the wire chambers will be high or low, or if, given the average resolution, say, of 18 percent envisaged for the calorimeter, it will actually be of 15 percent or 20 percent for that specific particle. And so it continues for all the other variables and characteristics of the first processed event. Then we will pass to the second, with a combination of values of the variables and the consequent responses of the detector, albeit always within the interval of the allowed values. And it still continues until all the generated events have been processed. In the end, we will get the simulated event distributions, which we will be able to compare with those obtained in reality with the experiment. This is illustrated in figure 15.12. In the abscissa, there is the energy of the electrons from muon decay obtained by a particular experiment. The unbroken line shows instead the distribution of the Monte Carlo–generated electrons, tracked and made to interact in the virtual detector. The points, with their respective statistical uncertainties, represent the experimental values. The agreement is excellent, which indicates that the simulation is correct and sufficiently accurate. The distribution of simulated data is also much smoother, with smaller statistical fluctuations. The reason for this is simple: it includes a large number of events. In effect, generating Monte Carlo events is undoubtedly more economical (and straightforward) than conducting a real experiment! Obviously, in the example given here, we

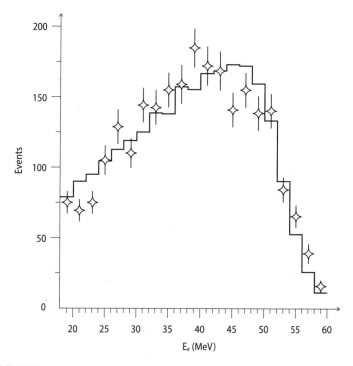

FIGURE 15.12
Example of a Monte Carlo simulation compared to the actual response of the detector. The diagram shows the distribution in energy of the electrons generated in the muon decay (experimental points) superimposed on the prediction of the simulation program (continuous line).

have examined the simulation and measurement of a well-known process. The objective was only to show how the real detector modified the ideal distributions of the events. In other cases, as mentioned before, Monte Carlo simulations are realized—relative to known physics, of course, but then they were confronted with those of the experiments to verify if any observed resonances or anomalies are due to new physics or simply to a particular response (usually an inadequate one) on the part of the detector. Analogously, it is possible to simulate processes and reactions never observed before, due perhaps to hypothetical new particles. The comparison with experimental data, therefore, could lead us to confirm their existence because the distribution of a particular physical quantity would end up being well explained by the predictions obtained by the simulation program.

Nowadays, with the enormous complexity of the experimental apparatuses, as well as that of the processes studied, Monte Carlo simulation programs have become very complicated and ingenious in translating reactions and detectors into numbers and images. They constitute an indispensable instrument, as important as the detector itself. They furnish algorithms that make it possible to predict very well the response of an apparatus with millions of readout channels, such as ATLAS or CMS—so well, in fact, that it almost seems unnecessary to do the actual experiment! Fortunately, however, this does not happen ... These programs can only simulate physics that are known or hypothetical. The real choices of nature only emerge from the experiment, and the experimental data will always have priority over any kind of simulations.

For now, however, it is opportune to concentrate again on the demand for simplification of descriptive and interpretative paradigms of nature that scientists have always developed. Somehow we have succeeded in doing this with elementary particles, passing from hundreds of examples to the Magnificent 12+12 fundamental fermions. So why not attempt a similar approach with regard to forces? Is it really necessary to have four different interactions in nature? Can we not conceive of a single superforce that, for some strange reason, might manifest itself to us in different ways according to the case in question? Yes, we probably can. So let's see if we can shed some further light on this scenario ...

16 UNITED WE WIN

> Every action in life is nothing but a slow process of unification.
> —Piero Martinetti

The fact that something was not altogether clear about weak interaction quickly became evident. You will recall that in chapter 10, we observed that the complex mathematical process of renormalization guarantees that Quantum Electrodynamics (QED) is a predictive theory, and it does not exhibit in its formulas infinite quantities capable of impeding the calculation of processes and reactions. The same happens for the other gauge theory, Quantum Chromo Dynamics (QCD). The mediator for both is a particle without mass—the photon in the case of QED, and the gluon for QCD. In the case of weak interaction, the extremely low intensity is due to its short range, which is in turn explained by assuming the exchange of mediators of very high mass. It can be demonstrated (though obviously we will not do so here) that a theory with mediators endowed with mass cannot be renormalizable. This in effect prevents us from correctly predicting the behavior of some reactions, since calculations bring incongruities such as probabilities higher than 100 percent and other such absurdities like mathematical divergences. This is a really tricky problem. It is, therefore, necessary to find a theoretical mechanism that makes it possible to eliminate such defects of the theory by finding a legitimate way of giving mass to the mediating bosons. We will see how this mechanism exists, provided that one introduces a new field—and its particle-quantum—which eliminates the infinites and, thanks to its interaction, generates the mass of the mediators and preserves the renormalizability of the theory, albeit with massive mediators. Once again, the introduction of a new, hypothetical particle makes it possible to solve the enigma, on the condition that the particle actually exists, that is! The model was first proposed in the 1960s by some brilliant theoretical physicists who came up with the idea of a field and its corresponding quantum—a field that would not only solve the problems outlined here, but also in passing provide mass to all fermions, not just to the mediator bosons of the weak interaction. But let's take one step at a time.

Physics has always discussed—or at the least dreamed of—the unification of forces. The first success in the line of theoretical research was achieved by James Clerk Maxwell, when with his glorious four equations he unified all the manifestations of electricity and magnetism into the coherent theory of electromagnetism. The electric and magnetic fields are transformed one into the other thanks to the relativistic Lorentz transformations. The price to pay for this, if we can put it that way, is that of introducing a fundamental constant in nature, the speed of light in a vacuum c, which is to be experimentally determined. This constant links together processes of an apparently different nature—namely, those related to electricity and magnetism.

In the case of the fundamental interactions, the unification acquires a precise meaning that is very clear. We have seen that the coupling constants of the forces are not really *constant* as a function of the energy; the electromagnetic one increases as the energy grows, while the strong counterpart diminishes. To speak of constants that are subject to variation may appear to be a contradiction, but what is meant by this is that to the couplings between mediators of forces and fermions, there are also contributions from Feynman diagrams of higher order, which themselves depend on the energy, hence determining so-called effective constants (as illustrated, for example, in figure 11.4 in chapter 11). The behavior of the coupling constants of the three fundamental interactions of the microcosm is shown in figure 16.1. Let's start with the strong force. Its intensity grows as the distance between particles increases, until it practically reaches the value 1 for $\sqrt{\alpha_s}$ for distances that are roughly equal to the size of the proton. The constant of the electromagnetic interaction $\sqrt{\alpha}$ diminishes slowly as the distance increases, always remaining much smaller than its strong sibling. The evolution of the weak coupling constant is peculiar; it is already virtually null for distances equal to a few percent of the proton radius; for still smaller dimensions, however, it increases significantly, becoming comparable with $\sqrt{\alpha}$ when the distance between interacting particles is only 10^{-16} cm, a length equivalent to less than a thousandth of the proton radius. As we already said, the reason why the weak force appears less intense than the others, therefore, is due to the large mass of the W and Z bosons that must be emitted in the interaction. In the light of the arguments already tackled

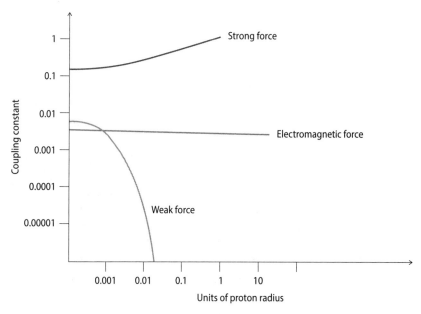

FIGURE 16.1
Behavior of the coupling constants of the fundamental forces as a function of the distance between the interacting particles, in units of proton radius (about 10^{-13} cm).

in chapter 8, virtual particles of such a mass can only be exchanged for an extremely short time, corresponding to a collision distance that is so small as to make it extremely improbable.

In 1968, Sheldon Glashow, Abdus Salam, and Steven Weinberg developed an aesthetically beautiful theory that predicted the unification of the electromagnetic and weak forces in one single comprehensive framework. An important effect of the unification is that, for sufficiently high energies, the respective coupling constants assume a very similar numerical value. The indications that the two forces were connected in some way were quite numerous. Let's consider in particular the charges of the electromagnetic and strong interactions. In the case of electromagnetism, we have the electric charge, and particles that have it can interact electromagnetically with each other. The electric charge that generates the electromagnetic force is a conserved quantity, an indicator of the existence of a symmetry of nature at the base of the mathematical structure of QED. Similarly, quarks are receptive to the strong force thanks to their three possible color charges. In this case also, the existence of a mathematical symmetry principle that generates the structure of QCD causes the color charge to be rigorously conserved in nature. None of this happens for the weak force. We did not show it yet, but in the case of the latter, there is nothing equivalent to the electric or color charge, indicating that if an eventual symmetry holds for very high energies it has to be broken, or if you prefer "violated" in the manifestation of the weak force in our laboratory experiments. Finally, we have observed that the mediating bosons of a force can themselves carry the charge in the interaction, just as gluons do with color charge. But think of the W^+ and W^- mediators of the weak force: they are charged indeed, but with the charge of QED—precisely a sign of some sort of commingling of the two forces. In short, when Glashow, Weinberg, and Salam made known their theory of electroweak unification, it became clear to everyone that a significant portion of the theoretical problems presented by the standard theory of the weak force could finally be resolved.

At the basis of the theoretical framework of the electroweak unification is the hypothesis of the existence of a sufficiently extended symmetry group that, starting from some high energy—or equivalently small distance, or short time after the Big Bang—could describe the characteristics of the unified model, which for lower energies could break generating the two distinct interactions. The theory predicts the unification of the two forces at around 100 GeV, an energy that is easily accessible with modern particle accelerators. Without getting into the detail of the complex mathematics involved, the model assumes that there exist—or, better still, existed in the remotest time immediately after the Big Bang—four bosons without mass (W^1, W^2, W^3, and B), mediators of a more general force than the electromagnetic and weak ones. The interaction is generated by a symmetry group that includes as subgroups those at the basis of the two sibling forces. Thanks to the fact that its bosons are massless, the theory is renormalizable, as was demonstrated by Gerardus 't Hooft and Martinus Veltman—a discovery that earned them the Nobel Prize in 1999. Two of the bosons without mass, with the diminishing of the energy and thanks to the mechanism of spontaneous symmetry breaking, transform themselves into the physical particles that intervene in the charged current interactions that we measure in nature today—that is, the W^+ and the W^-. The latter do not have null mass, but the model of symmetry breaking guarantees that renormalizability is maintained. The other two bosons, the W^3 and the B, remain massless and mix with the two electrically neutral physical states, the Z and the photon, just as happens for the neutrino flavor and mass eigenstates. The result is that the photon, the quantum of QED, remains massless, while the Z acquires a large mass of about 90 GeV/c^2, just as we observe experimentally!

The mixing of the mediator bosons happens thanks to a 2×2 rotation matrix through an angle called θ_w—where W stands for weak, or Weinberg, who made a substantial contribution to the development of the theoretical formalism of the model. The Weinberg angle, along with the electron charge, is the theory's only external parameter to be determined experimentally, just as the speed of light in a vacuum c is the only constant of electromagnetism that unifies electricity and magnetism. Here, θ_w enters into the various quantities that express the unification theory, as well as into the vertices of the Feynman diagrams that describe neutral current's electroweak processes. In particular, through its cosine, it links the masses of the electroweak bosons, while its sine defines the relation between the electromagnetic and weak coupling constants, ultimately making it possible to predict with excellent accuracy the masses of the physical mediators, as shown in the following three relations:

$$M_z = \frac{M_w}{\cos\theta_w} \quad \sin\theta_w = \frac{e}{g} \quad M_w = \frac{37.4\left(\frac{\text{GeV}}{c^2}\right)}{\sin\theta_w}. \tag{9}$$

The Weinberg angle was measured with great precision between the 1970s and 1980s by studying various electroweak processes and comparing the results with the predictions of the theory, which are a function of the angle. In many cases, reactions using neutrinos were used, and the reason for this is simple: given that the neutrino interacts only weakly, the comparison of the experimental results with the theory makes it possible to determine θ_W quite directly, without having to subtract the contributions from other interactions. It should be noted in passing that the neutrinos are assumed to be massless in the electroweak theory. Hence we would have LH neutrinos and RH antineutrinos, while RH neutrinos and LH antineutrinos are simply assumed not to exist. In the wake of the discovery of neutrino oscillations, however, the scheme is no longer rigorously valid. Even though it was at a very low level, by virtue of the minimal mass of our fleeting particles, there exists a minuscule component of neutrinos of wrong helicity. Today, this is considered to be the first experimental evidence of new physics beyond the electroweak theory, and it inspires the theoretical and experimental attempts to extend the original unification model. Finally, it should be stressed again that while weak charged currents only couple to the LH (RH) components of particle (antiparticle) wave functions, neutral currents occur indifferently for both helicity states.

At the beginning of the 1980s, I found myself participating in one of the projects for measuring θ_W: the CHARM II experiment at CERN, the apparatus

FIGURE 16.2
The CHARM II experiment in the process of being built at CERN in 1984. Courtesy of CERN.

of which is shown in figure 16.2. CHARM II consisted of a detector that was of remarkable size for the time—more than 40 m long and weighing about 1,000 tons—but that was conceptually simple. Its design featured a huge sandwich of glass plates for the interaction of neutrinos, interspersed with large planes of wire detectors to measure the tracks of the particles produced in neutrino collisions. The goal was in effect the precise measurement of the square of the sine of θ_W by studying neutral current scattering of muon neutrinos and antineutrinos off electrons, à la Gargamelle. The experiment yielded excellent results, detecting the rare interactions and identifying 1,000 times more events than had been detected a few years earlier by Gargamelle, thereby making it possible to achieve high statistics for a precise measurement of the Weinberg angle. With CHARM II, my own neutrino adventure began. As a young researcher, I spent several years concentrating on the design and construction of the experiment—and afterward on the gathering and analysis of the data. For many months between 1984 and 1985, I oversaw the installation and the putting into operation of the 440 large wire-chamber planes that made up the heart of the detector. Beginning at 8:30 a.m., I would don the overalls of a mechanic and, together with a small group of technicians, I would clamber over the structure of the apparatus, tighten screws, pull and push, wire, test, assemble, and sometimes disassemble. In the afternoon, my team would put down the tools and I could finally head to the office to be a physicist, working at the computer. I mention this to recall good times past, but also to point out to potential scientists of the future that our field is wonderful, but that it also demands dedication, hard work, and sometimes sacrifice. As the late Giancarlo Piredda, a colleague of mine who also worked as a "mechanical worker" on CHARM II, used to say: "You cannot cut corners in physics!" How very true that is.

Today, we know that θ_W is approximately equal to 28.5 degrees and that its sine is about 0.5. For this reason, the second of equations (9) introduces, in effect, the unification of the electromagnetic and weak forces, resulting in two coupling constants of the same order of magnitude. Actually, the value of the g constant at the scale of the unification energy (100 GeV) turns out to be approximately double that of e, which in turn is equivalent to $\sqrt{\alpha}$. Thus, we find confirmed that the apparent feebleness of the weak force is only due to the large mass of the W and Z bosons that must be exchanged in the interaction. The electroweak unification theory is undoubtedly an elegant and successful one—not just because it allows us to reclaim the phenomenology that we know, compatible with the existence of weak charged currents involving W^+ and W^- bosons, but because it also anticipates two truly unexpected results. On the one hand, the theory predicts the existence of neutral currents (mediated by the Z) that were discovered in 1974, thus furnishing strong proof of its correctness; on the other hand, it indicates that neutral currents, which are exquisitely weak, mix with electromagnetic reactions mediated by the photon. The hypothesis that the W^+, W^-, and Z^0 mediators exist as real particles remained for many years based on just indirect evidence. The fact that processes such as those illustrated in figure 16.3 really happened was obviously in keeping with the electroweak theory, but in those diagrams, the bosons are virtual particles

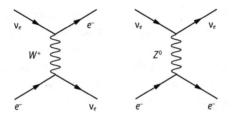

FIGURE 16.3
Feynman diagrams describing ν_e scattering off an electron through both charged and neutral current reactions. In both cases, the mediating bosons are virtual particles. The energy of the colliding neutrino is not generally sufficient to produce a "real" W or Z, given their large mass (of about 80 and 90 GeV/c^2, respectively).

and not directly produced in the experiments, due also to the technical limitations determined by their very large mass. To actually produce and therefore discover them, it was required to have at our disposal an energy that was correspondingly huge.

On account of this, toward the end of the 1970s, a remarkable scientific endeavor was launched at CERN with the aim of making the direct discovery of the electroweak bosons and verifying their physical properties as predicted by the unification theory. The preexisting proton accelerator of CERN, the Super Proton Synchrotron (SPS), was modified so that it would function as a collider, accelerating and causing protons to collide with antiprotons at the very high energy of (270+270) GeV, which was considered to be sufficient to produce real (and not only virtual) particles with a very high mass, and hopefully to verify in the laboratory the unification of the electromagnetic and weak forces. The mass of the electroweak bosons, on the other hand, had already been predicted with a good accuracy, thanks to the measurement of the Weinberg angle. Similarly, the value of the production cross section was also well known, matching quite well with the design performance of the accelerator. With this winning scientific setup, CERN played with a topflight team, spearheaded by Carlo Rubbia and Simon van der Meer (figure 16.4). Needless to say, their role in the development of the proton-antiproton collider (van der Meer), as well as the construction and direction of the UA1 experiment (Carlo Rubbia), was recognized with a Nobel Prize in 1984, immediately after the discovery of the W and the Z. It should be added that for the sake of confirmation, another independent experiment was constructed and successfully operated, UA2, based on a different detection technique. This produced the same spectacular results as UA1.

FIGURE 16.4
Simon van der Meer (left) and Carlo Rubbia, recipients of the Nobel Prize in 1984 for the discovery of the W and Z electroweak bosons.

The principle of the measurement is illustrated in figure 16.5. In this example, a d quark of a proton collides with a $\bar{u}$ of an antiproton (the process would work with a $\bar{d}u$ collision as well). Regarding this objective, we should remark that the possibility of having very intense beams of antiprotons—thanks to the groundbreaking ideas of van der Meer on the so-called stochastic cooling—was one of the principal reasons for the success of the whole enterprise. If the momentum of the quarks is sufficiently high, a W^- boson can be created through electroweak interaction. Given that the W mass is nearly 80 GeV/c^2, it is necessary for the total energy of the collision between the quarks to be at least as high. At this point, however, a brief digression is in order. When we accelerate a pointlike particle, such as an electron or a muon, it acquires the total amount of energy supplied. For this reason, by energizing electrons and positrons (say, to 5 GeV), in their head-on collision we have 10 GeV at our disposal for the creation of new particles in the final state. The situation is different if we accelerate protons and make them collide with antiprotons. If their momentum is low, we know that we will not be able to evidence their internal structure made up of quarks and gluons; therefore, the particles appear to be pointlike, just like

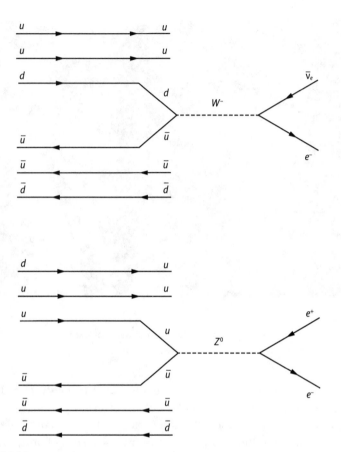

FIGURE 16.5
Diagrams of the production of the W^- and Z^0 at the SPS collider at CERN, in which protons collided with antiprotons. Among the various types of W and Z decays, two of the clearest from the experimental point of view are shown; they were exploited for the discovery of the particles with the UA1 and UA2 experiments.

the electron and the positron: in this case, the collision happens between a proton and an antiproton as a whole. If we place ourselves in the conditions of the SPS collider, however, with two beams of high energy (each of 270 GeV), the collisions likely take place at the level of the constituents of the two hadrons. Now, you will recall that a proton (or an antiproton) has an internal structure that is very complex (see figure 12.10 in chapter 12), with three valence quarks and a sea of virtual quark-antiquark pairs and gluons. Consequently, the proton's 270 GeV/c of momentum must be subdivided among all the internal constituents. We cannot help but observe, for the sake of completeness, that we should always speak of momentum to characterize particles extracted from an accelerator such as the SPS. That said, given that 270 GeV/c is a very high value compared to the proton mass of 1 GeV/c^2, on the basis of equation (1) in chapter 3, we are not making a major error when talking similarly of an energy of 270 GeV. For a given proton, the manner in which the momentum is distributed between the constituents at the precise moment of the collision is wholly random, in keeping with the microcosm's probabilistic laws. What we can do, however (thanks to QCD), is to calculate the statistical distribution of the momentum of quarks and gluons, the so-called density or structure functions, and then to measure them in the experiments. Hence, we know that approximately 50 percent of the total momentum of the proton is taken by the sea of virtual particles, including gluons, while the remaining 50 percent is subdivided among the three valence quarks (u, u, and d). This, I repeat, is valid only when averaging over a large number of events. For each individual collision, ample fluctuations of the quota of momentum carried by the colliding quarks are possible. The moral of this tale is that although the total 540 GeV/c is more than enough to produce the W or Z mass, most times the interacting quarks do not possess the necessary momentum in the collision, and our mediator bosons are not created as real particles.

Complicating life further for physicists, W production can occur only if the colliding quark (coming from one direction) is left-handed (LH, spin = –1/2, with respect to the direction of the momentum) and the antiquark (from the opposite direction) is right-handed (RH, spin = +1/2), to build up a W with spin equal to 1. This requirement does not apply to Z production: as we have already noted, unlike charged currents, the neutral ones occur indifferently for both LH and RH particles, provided their spins add up to make that of the Z equal to 1. Nevertheless, the Z production cross section is appreciably smaller than for the W. Last but not least, the colliding quarks, as shown in both diagrams of figure 16.5, must have corresponding color and anticolor, to ensure that W and Z are both white (no strong color charge). These further requirements make the production of electroweak bosons even less probable. At this stage, if the creation of W and Z manages to occur regardless, the choice of how their decay is to be detected is purely a matter of considerations of an experimental kind. It is understood that given the extremely short lifetime of the electroweak bosons, the only way of identifying them consists of reconstructing and studying the particles produced in the process of their decay. In figure 16.5, we can see two relevant channels employed by the UA1 experiments for the discovery of W and Z, in each case involving electrons. In fact, the decay channels into hadrons are largely affected by backgrounds due to other processes with analogous final states.

The UA1 experiment was based on a typical onion-type detector, like those described in the previous chapter. The experimental apparatus, of considerable size and complexity (as shown in figure 16.6), was equipped with an excellent detection system for high-energy electrons (wire trackers and calorimeters) and was above all a hermetic detector—an innovative characteristic that turned out to be of fundamental importance. In the case of the production of a W^-, its 80 GeV/c^2 may be subdivided into the energy and momentum of the electron and of the electron antineutrino. To conserve momentum, the two particles must be produced in opposite directions in the reference frame where the W is at rest. The identification of the W is therefore translated into detecting an electron of high energy (around 40 GeV), for which there is no appreciable corresponding energy release in the opposite direction due to the escaping antineutrino. We speak of "missing energy" in such cases. The detector being sufficiently compact and hermetic, the sum of the energy of all the detected particles must be compatible with the initial 540 GeV, apart from the energy of the particles

FIGURE 16.6
The UA1 detector at CERN that in 1983 contributed to the discovery of the W and Z electroweak bosons. The detector is installed underground in one of the collision points of the proton–antiproton accelerator. Courtesy of CERN.

produced at small angles that in practice do not escape the vacuum tube of the accelerator. To be more precise, one generally uses the so-called transverse variables for which only the components of momentum and energy in the directions orthogonal to that of the beam are considered. In the end, the apparent missing detected energy in the opposite direction to that of the electron is a strong indication of the presence of an antineutrino that carries with it the lost energy—remember the exposition at the end of chapter 15. This constitutes an indirect detection of a high-energy neutrino that is clever indeed.

The method used for the detection of the Z was analogous to that described for the discovery of the J/Ψ meson. The measurement of an electron and of a positron (or of a positive muon and a negative one), emitted back-to-back in the reference frame where the Z is at rest and with an invariant mass equal to approximately 90 GeV/c^2, is a clear signature for the production of the particle. The subsequent resonance is not appreciably affected by background processes, contrary to what would have happened for hadronic decays of the Z. In figure 16.7 we have one of the events attributed to the production of a Z in UA1, as reconstructed by the computer, in which the boson decays into an high-energy electron-positron pair: the two leptons are clearly discernible and the invariant mass of the pair is compatible with that expected for the Z.

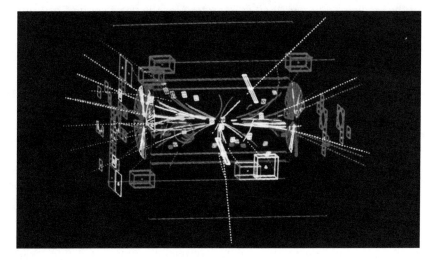

FIGURE 16.7
One of the UA1 events compatible with the production and subsequent decay of a Z boson into an electron–positron pair. The negligible curvature of the tracks is an indication of the high momentum of each of the two particles. Courtesy of CERN.

From the measurement of the W and Z masses, it became possible to determine the value of the Weinberg angle, thanks to the first formula of equation (9). The value derived from this turned out to be in perfect agreement with the one obtained by the lower-energy experiments conducted with neutrinos. The discovery of the electroweak bosons and the proof of the correctness of the Glashow, Weinberg, and Salam theory led to a long series of experiments that have verified in detail the predictions of the model. The electroweak theory, together with QCD, which describes strong interactions, constitutes what today is universally known as the *Standard Model of elementary particles and of fundamental interactions*. This title, while somewhat pompous, seems nevertheless quite fitting for the formidable theoretical framework, supported by so many experimental results that were perfectly predicted by the theory.

In the 1990s at CERN, a large accelerator collider of electrons and positrons was realized—the Large Electron–Positron (LEP) collider—a circular machine with a 27-km-long vacuum tube and situated underground in the tunnel now occupied by the LHC. The primary task of LEP, and of its four large experimental apparatuses, ALEPH, DELPHI, L3, and OPAL, was to measure the Z and its electroweak couplings to all the fermions, with much larger statistics than that of UA1 and UA2. The result of LEP being in operation for so many years may be summarized by observing the behavior of the predicted and measured cross section for the production of hadrons in electron-positron collisions as a function of the energy (figure 16.8). The huge peak, at about 90 GeV, is due to the electroweak creation of the Z. The latter produces a marked resonance, which is very broad due to the numerous final states that are possible for its decay, and hence with an extremely short lifetime: an outcome that had once again been quantitatively predicted by the electroweak theory.

From the study of the Z conducted through the experiments at LEP, many predicted features of the unification theory were accurately confirmed, in particular those related to the coupling of the boson to quarks and leptons. But another extremely important result ensued with respect to the counting of the neutrino species existing in nature. Remembering what has previously been stated in relation to the symmetry between the three fermion families (see figure 12.9 in chapter 12), and about the existence of only twelve of these elementary fermions, we may ask ourselves if replicas of these three families can exist with quarks and leptons of higher mass: two new quarks, a new charged lepton, and consequently another neutrino into which the new lepton can reciprocally transform itself via a charged current reaction. The main avenue for obtaining an answer is to conduct

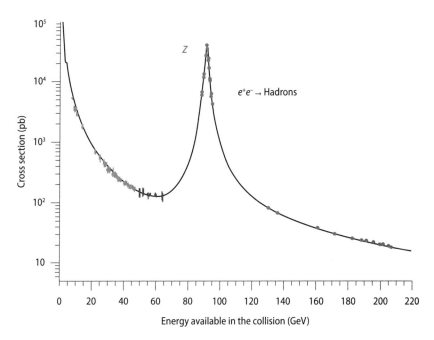

FIGURE 16.8
Distribution of the cross section for the production of hadrons in electron–positron collisions as a function of the available energy. The circles represent the experimental points, and the continuous line the prediction by the model of electroweak unification. The broad peak at around 90 GeV corresponds to the production of the Z and its subsequent decay into hadrons, as studied by the LEP experiments at CERN. The experimental data for energies below 80 GeV were obtained with other accelerators that preceded LEP.

experiments at ever-higher energies and to see whether these hypothetical fermions with large mass are in fact revealed, precisely in the way that the tau lepton, the *bottom* quark, and the *top* quark were discovered. In reality, there is also a highly refined method that was used at LEP, which restricted itself to the energy necessary to produce an actual Z, which is to say only about 90 GeV. As we have seen by now, the energy width of an unstable particle or of a resonance depends on the number of possible decay modes. Take the Z, for instance. It can decay through a remarkable number of channels—into quark-antiquark pairs, to be precise: *up-antiup, down-antidown, strange-antistrange,* and *bottom-antibottom,* and then into lepton-antilepton pairs: e^+e^-, $\mu^+\mu^-$, $\tau^+\tau^-$, $\nu_e\bar{\nu}_e$, $\nu_\mu\bar{\nu}_\mu$, and $\nu_\tau\bar{\nu}_\tau$. Decay into *top-antitop* quarks is impossible because the mass of that pair is around $(173+173)$ GeV/c^2, significantly

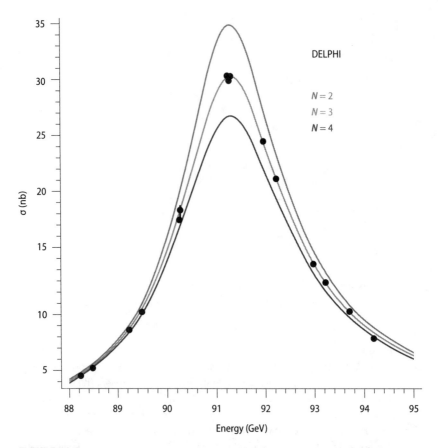

FIGURE 16.9
Precise measurement of the width of the Z conducted at LEP with the DELPHI experiment. The experimental points are indicated by black circles. The three curves represent the prediction of the Standard Model in the hypothesis of two (top curve), three (central curve), or four (bottom curve) conventional neutrinos, respectively, in nature (i.e., with a mass lower than 45 GeV/c^2—half the Z mass). The agreement with the hypothesis of only three neutrinos is excellent.

larger than that of the Z "parent." Instead, each of the decays mentioned here is possible because the mass of each of the particles of the various pairs is less than the Z mass divided by 2 (45 GeV/c^2). At the time of LEP, it was already known that there was no evidence for new elementary fermions, up to a mass of over 100 GeV/c^2, but nothing would have prevented the existence of a neutrino (say, of 10–20 GeV/c^2), partnered by a still heavier charged lepton with a mass outside the observational limit of LEP—perhaps 500 GeV/c^2 or more. The existence of a new hypothetical neutrino—and of the corresponding antineutrino—would have added a pair to the list of possible decays, making the width of the Z shown in figure 16.8 a little larger.

The resonance width was measured with great precision by the experiments at LEP and compared to the theoretical predictions obtained by assuming the existence of two, three, or four active neutrinos capable of coupling with the Z, and consequently of having a charged leptonic partner. The result was momentous: in figure 16.9, one of the measurements is recorded. The experimental points (black circles) are in excellent agreement with the existence of three and only three neutrinos with a mass that is less than half that of the Z, while the predictions corresponding to two or four neutrinos are clearly disfavored by the experimental data. At this point, it is understood that if the number of neutrinos (let's say the standard ones) has to be three, then we have reasonable grounds to assume (though caution is required …) that there are also three families of fermions. But we must be very clear on this point: it does not constitute a proof, only a strong indication of plausibility. Further, in the final analysis, to know that in nature there are eight quarks, and only three neutrinos, would not satisfy anyone …

Be that as it may, and despite a long series of successes, physicists have for many years looked for potential signs of weakness in the Standard Model, secretly hoping to find them in order to obtain some kind of evidence for new physics. As is well known, physicists are rather like children: having built a tower of cards through strenuous effort and application, they destroy it the moment that it's finished, whether for the pleasure of making another one, seeing how it collapses, or discovering what the structural weak points are. Until now, the only crack in the edifice of the Standard Model has been the discovery of neutrino oscillation in 1998; in the theory, the neutrino is a particle assumed to be massless and therefore unable to oscillate. Apart from that, there was absolutely nothing firmly established, apart from a few "tensions" here and there, in particular in relation to some anomalies in the decay of B mesons (composed of a b quark), which in the mid-2010s began to intrigue particle physicists. The model functions very well—almost tediously so. It is no wonder, therefore, that physicists concentrated in the end on finding the architrave of the Standard Model structure, the conclusive and decisive proof of the mechanism that preserves the renormalizability of the electroweak theory through the spontaneous breaking of the original symmetry. This physical-mathematical mechanism that supplies mass to three of the electroweak mediating bosons was proposed in 1964 by Robert Brout, François Englert, and the British physicist Peter Higgs—from which the model takes its name—and independently by Gerald Guralnik, Carl Hagen, and Tom Kibble. Their theory implies the existence of the Higgs field and of the respective boson, the famous God

Particle (from the title of a book by Leon Lederman), borrowed so as to underline somewhat the very special role that this field plays in pervading the entire volume of our universe, and through its interaction, in the attribution of mass to all the other elementary particles (leptons and quarks). A curious fact: it seems that Lederman's original definition of the Higgs boson had been *The Goddamn Particle*, so called because the cursed thing had evaded the research of physicists—but perhaps for editorial reasons, it was modified to a title that certainly had a much bigger and wider impact.

Returning to the theory, the discussion of the spontaneous symmetry breaking and the way in which the Higgs field generates the mass of the electroweak bosons and, as a collateral effect, of all the fermions, is very complex and requires no ordinary range of knowledge. To describe them, it is sometimes customary to resort to simple metaphors, but in my opinion, these are ultimately not much help. One anecdote can serve to illustrate this point: I guarantee that every time a group of physicists happens to be gathered around a well-laid round dining table, with multiple glasses and cutlery and a little plate for the bread, not even a minute after everyone has taken their seats, the most extroverted member of the group will say: "Now we must break the symmetry, choosing whether the plate for the bread is on the left or the right!" Apart from the fact that etiquette tells us that the correct plate is on the left, it remains true that the perfect symmetry of the setting is broken the moment the first guest chooses which of the small plates to use, and then all the others must follow suit. The symmetry is broken in the wake of the action of one of the physicists. This is all well and good, but we must go quite a way further to explain the existence of the Higgs field. Nevertheless, we can attempt to provide some interpretation that may help us to at least grasp the essence of the theory, putting to one side the technical explanation of the symmetry breaking mechanism.

The first thing that needs to be understood is the concept of field, extended to all the elementary particles and not just to the mediators of the forces. This concept forms the basis of quantum field theories such as QED and QCD. Until now, for the creation of elementary particles, we have mostly considered the energy side of the issue. I accelerate a pair of electrons and positrons and make them collide with each other; from their total energy, new particles materialize in honor of $E=mc^2$, always respecting the conservation principles and the probability laws defined by the particular type of interaction. We have also learned that speaking of an interaction mediated by a force field is wholly equivalent to assuming that the particles interact by exchanging quanta of the corresponding field. The photons exchanged between two electrically charged particles are nothing other than the quanta of the electromagnetic force. Extending these concepts to encompass our elementary fermions, quarks, and leptons, we can reason about this in the following way. There exists, for example, the field of the muon, which unfolded and filled the entire universe a little after the Big Bang. Today, this field is everywhere, but in the current, cold cosmos, which is not sufficiently hot even in the interior of the stars, in the cosmic vacuum of practically null energy-temperature, it has an intensity equal to zero. The only way to reawaken the muonic field from its torpid state of zero energy in the vacuum is to apply energy so as to excite it. This is what happens in our accelerators. At the spacetime point where the electron-positron or proton-proton collision occurs, the enormous energy density that is concentrated there—even equivalent to that of the superheated universe when less than one picosecond old—produces an excitation of the field, like a dropped stone disturbing a calm lake. This disturbance materializes in a quantum of the muon field (i.e., in the physical muon particle that begins its own autonomous life in spacetime), propagates, and then decays into an electron and a pair of neutrinos, finally disappearing again. The case is different for stable particles that cannot decay into other lighter ones without violating conservation principles: the electron, the u and d quarks, and the three neutrinos of the electron, the muon, and the tau. Today, we find all these particles to be free in the universe—and the first three of these constitute the ordinary matter that fills it. The other exotic fields—muons, taus, heavy quarks, etc.—have instead a null value in the vacuum until someone arrives and produces a local excitation, supplying energy and creating a quantum that is observable as a real particle, even just for a short while.

The Higgs field also spread into all corners of the universe just ten picoseconds after the Big Bang, when the quark and lepton fields had already been created. It is like an ocean filled with water generated by an indistinct and rapidly condensed aqueous vapor, that rushes precipitously in to fill an arid land in the blink of an eye. The condensation of water happens for temperatures that are lower than 100 degrees centigrade, while the Higgs field condensed when the temperature dropped below 10^{15} degrees! Before this, the vapor-Higgs field had a very small value; all our elementary particles (particularly the electroweak bosons) were without mass. A perfect symmetry existed, therefore, between the *weak bosons* and the photon, given that they were each strictly massless so as to render equivalent (unified) the weak and electromagnetic interactions. It is important to emphasize that the relevant parameter when we speak of the unfolding of the fields of fermions, of the action of the Higgs field, and of all the fantastic occurrences that took place after the Big Bang, is the energy density or, if you prefer, the temperature of the universe. The latter diminished extremely rapidly after the initial moment, due to which we often use time to characterize the events that happened, one after the other. Therefore, the most incredible things took place in an extremely short time just because the energy density diminished very rapidly indeed in that brief interval due to the explosive expansion of the cosmos.

In any case, from the moment that the universe was filled by "the water of the sea of Higgs," so to speak, all particles were influenced by it and began to interact with the newborn field that, upon cooling, had acquired a very high value. In a similar fashion to the force that an electric charge feels when moving into the electric field generated by another, the leptons, quarks, and electroweak mediator bosons became subject to the Higgs field, which acted like a dense medium that causes friction (and therefore inertia) to anything passing through it. A classic example is that of a tea cup. Imagine that you are blindfolded and that you are stirring a teaspoon inside an empty cup. In moving the teaspoon, you do not feel any force, so you rightly assume that it is very light—we might even say weightless, or massless. If someone suddenly pours honey into the cup, you will feel the effect of its viscosity as a medium in which the teaspoon is now moving. With equal justification, you will now think that the teaspoon has become heavy and that it has acquired a mass. Good. The internal volume of the teacup is the universe, the teaspoon is one of our elementary particles (previously massless), and the honey is the Higgs field that fills and pervades everything from the beginning of time and for the entire foreseeable future life of the universe, dynamically generating the mass of the other particles through its interaction with them. The elementary particles experience this friction; the photon doesn't, and so it remains without mass; meanwhile, the quarks and the leptons do so in their own way. The masses of the particles originate according to the particular attitude that they have in relation to the omnipresent Higgs field, or more particularly, according to the coupling constant between the particle and the field itself: the stronger this coupling, the greater the mass that the particle acquires. W and Z, in contrast to the photon, got a notable mass; then the electroweak symmetry broke, and from then on, the daughter interactions appear so different. In that moment, a kind of new interaction was generated, with an extremely short radius—the Higgs force. The most important point here is that there is a profound difference between the behavior of the fields associated with the various elementary particles—elementary fermions: quarks and leptons—and the Higgs field in the act of its condensation. In the first case, at very high energy, the field is excited and has a relatively high value. As the energy diminishes, the intensity of the field decreases until it becomes null when the energy of the system is zero (i.e., the *stable* state of the vacuum, the state of minimum energy). Thus, there is no energy, and consequently there is a null value for the field of the fermions: a vacuum full of lazy, dormant fields.

The Higgs field behaves in a completely different way. It has an almost null value in the very hot and young universe. With its cooling down, the relation between energy and field intensity becomes that shown in figure 16.10. The Higgs field starts in a position of unstable equilibrium, corresponding to the peak of its curve, with an intensity centered around zero; then, thanks to quantum fluctuations, its magnitude could grow, bringing it toward a new position of stable equilibrium corresponding to $E=0$, in the vacuum state, in the valley of the curve in the illustration. The stable point of minimum energy, however, no longer corresponds to a null value

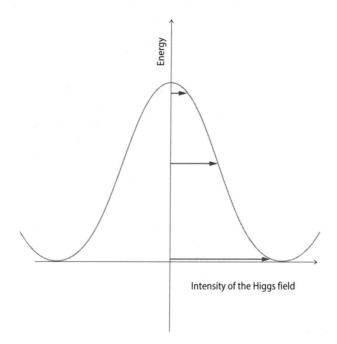

FIGURE 16.10
Dependence of the value of the Higgs field (indicated by the arrows), on the value of the energy of the field itself. Following the symmetry breaking (Higgs condensation) the Higgs field originally of null intensity, thanks to the little "kick" provided by quantum fluctuations, acquires a finite, nonvanishing value in the vacuum, namely in the state defined by $E=0$.

of the field! In such a vacuum, there is something active, like a sea that is calm, still, and silent—but it nevertheless causes friction with the fishes that swim through the water and boats that travel on top of it—our elementary particles. In the end, therefore, we can say that the entire volume of the universe is filled by the Higgs field, which possesses a value that is different from zero in the vacuum. But there are also fields (some of them dormant) of leptons (6); of quarks, each one with three possible colors (18); of gluons (8); and of electroweak bosons (4). Thus, there is a total of 37 fundamental fields. This is an alternative, but equivalent, way of classifying the elementary particles to the one we have had occasion to discuss at numerous points in this book.

I invite the reader to reflect here on another crucial issue. We have said that the moment the Higgs field condensed the original symmetry was broken. This assertion can be understood by continuing the analogy with aqueous vapor and water. Even if it seems counterintuitive, vapor possesses a greater level of symmetry than water. Its molecules can be rotated and shifted in any possible way, and the system will not change—think of the rotation of a vase around its axis of symmetry. In liquid water, on the contrary, each of the molecules of H_2O must adhere to the conditions of its bond with other molecules. These bonds reduce its degree of symmetry (i.e., its capacity to remain identical after a certain operation—rotation, reflection, or translation). For this reason, the breaking of the electroweak symmetry occurs precisely in the act of the phase transition of the Higgs field—in other words, at the moment of its condensation. We can also take into account a further, very intriguing consideration. With a little imagination, we can assert that with the Higgs field, the concept of ether that at the beginning of the twentieth century had been marched out of the door had come back in through the window at the start of the new millennium. After all, to think that every point of spacetime is replete with the "honey," or of the "sea" of the not null Higgs field, is something that leaves one perplexed and poses a series of questions. Could it be the new ether? To reassure the reader, we should say that the only point in common between nineteenth-century ether—which does not exist!—and the Higgs field—that does exist (and how!)—is that they are both assumed to be ubiquitous.

When asserting that the Higgs field generates the mass of all the particles, and consequently of the entire universe, we also need to be cautious. In chapter 12, we observed that nearly 98 percent of the hadrons, and therefore of the neutrons and protons that make up the atomic nucleus, are generated by the interaction energy between quarks and gluons. And the atomic nuclei constitute almost all the ordinary matter of which the stars, planets, galaxies, and everything else is made. The electrons, and even the quarks contained in the protons and neutrons, have entirely negligible mass. The moral of all this is that while it is true that the Higgs field created the mass of the fundamental fermions that constitute the universe (electrons and quarks), this mass is itself rather insignificant compared to that generated by QCD, from the energy of the color field. But this is not all. Today, we know that the sum of the masses of all the stars, planets, and intergalactic gas in the visible cosmos (i.e., what we call "all that surrounds us"), made up of our elementary fermions and of the dominant color field energy, amounts to only 5 percent of the overall energy-mass balance of the universe, as determined by independent measurements. Approximately 27 percent of such energy belongs to a form of nonvisible matter, hence called *dark*, which is still unknown—and measured using the observation of its gravitational effects on the rotation of galaxies. Even though it is dark, it is matter, and therefore has a mass that affects the motion of large massive objects such galaxies. What's amazing is that the remaining 68 percent of the universe consists of something we know even less about. I refer to *dark energy*, which, among other things, is responsible for the fact that for a few billion years, the space of the universe has even begun to accelerate its expansion, overcoming the attractive force exerted by matter and dark matter. It's a bit like the difference between a freewheeling car that moves in neutral without needing to use gasoline, proceeding at constant speed (neglecting frictions), and another vehicle in which, by pressing the throttle pedal and therefore consuming more gas (energy), we can obtain an accelerated motion. Dark energy is precisely the cause of the acceleration of the objects that fill the universe. To be precise, we are dealing with the expansion of the same space in which stars and galaxies are situated. The evidence for the existence of dark energy arrived at the end of the 1990s. Accurate measurements of the motion of the most distant galaxies showed that they are moving away from all the others, and thus from us, at increasing speeds, and therefore in an accelerated manner. This outstandingly important discovery earned a Nobel Prize in 2011 for Saul Perlmutter, Brian Schmidt, and Adam Riess. One of the primary objectives for future research in astrophysics and cosmology will be to understand the nature of this mysterious energy. There is, for instance, the EUCLID experiment of the European Space Agency, which, via complex equipment placed on a satellite, should make it possible to solve this fascinating problem through the observation and study of a considerable number of galaxies—several billion of them, in fact!

Now, a further consideration arises, also of a counterintuitive kind, and one that has extraordinary consequences. While expanding at increasing speed (being accelerated), space drags with it the matter (galaxies) with which it is filled. Then, such an expansion of the universe's space can take place at speeds that are even higher than the speed of light. Those galaxies that are most distant from us, therefore, can move away at superluminal speed, but nonetheless without violating Einstein's laws of relativity because the expansion refers to the space itself on which they sit. Some of the fastest galaxies are now emitting the last light pulses capable of reaching us when, in some billion years, they will have traveled the enormous distance to us. But when their speed becomes faster than that of light, the luminous signals produced by their innumerable stars will no longer have any chance of reaching us. If we assume that the accelerated expansion will proceed indefinitely, or at least for an extremely long time, all galaxies will become progressively invisible to us. Eventually they will vanish from our horizon forever, and the two respective parts of the universe will be eternally unobservable to each other. This fact leads us to believe that the cosmos that is observable today is in fact only a tiny part of what has been produced and developed during the course of 13.8 billion years. We can even foresee that if our galaxy survives long enough, then its inhabitants might one day believe that it constitutes the entire cosmos—a single galaxy, the Milky Way, in what would effectively be a rather small universe. The hope is that our great-grandchildren may still have at their disposal the beautiful photographs taken with the Hubble telescope—unequivocal proof of the existence of billions upon billions of other galaxies, sisters of our own (figure 16.11).

Returning to the Higgs field, for now, we will point out the possibility that it is not connected to dark energy at all. And because we don't know if dark matter is itself generated by the Higgs mechanism, we can only say cautiously that the Higgs field is only responsible for the creation of a very small fraction of the energy-mass of the universe. But there is a last crucial aspect of the Higgs model closely connected to our elementary particles. Just as the quantum in the electromagnetic field is the photon, the Higgs boson (indicated by H) is the quantum of the Higgs field—a particle with null electric charge and with spin equal to 0, in contrast to the photon, which has a spin of 1. This characteristic is reflected in the different properties of the two fields. Let me explain: the electromagnetic field, mediated by the photon with unit spin, is mathematically a vector mathematical entity defined by its magnitude and direction—as illustrated, for example, in figure 8.3 in chapter 8. The Higgs field, instead, is a scalar quantity completely determined by just the numerical value of its intensity (magnitude), a little like the temperature field, defined at every point of the volume of a given

FIGURE 16.11
Photograph by the Hubble telescope that shows a large number of the universe's far-off galaxies, each one containing billions of stars, which in turn may credibly be endowed with planetary systems such as ours.

ambient by a number. The H particle may be detected experimentally by means of high-energy interactions, such as those that take place in the LHC at CERN, exciting the Higgs field, which is present at the collision point between the beams of protons, just as for the muonic field in the previous example. To better establish the difference between the Higgs field and the Higgs boson H, we can use an illustration devised in 1993 by the English physicist David Miller in response to a request from the ministry in charge of scientific research for an explanation of the mechanism that would be accessible to all. I take the liberty of offering a slightly modified version of the illustration, but the underlying substance is the same.

Let's imagine that we are at the coffee break during an important physics conference dedicated to elementary particles, with colleagues crowded together in a hall discussing the latest findings that have been presented at the event. Suddenly, from a side door, who should appear but Albert

FIGURE 16.12
Cartoon illustrating the Higgs field mechanism: (a) and (b), and the Higgs boson: (c) and (d).

Einstein himself (figure 16.12a), probably an avatar from a parallel universe, having pushed himself through a spacetime wormhole … The physicists closest to him surround him immediately, creating a phalanx. Einstein attempts to move from one side of the hall to the other, but he finds it difficult to do so because he is encircled by admirers drawn to his astonishing, unexpected presence. It is as though the inertia of his motion were increasing—and in effect, this is how his mass grows. He finally manages to reach the door at the other side of the hall (b). Upon hearing the news, as I am also present at the conference, I enter the hall and manage to cross it easily, going quickly past my gathered colleagues to reach the door from which Einstein has exited. Those colleagues, rightly, pay no attention to me. At this point, a physicist appears at the first door (c), announcing that Einstein will shortly be holding a press conference to explain to everyone the correct interpretation of the recent scientific results that nobody has been able to understand. The news causes a dense group of physicists to form, and to discuss the news in an animated way (d). So what is the point of this scenario? It's simple. The initial crowd of physicists at the coffee break represents the barely agitated, almost dormant Higgs field, which fills the entire volume of the universe—the coffee break hall, in this case. At its arrival, a particle without mass—Einstein's avatar—acquires inertia in its movement and becomes endowed with mass by interacting with the Higgs field. Given the fame of this eminent scientist, the acquired mass that provides resistance to his motion is considerable. Meanwhile, I, passing through the room undisturbed (ignored) from one side to the other, represent a photon without mass at the outset, which remains without mass after passing through the Higgs field. The Higgs boson is the excitation of the field, or the gathering of the physicists that regroups upon hearing the announcement of the press conference. The announcement shouted by the news-bearing physicist corresponds to the energy of the collision in the LHC, capable of creating the H particle. Given that this localized excitation also moves with significant inertia, the Higgs boson also acquires a mass from the Higgs mechanism.

On a more serious note, the experimental search for the Higgs boson has constituted a titanic enterprise, a feat that has required over twenty years of work on the part of thousands of physicists, engineers, technicians, and students dedicated to creating the LHC, the gigantic and complex detectors associated with it, and the software programs necessary to operate the detectors and analyze the data. This is one of the most arduous and costly scientific enterprises in the history of humanity, which has been possible to carry through only thanks to the coherent efforts of the entire international community of scientists and with the continuous support of society, through the various funding agencies for scientific research. From this point of view, the LHC represents an exemplary model for future projects due to its farsightedness, its efficiency, and above all, its undeniable success. Despite all this, the media clamor generated around the project when the machine was put into operation has caused some insidious collateral effects. As had already happened with other accelerators, there were those who, if not exactly dilettantes, were at best not sufficiently qualified (in technical and scientific terms), sounding alarm bells by claiming that in the collisions that it engineered between high-energy protons, the accelerator would create black holes. These would then immediately swallow the experiment, CERN, the city of Geneva, and in an apocalyptic crescendo, the whole of Switzerland itself, followed by Europe and the entire planet! In support of this prophecy, of course, there was the inevitable solid theory, evidently correct and certified, which foresaw such a catastrophe with absolute conviction ... without the least glimmer of doubt. Perhaps you are aware that the search for micro–black holes is one of the current objectives of our experiments at the LHC, and their discovery will represent a sensational but fortunately innocuous event, given that the entity and the life span of such microscopic objects would be altogether too small to produce anything damaging to physicists, being hardly detectable by them. With near-categorical certainty (i.e., with a probability so high as to be tantamount to *certainty* for scientists), the resulting minuscule black hole would produce nothing more dangerous than joy for the researchers. Instead, our so-called experts were guaranteeing a catastrophe, advancing their claim with a degree of militancy frankly unjustified by actual competence in such matters. So, what is the outcome of all this? The physicists were obliged to intervene, to reassure the public—or rather a few of them, fortunately. A flawed understanding of the concept of probability, a tendency to always fear the worst, and above all an inclination to give credence to those not sufficiently knowledgeable in scientific matters conspired to cause some to wait with bated breath until it was proved (experimentally—hurrah!) that in fact, they would not be swallowed by an artificially created black hole.

To return to the Higgs boson, its mass is not predicted by the theory, unlike its other characteristics, such as the intensity of its coupling to the various fermions, which is effectively proportional to their mass. For this reason, in designing the LHC, it was determined to reach the highest energy that could be technologically obtainable to make protons collide with other protons and detect the fruits of their interactions. With the LHC, we succeed in attaining an available energy of up to 13 TeV in the collision—that is, 13,000 GeV. The discovery of the Higgs boson was announced at CERN on July 4, 2012, from the two scientific collaborations of the ATLAS experiment—of which my own group from Bern is also part—and of CMS. The LHC had begun to function only relatively recently, at a reduced capacity, producing collisions of only 7 or 8 TeV of energy in total. Nevertheless, the signal of the production of the new particle rapidly and unequivocally emerged. The following year, Englert and Higgs (figure 16.13) received the Nobel Prize for the experimental confirmation of their theory, after a wait of fifty years. Unfortunately, poor Brout was unable to share the celebration with his colleagues, having passed away in 2011. Like other occurrences and recurrences of history, the Higgs boson was observed through a resonance in the distribution of the invariant mass of pairs of high-energy photons, according to the reaction: $H \rightarrow \gamma + \gamma$. However, this signal overlaps with the concomitant background events which have also with two photons in the final state. This result is illustrated in figure 16.14; it involves

FIGURE 16.13
Peter Higgs, the recipient with François Englert of the Nobel Prize in 2013 for the discovery of the theoretical mechanism for the generation of the mass of the electroweak bosons and of the elementary fermions.

the original graph produced by the ATLAS collaboration and presented on the day that the discovery of the Higgs was announced. In the upper histogram, the distribution of the invariant mass of high-energy photon pairs is shown, essentially constituted by background events; for those events, two uncorrelated protons, identified in the ATLAS detector, are combined to reconstruct their invariant mass. This background slowly diminishes as energy increases. At approximately 125 GeV/c^2, we observe a small but significant resonance due to concurrent Higgs decays, representing a few signal events added to the background. As explained in chapter 4, given that the probability of simultaneous fluctuation of background events required to generate the observed peak is calculated to less than one in a few million, we can without doubt speak of a discovery. In the lower histogram, the background events are statistically subtracted from the original distribution to better show the peak of the resonance due to the Higgs production and subsequent decay into two photons. These few events recorded by the ATLAS and CMS experiments constitute, without fear of contradiction, one of the greatest discoveries of modern science. Imagine that in order to find one Higgs particle, it is necessary to analyze 10,000 billion proton-proton

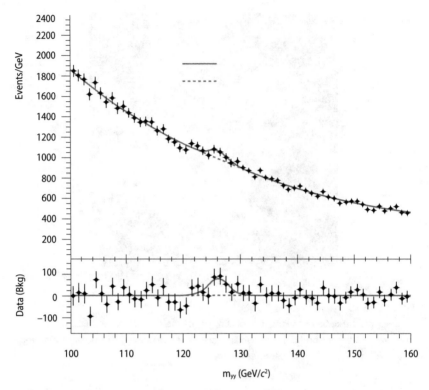

FIGURE 16.14

Distribution of the invariant mass of pairs of high-energy photons presented by the ATLAS collaboration of CERN on July 4, 2012. The peak, at around 125 GeV/c^2, corresponds to the resonance due to the creation and decay of the Higgs boson.

collision events, produced by a machine such as the LHC, which in the ATLAS experiment generates approximately 1 billion events per second, of which only a few hundred per second are interesting for physicists!

Naturally, the $H \rightarrow \gamma + \gamma$ decay is only one of the allowed decay modes. Similarly, there are many ways of producing the particle, each with a cross section that is generally a function of the energy. Figure 16.15 shows some of the notable production processes of the Higgs boson. In figure 16.15a, two gluons emitted by the colliding protons fuse, giving way to a virtual quark-antiquark pair loop, which in turn couples with a Higgs boson. Because the coupling of the Higgs with the fermions is proportional to their mass, the process is all the more probable the greater the mass of the quarks involved, as is the case for *top* or *bottom* quarks. In figure 16.15b, a quark

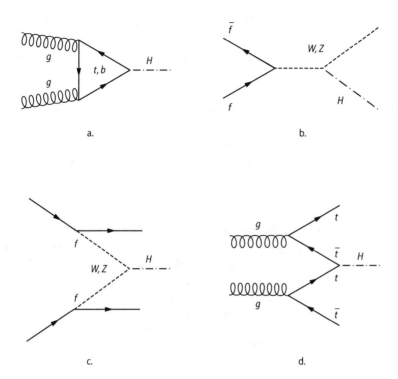

FIGURE 16.15
The main Feynman diagrams for the production of the Higgs boson at the LHC.

and an antiquark, the latter coming from the sea of one of the two colliding protons, interact via the electroweak force, producing a W or a Z according to whether it involves a quark-antiquark pair of different or of the same flavor. The virtual electroweak boson in turn emits a real Higgs. Figure 16.15c illustrates the radiation of two bosons from the colliding quarks; the W or Z pair fuses and creates the Higgs. Finally, figure 16.15d shows the production of a Higgs through *top-antitop* quark fusion. Given that the *top* is the quark with the highest mass, the coupling with the Higgs is maximal.

Similar arguments apply to the decays of the Higgs particle. Once it has been produced, it decays extremely quickly (in about 10^{-22} s). Given that its coupling is stronger with the particles of higher mass, we are not surprised to find heavy quarks (t or b) in the decay diagrams. Nevertheless, as the reader is already well aware, hadronic decays are more difficult to study experimentally. The quarks fragment, giving way to jets of hadrons that saturate our detectors, making more complex the reconstruction of the events and the extraction of the signal of Higgs production. Hence, it is better to rely on decay channels with smaller cross sections but greater clarity from an experimental point of view, such as the aforementioned decay into two photons, or that into leptons. These two processes are illustrated in figures 16.16a and 16.16b for a Higgs produced from a virtual quark loop. As we have seen, the photon does not couple to the Higgs, and as a result, the first decay proceeds indirectly, and for the most part by way of the primary coupling of the Higgs with another virtual *top* or *bottom* quark-antiquark pairs. The quarks can radiate two real photons, the invariant mass of which is equal to that of the Higgs boson. In figure 16.16b, the Higgs decays into a Z pair that in turn produces four charged leptons (e.g., four muons, or a muon-antimuon pair and an electron-positron one). The detection of charged leptons is notoriously less challenging. In this case too, the invariant mass of the four particles coincides with that of the Higgs. An event that is potentially due to the production and decay of a Higgs particle is shown in figure 16.17. I say

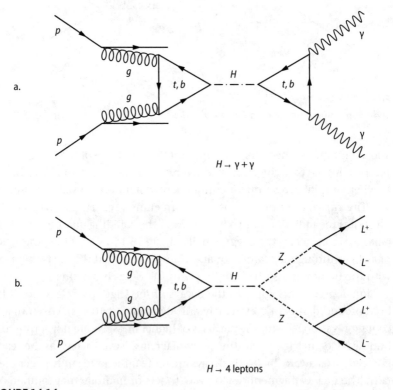

FIGURE 16.16
Diagrams of Higgs boson production and decay of particular experimental interest.

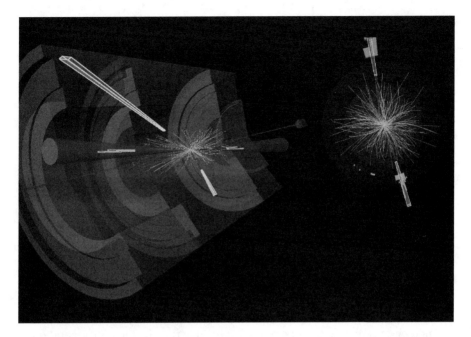

FIGURE 16.17
Visualization of an event of the ATLAS experiment—in both longitudinal and transverse projections—that can be attributed to the decay of a Higgs boson into two high-energy photons. The latter are identified by the two electromagnetic showers that they generate in the calorimeter at rather large angle with respect to the beam direction.

"potentially" because the event can be (principally) caused by one of the background processes with a similar signature, as depicted in figure 16.14. It involves the decay in two energetic photons with an invariant mass of about 125 GeV/c^2, as registered by the ATLAS experiment.

Before concluding this chapter, let's return to the initial theme of the unification of the fundamental interactions. Having achieved the unification of electromagnetic and weak forces, the ambition of physicists naturally turned to considering the possibility—and why not—of unifying the electroweak interaction with the strong force, certainly for a much higher energy and corresponding to a time still closer to the Big Bang. It is fascinating to think that an instant after the beginning of everything, there existed a "Superforce" that included the three most intense interactions that we observe today, differentiated now in a universe that has become old and cold ... In tackling this subject, we must first make a few preliminary considerations. The concept of unification of forces means that the same interactions are strongly related from a formal perspective, under a common umbrella represented by a mathematical symmetry paradigm that generates the characteristics of the hypothetical Superforce. We have frequently used the example of electromagnetism. In the light of the Maxwell equations, though, it makes no sense to speak of the electric and magnetic forces separately—they are closely connected, even if in practice, we can readily observe effects that may be ascribed to only one of the two types. In the electroweak interaction, the initial forces are only partly unified. The electromagnetic and weak forces mix, but they maintain their respective coupling constants, relatively independently and with a different behavior as a function of the energy. The Higgs field, on the other hand, does the work of breaking the original symmetry by creating massive electroweak bosons and leaves the photon without mass, while still guaranteeing the calculability of the various Feynman diagrams. The case is quite different for a real Grand Unification Theory (GUT), based on the symmetry laws of the three forces present in our world—the electromagnetic, the weak, and the strong—together with their coupling constants, which are, as we said, functions of the energy. In this case, we look again for a more extended symmetry group that on the one hand can include the offspring forces, and on the other "unpacks itself" into the three interactions at lower energy. Such a group must ultimately generate a new interaction at extremely high energy, with a single, shared coupling constant and with new mediating bosons.

From this perspective, we begin from the study of the three daughter interactions, in order to understand the dependence of their coupling constants from the energy. We have already seen that the intensity of the electromagnetic force grows with the increase in energy ($\sqrt{\alpha}$ grows), while that of the strong force decreases markedly ($\sqrt{\alpha_s}$ diminishes). From precise measurements made in the 1990s at the LEP accelerator at CERN, it also follows that g, the weak coupling constant, also decreases as the energy grows. The reason is similar to that which determines the dependence on the energy of the electromagnetic and strong constants. The contribution of higher order diagrams creates a general effect similar to that of the anti-screening of the strong force discussed in chapters 10 and 11. In general, every additional diagram that is considered can bring something different, increasing or diminishing the effective value (strength) of the coupling. At this point, it is natural to ask if the respective progressions of the three constants can lead to a point of union, perhaps at very high energy, in which the three may acquire the same value and therefore indicate, without proof, the existence of the hypothetical unified Superforce. This problem in particular was tackled in 1991 by Ugo Amaldi, together with his collaborators

Wim de Boer and Hermann Fürstenau. Utilizing the experimental data at their disposal and extrapolating the behavior of the electromagnetic, weak, and strong coupling constants, using the respective (Standard Model) theories, the three physicists obtained the result shown in figure 16.18. On the x-axis, there is the energy that reaches up to 10^{18} GeV—an extremely high value, corresponding to the energy density of the universe as it was just 10^{-36}–10^{-37} seconds after the Big Bang—a truly small "fraction of nothing." In the ordinate, the inverse of the values of the coupling constants are plotted; the more intense the force, the lower the numerical value in the graph. The three constants begin from the experimental value of low energy that was explored at the time (about 100 GeV) and are then extrapolated

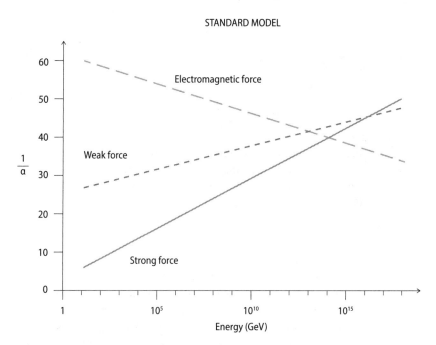

FIGURE 16.18

Behavior of the inverse of the three coupling constants of the fundamental forces: electromagnetic, weak, and strong, as a function of the energy. The inverse of the couplings is shown to more readily give evidence of an eventual point of encounter—which unfortunately does not appear. What is shown here is that at around 100 GeV, the weak coupling constant is approximately double that of the electromagnetic one, as follows from the theory of electroweak unification. The constants are extrapolated at very high energy, assuming the validity of the Standard Model for the whole energy interval.

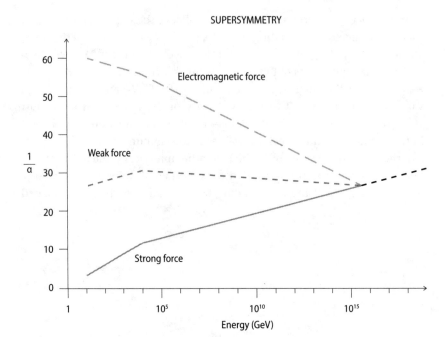

FIGURE 16.19
The counterpart of the graph in figure 16.18, obtained by assuming the existence of a particular Supersymmetry, generating the Grand Unification of the forces at extremely high energy. In this case, the constants of the fundamental forces meet at around 10^{16} GeV.

following the theoretical predictions. We observe that at around 10^{-14}–10^{-15} GeV, there are encouraging signs of unification, but the constants miss their appointment emphatically. Nevertheless, we should add that the mathematical symmetry model used for the extrapolation at high energy was quite simple. The negative result showed unequivocally the need to resort to a more complex symmetry structure as a basis for the Superforce hypothesis. Because of this, the three physicists formulated a bold hypothesis. Their conclusions are illustrated in figure 16.19. As if by miracle, the three constants meet, and this fact permits us to hope that the Grand Unification can really exist—or rather can really have existed for a brief instant, up to 10^{-35} seconds after the Big Bang. But what kind of magic did Amaldi et al. manage to invoke? Nothing less than the existence of a Supersymmetry, obviously! We shall be discussing this at more length in the next chapter …

17 DREAMS, THE UNKNOWN, ADVENTURE

> The task of science is not to open a door onto infinite knowledge, but to put up a barrier against infinite ignorance.
>
> —Bertolt Brecht, *The Life of Galileo*

We are reaching the end of our journey through the world of elementary particles. We have surveyed the history of approximately 100 years of theoretical, experimental, and technological efforts and discussed scientists, experiments, pointlike and composite particles, and interactions. We have attempted to frame this narrative within its historical contexts, and as promised, we have tried to organize the observed phenomenology into appropriate interpretative frameworks. Now we are on the verge of taking the last step: looking ahead to what tomorrow might bring, and searching into the unknown, theories and dreams leading to an adventure that will take us who knows where. But let's be careful. To indulge in predictions in science, especially in particle physics, is a difficult and perhaps ultimately nonsensical activity. The most widely supported theories can evaporate when confronted with new experimental evidence, expected results may not materialize and, vice versa, great problems, the solution of which was expected to take years, are solved quickly, probably due to a nudge on the part of nature itself. We shall see. In any case, the point of departure from which to speak about the future is today. Let's return to the point where we interrupted our narrative in the previous chapter, in order to see if there is any possibility that the Grand Unification is real, or whether it is just a desirable outcome with no empirical grounding in reality.

You will recall that the two principal families of particle-fields are fermions, with half-integer spin that compose the visible matter of the universe, and bosons, the particles with integer spin that include the mediators of the fundamental forces. Between them, there is a profound difference, which is also reflected in the diversity of behavior in relation to their statistics: two identical fermions cannot both stay in the same quantum state, due to the Pauli exclusion principle, while the many bosons can coexist in the same state. This difference also has notable consequences at the macroscopic and cosmological scale. Fermions are subject to conservation principles: those of the baryon and lepton numbers. The consequence of this is the stability of matter, expected to hold well beyond the present age of our universe. Nothing similar applies for bosons, which can be created or destroyed with impunity.

The need for very general and extended symmetries, which can explain the nature of the fundamental interactions and guarantee their unification, leads us to determine whether any of these really have any significance in nature, rather than being just an interesting mathematical exercise. Why not think, then, that at an arbitrarily high energy, a symmetry might exist, which we shall call a *Supersymmetry*, precisely between fermions and bosons? In such a paradigm, every fermion will have a corresponding boson partner. Therefore, there will be an electron of integer and a photon of half-integer spin. Given that despite all our best efforts, they have yet to be observed, we must assume that such superpartners of ordinary particles, if they were in reality to exist—that is to repeat, if Supersymmetry were valid—would have a very high mass, beyond what we have explored to date with the Large Hadron Collider; let's say a mass of 1,000 GeV/c^2, or perhaps even more. This fact implies that Supersymmetry itself must also be spontaneously broken. In effect, at best, each supersymmetric particle would have a mass many times larger than that of its counterpart as it is already known to be in nature.

There are various theoretical models capable of generating the zoology of Supersymmetric particles. The original ideas go back a long way, to when the theory was seen above all as an aesthetic model of mathematical beauty that nature might possibly have obeyed. The major contributions in this area came from Yuri Golfand and Eugeni Likhtman, Dmitry Volkov and Vladimir Akulov, Jean-Loup Gervais and Bunji Sakita—as well as, finally, from Julius Wess and Bruno Zumino. The convention that physicists have adopted to designate the boson partners of the fermions is to prefix an *s* to their names. Hence, we have the *selectron*, the *smuon*, and the *up* and the *top squarks*. Similarly, the fermions corresponding to the bosons of the Standard Model take the names *gluino*, *photino*, *Wino*, *gravitino*, and so forth, identified by the suffix *ino*. Supersymmetric particles partners of those experimentally known are given a tilde accent (~) above the symbol, as shown in figure 17.1. So far, so good. It's an elegant symmetry, in the end,

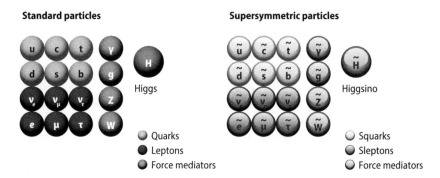

FIGURE 17.1
The elementary particles of the Standard Model and their hypothetical supersymmetric partners.

and one that's undoubtedly extensive and inclusive. If it were verified, we would acquire at a stroke thirty-six fundamental fields, plus five *Higgsinos* in the most simple Supersymmetry model, bringing the family of elementary particles to seventy-eight members—and possibly more, depending on the specific theory. Here we go again … farewell to simplicity and reductionism! But the question that we are obliged to pose, and that we have already raised implicitly, is this: is there an experimental verification, or at least an indication of plausibility? While there is no direct proof as yet, there are nevertheless good reasons to suspect that Supersymmetry might be valid. The first of these has just been pointed out. In figure 16.19 we see that the coupling constants of the interactions meet "astonishingly" at an energy of approximately 10^{15} GeV, assuming that the simplest of all the supersymmetric models is valid—the so-called *Minimal Supersymmetric Standard Model* (MSSM). On the basis of what has been explained in chapters 10 and 11, it is clear that the "running" of the coupling constants with the energy is due to the contributions of higher-order Feynman diagrams. These can include virtual particles of various kinds, as illustrated in figure 17.2 for an electromagnetic interaction; similar considerations apply to the weak and strong diagrams. Naturally, the contributions of virtual particles with a larger mass become relevant at higher energies (or for shorter life spans), to respect Heisenberg's uncertainty principle. Each diagram can contribute in the direction of either increasing or decreasing the effective value of the coupling constant (remember figure 11.4 in chapter 11). In the hypothesis for the validity of Supersymmetry, the number of fermions potentially available to create loops, such as those shown in figure 17.2, doubles. And the effects will generally vary from diagram to diagram. The result of this is

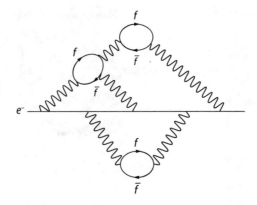

FIGURE 17.2
An electron continuously emits and reabsorbs virtual photons. These can be transformed into so-called loops of fermion-antifermion pairs. All the existing charged fermions (quarks and leptons) contribute to the loops. The existence of other particles, such as those generated by Supersymmetry, creates a series of additional contributions that influence the value of the effective coupling constant—in this case, the electromagnetic one.

impressive: the meeting of the constants occurs then, thanks to the contributions of diagrams involving both the Standard Model particles and their supersymmetric siblings.

A typical Feynman diagram involving particles generated by Supersymmetry is provided in figure 17.3, as an example. Let's try to interpret it. Two quarks collide, coming perhaps from protons energized in a high-energy accelerator. Through the creation of a gluino—the supersymmetric partner of the gluon—two squarks are produced (the flavor is irrelevant in this example). Both undergo the same decay into a quark and a *chargino*. The latter is a state originating from the quantum mixing (remember the neutrinos?) between the Wino and the Higgsino, while the neutral chargino, known as the *neutralino*, is a combination of the neutral partners of the electroweak bosons and the Higgsino. It all seems very complicated, and perhaps it is, but as I say to my students: all the easy things have already been discovered ... Returning to the diagram in figure 17.3, the chargino $\tilde{\chi}^{\pm}$ decays into a neutralino $\tilde{\chi}^0$ and a standard W. The latter converts, conventionally, into a charged lepton and its neutrino. The neutralino instead is stable, being the lightest of the supersymmetric particles. In this case, we assume the validity of the conservation of the total number of supersymmetric particles, just as happens to baryons and leptons. Such a conservation principle may or may not be a characteristic of Supersymmetry—but if it were to be, really interesting scenarios would open up. Remember dark

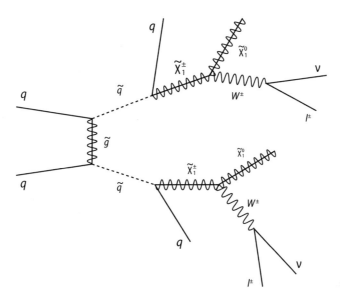

FIGURE 17.3
A tangled example of a Feynman diagram involving Supersymmetry. Supersymmetric partners of bosons are indicated by a continuous segment superimposed to a wavy curve, while squarks are shown as broken lines.

matter? More than a quarter of the energy-mass of the universe is made up of invisible (that is, in effect dark) matter. It emits no electromagnetic radiation and remains concentrated around the galaxies due to gravitational attraction. Until now, we have detected its effects only indirectly. Given that we are dealing with matter that has mass, however, it enters into the balance of the galaxies, and their rotation speed will depend on it. Measuring, for example, the rotation velocity of a given galaxy in relation to the one that it should have had with only its visible matter, made up of stars and intergalactic gas, allows us to estimate the quantity and other characteristics of the cloud of dark matter that was trapped by gravity.

An obvious candidate for dark matter is the neutrino. We have seen that our elusive particle has a mass, which, though certainly minuscule, when multiplied by the immense number of neutrinos that wander about the universe, could be relevant to solving the mystery. This, however, is sadly not the case. The total estimated mass of neutrinos in the cosmos does not appear to be sufficient for this purpose. A variety of other potential candidates exist, some of which are of a really fantastical kind. One possibility could be that there are a large number of so-called primordial micro black holes—as previously discussed in relation to the LHC—still present in the universe. They might not have come from the death of stars, but were directly produced soon after the Big Bang. They might be present in the universe in a quantity sufficient, even partly, to explain the dark matter. This hypothesis is certainly fascinating, but totally lacks verification. In short, we are still groping in the dark (!). Nevertheless, it is an interesting fact that if Supersymmetry exists and is conserved (i.e., if the number of supersymmetric particles is constant before and after any reaction in which they might be involved), then the lightest supersymmetric particle cannot decay into others—and the neutralino is an excellent candidate for constituting the dark matter of the universe. This is one more good reason to bet on the existence of Supersymmetry. The free neutralinos in the universe, trapped as in a halo in the gravitational field of our galaxy, could be directly identified by terrestrial detectors that, together with the Earth, move through this halo, generating potential interactions. It is a little bit like the windshield of a speeding car colliding with flies on a highway. Unfortunately, however, the very small cross section of neutralinos with ordinary matter (and with our detectors), as well as the ambiguous signature of the eventual interaction, makes their direct detection a formidable experimental challenge—and one that's still unresolved, despite the best efforts of researchers. On this point, in figure 17.4 we show the XENON detector in the underground laboratory of Gran Sasso, which is searching for evidence of these rare events, well screened from concomitant background reactions principally induced by neutrons and cosmic rays. The detector is comprised of 1 ton of liquefied xenon gas contained within the volume of a special detector.

The third reason to hope that Supersymmetry exists is still more technical and challenging, but it happens also to be the most important one. In chapter 10, it was shown that the real mass of the electron is arrived at by taking into account all the possible diagrams of self-interaction for the particle. An example of this is given in figure 17.2. This mass, thanks to the complex mechanism of renormalization, is substituted for the naked one, which would lead to the noncalculability of the Feynman graphs. So far, so familiar. What's different and new, however, is that the Higgs particle, always subject to the uncertainty principle, can itself produce virtual particle-antiparticle pairs during its brief life, as a result of the excitation of the Higgs field (figure 17.5). While it is true that these virtual pairs live for next to no time, as we have already learned, they nevertheless contribute to the real value of the mass of the particle that produces them. Now let's take all the possible diagrams of self-interaction of the Higgs. To our surprise, the value that we obtain for its mass is not at all reasonable, as happens in the case of the electron. Rather, it is shamefully large—huge

FIGURE 17.4
XENON, the hunter of dark matter at the Gran Sasso underground laboratory. On the left is the great cryostat containing the detector based on liquid xenon—depicted in the poster pinned to the detector; on the right, the control rooms and the spaces housing the electronics for the data acquisition system. © INFN Gran Sasso National Laboratory/XENON Collaboration.

in fact. To be precise, it is around 10^{16} GeV/c^2, which is actually similar to that required for Grand Unification. How do we lose fourteen orders of magnitude to arrive at the 125 GeV/c^2 that we have effectively measured with the LHC? It's simple! A natural way of solving the conundrum is provided once again by Supersymmetry. If we include the new particles, these will also come to be part of the loops shown in figure 17.5. As if by magic, the contributions of these new diagrams cancel the hypertrophic growth of the Higgs mass, and bring it down precisely to around 125 GeV/c^2. All this occurs on the understanding that the new particles (currently unknown …) have masses that are smaller than 1,000–2,000 GeV/c^2. This strongly motivates us to look for proof of the existence of supersymmetric particles, exploiting the LHC.

Nevertheless, it should again be stressed that we must not become too enamored of theories, even when they promise to solve such problems and

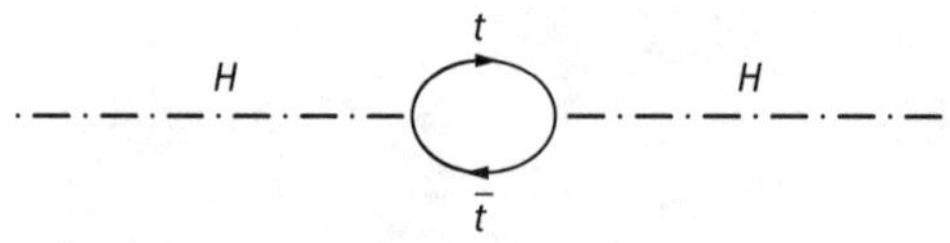

FIGURE 17.5
A typical process of self-interaction of the Higgs, in which a virtual *top-antitop* quark pair is produced and then quickly reabsorbed. All the possible diagrams of this kind make the value of the Higgs mass enormous—far larger than the one measured experimentally.

incongruities as the persuasive Supersymmetry does. If we were to find no evidence to support it at the LHC, then we could conclude that (1) it manifests itself at an even higher energy—and at this point, perhaps, it would not be of much use to us; (2) we would find ourselves, as in the case already mentioned, of having an indication that "pigs don't fly." Here's an awkward situation. For every theory that fails, however, the ingenuity of physicists rallies to produce alternative solutions. And in the end, in the great Galilean tradition, only experimental data will lead us to the truth.

At this point, we could remain satisfied with what we have discussed of the physics of elementary particles and of their interactions, even though we have been obliged to pass over many interesting subjects. One of these is the production of the so-called plasma of quarks and gluons. In general, when we say "plasma," we mean an ionized and energetic gas. Without getting bogged down in the details, imagine that we induce accelerated nuclei of heavy elements to collide, such as lead against lead or gold against gold. To make this happen, we currently use the LHC again, this time in a slightly different way. The theoretical idea is that a little after the Big Bang—and the reader knows by now that "little" means very little indeed—as soon as quarks and gluons were formed, they filled the universe with an energetic "boiling soup". They possessed so much energy that the hadrons, including the protons and neutrons, could not yet form as bound states. Therefore, quarks and gluons were free to wander through the newly born universe, constituting a special state of aggregated, superhot matter, at a temperature higher than around 10^{12} degrees centigrade—precisely the plasma of quarks and gluons. In order to study this exotic state of primordial matter, we make heavy atomic nuclei collide at extremely high energy. The hope is that in the ensuing "supercrash," we will succeed in rupturing the virtual containers constituted by protons and neutrons and put all the individual particles together (quarks and gluons), creating in the process a small, hot, and dense volume full of plasma (figure 17.6). The main issue is to understand whether

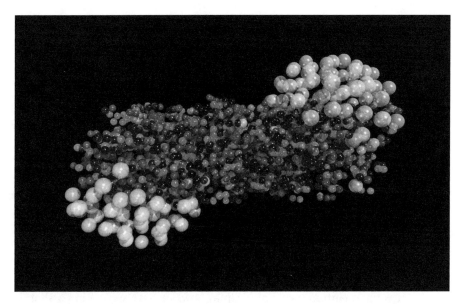

FIGURE 17.6
An artist's impression of a collision between two heavy nuclei, in which individual protons and neutrons are shown in white, giving place to an indistinct plasma of quarks and gluons.

the freeing of quarks and gluons actually happens, and what the indicators are of the formation that has occurred. What's clear is that the plasma, once it has cooled, causes the quarks that until then were free to recombine to create new mesons and baryons, conserving a memory of their former freedom. The investigation—and the explanation of it—are again very complex. The available experimental indications are nevertheless very encouraging, and it is anticipated that the coming work, in particular with the LHC, will provide us with some big surprises. We are waiting impatiently.

For the moment, however, I am sure that some of my readers will have asked "The Question": assuming that Grand Unification is something given, or at the very least that we will succeed in having sufficient indirect proof of its validity, can we then dream of unifying the Superforce, with its single coupling constant, with gravity as well—thus fulfilling the ambition of generations of scientists, thanks to the hypothesis of the Theory of Everything? Dreaming costs little, and physicists have already tackled the subject with an approach that was anything but detached—and which, at least on the part of the theorists—has yielded notable results. The experimental physicists, unfortunately, are virtually watching from the sidelines.

Once again, we find ourselves at a point in the history of science where we are in the embarrassing position of risking not having any predictions to verify, or phenomenology to study, with the experimental instruments that are available. We are talking about energies completely beyond our current capabilities! The methodological and philosophical problems concomitant with this considerable handicap are serious, and in a different context it would be timely to discuss them. But because what's at stake, as well as a dream, is a significant endeavor on the part of scientists, let's try to explain the situation as it actually is—and how it is projected to be in the near future. Before elaborating on what we could name Hyperunification, we would do well to take stock of what we have obtained in the last 100 years or so with regard to the formulation of a potential Theory of Everything. The logical course of this long historic and scientific evolution is mapped in figure 17.7. The last box in this figure perfectly targets the crux of a unifying theory, a problem of such weight that it drags with it two of the most beautiful and powerful theories developed in the twentieth century: general relativity and quantum mechanics. Up to now, we have spoken in passing about Einstein's general relativity, the best theory that we have to describe

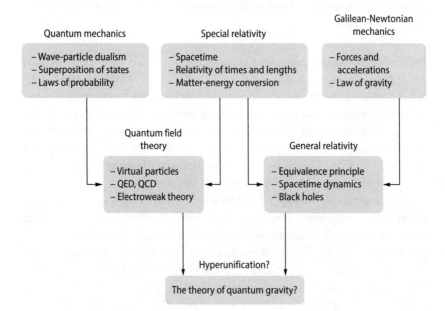

FIGURE 17.7
The development of Hyperunification and of the intermediate steps based on discoveries and theories.

gravity and its effects. We can simplistically say that the theory reinterprets the gravitational force as generated by disturbances of the spacetime fabric, which in turn are caused by the mass of bodies. A satellite is attracted to the Earth because it falls into the spacetime hole that is created by the large terrestrial mass (figure 17.8). In addition to this, Einstein's theory predicts the equivalence of accelerations and gravitational fields from the point of view of the effects that they cause. The force that we feel due to a gravitational field, therefore, is altogether equivalent to that which is perceived in a noninertial frame (i.e., in accelerated motion). The interpretation of gravity as a modification to the spacetime structure of the cosmos implies, for example, that photons—despite being massless particles—can be subject to the force of gravity. Like all particles in motion in the space that is curved by the presence of large masses, they do not travel in a rectilinear fashion, but precisely follow actually curved trajectories, just like the satellite in figure 17.8. The first verification of the predictions of general relativity dates all the

FIGURE 17.8
The interpretation of gravity according to general relativity. The mass of the Earth curves spacetime (for simplicity's sake represented in two dimensions rather than four), and the satellite "perceives" the hole, going into orbit around the planet or falling on it.

way back to 1919, just a few years after Einstein formulated it, to a solar eclipse during which Arthur Eddington measured the curvature caused by the mass of the Sun to a beam of light (photons) originating from distant stars. However, the most spectacular confirmation of Einstein's theory was the discovery of gravitational waves in 2015 for which Rainer Weiss, Barry Barish, and Kip Thorne, received the Nobel Prize in 2017. With that result, a new window opened on the unknown, through the "multimessenger" approach: namely, to look at the far universe not exclusively with electromagnetic radiation (visible light, X-rays, etc.) but also using neutrinos and, by now, gravitational waves. Let's also remember here the abovementioned detection of extragalactic neutrinos by IceCube, in coincidence with optical and γ-ray signals.

Another significant aspect of general relativity is that it is a classic, deterministic theory that acts in nature with continuity—and evidently not through quanta, as it deals with macroscopic objects. I am here now, and at the point in which I find myself, I feel the effect on my mass of the force exerted by the mass of the Earth. The eclipse of the Moon will begin at the precise moment foreseen by the solution of the equations of motion of the Earth-Moon system. Nevertheless, as the intensity of gravity increases as the distance between particles reduces, one encounters mathematical problems (infinities) for the typical distances when the hypothesized Hyperunification was possibly achieved in the very beginning of the life of the universe. Note that we are speaking of energies of around 10^{19} GeV, corresponding to a time of only 10^{-43} seconds after the Big Bang. This is completely outside the sphere of influence of Newtonian-Einsteinian gravity, but we formally have the right (nay, the duty) to consider this eventuality. As if this were not enough, the continuity of the structure of spacetime in general relativity clashes with Heisenberg's uncertainty principle, the intrinsic indetermination of which is also applied to the gravitational field. At extremely small distances, the latter would begin to fluctuate chaotically, and it takes only an instant to reach the incomprehensible assumption that the same spacetime becomes quantized. In practice, that means that between two points in space, there could be no space in the middle ...

Finally, there is the issue of pointlike particles, a sort of "stone guest" for particle physics. Until now, in agreement with the experimental measurements, we have taken quarks and leptons to be pointlike—and without any inner structure—independent of their mass: a concept that on the one hand is difficult to digest, and on the other hand is a harbinger of problems when we consider the gravitational interaction between extremely close and enormously energetic particles—a source of a real headache for physicists. Then again, not even quantum mechanics operates comfortably at cosmological scale. The self-interaction of the fields and other similar attributes create absurd behaviors at large scales, which are not at all supported by experimental results. Must we conclude, then, that two of the most beautiful and successful theories in physics fail when describing the universe, each at one of the two extreme length and time scales? The question is how to combine them, extending and unifying the two models into a completely different paradigm. Resorting to an example that has been used many times before, it is as if we were developing special relativity so as to incorporate classic mechanics without, nevertheless, canceling the latter from the domain in which it is more than valid (i.e., where it explains the experimental facts well, such as in everyday life). As you will have gathered, this involves very complex problems indeed. The extremely close connection with the world of the elementary particles obliges us as a result to consider here at least the essential aspects of the coexistence of quantum mechanics and gravity in the microcosmic realm.

One possible solution for this infinite series of problems is string theory, which in principle is capable of constructing a reasonable scenario of Hyperunification. The strings are hypothetical one-dimensional objects, infinitely thin threads with a length that is infinitesimal but more than nothing (10^{-33} cm), which are closed on themselves like rings, or else open. These replace the pointlike particles at a scale smaller than the microscopic, reducing at least the mathematical problems generated by the assumption that particles—the sources of fields, subject to fundamental interactions—have no spatial extension (figure 17.9). Quantum strings continuously vibrate like the strings of a violin. Their vibration modes correspond to the various particles, each with its own sound. So we have a real quantum concert, but an extremely quiet one. There is something magical about the minuscule length of the strings—of the order of 10^{-33} cm. Also known as the *Planck length,* it is the smallest dimension possible and imaginable, 10^{20} times smaller than the size of a proton. To give an idea of what we are talking about, think that if the Planck length were equal to 1 meter, the diameter of the proton would be 10,000 light years—equal to a tenth of the diameter of our entire galaxy! The Planck length (l_P) is obtained by combining fundamental constants as follows:

$$l_P = \sqrt{\frac{\hbar G}{2\pi c^3}} = 1.6 \times 10^{-33} \text{ cm.}$$

In the formula, you will find several well-known quantities, together with G, Newton's gravitational constant. The Planck length constitutes the limit

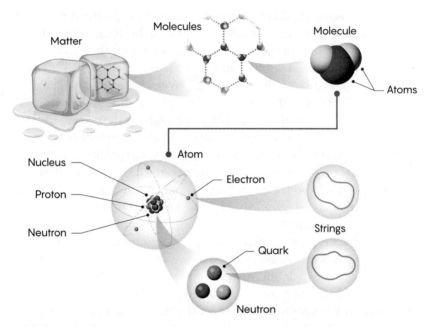

FIGURE 17.9
According to string theory, it is these strings that constitute the intimate structure of particles, of those that appear to be pointlike, such as the quarks and the leptons, at a scale much larger than that of the strings.

of what could be hypothetically measured. Below such a dimension, it makes no sense even to conceive of a length measurement because spacetime quantum fluctuations would be larger than the distance to be measured. We are once again running the risk of a serious headache ... But let's return to strings. It seems almost obvious that every time they are observed from the abyssal distance of the size of a proton (sic!), our beautiful oscillating strings appear as points, returning us to the approximation of pointlike particles. If we get close enough, however—if the energy increases or, equivalently, if the time after the Big Bang decreases—the strings show all the beauty of much-less-than-nano musical instruments, their continuous vibration generating the "notes," which are nothing other than our beloved elementary particles. String theory emerged around 1985, characterized by a pronounced mathematical complexity. After various vicissitudes, the American theorist Edward (Ed) Witten (figure 17.10) produced in 1995 a formulation of the theory that in a way summarized and encompassed all the preceding versions, reaching a theoretical scheme called *M-Theory*,

FIGURE 17.10
Edward (Ed) Witten, one of the major authors of string theory.

which was comprehensive and coherent. Hyperunification seems possible, the conflicts between quantum mechanics and general relativity can be avoided, and the phenomenology of the particles observed in nature could be explained as well; and all of this occurs at the cost of a "small" complication—the need to have a nature that is no longer in just three spatial and one time dimension, but that has (pay close attention) ten spatial dimensions and a temporal one. This may strike some as a step too far, but let's assess this new scenario objectively, with the help of a little logic.

In principle, the idea that there can be more than three spatial dimensions should not astonish us too much. From the mathematical point of view, there is no problem, apart from having to make the equations that describe the motion of particles and their interactions more complex. From the physics point of view, a few timely observation are in order. Imagine that you have some two-dimensional creatures, like extremely thin Dover soles confined to the surface of a planet—why not a two-dimensional planet, as well, like the one postulated by flat earthers ... These creatures will spend their lives moving on the surface, meeting and interrelating in a world of just two spatial dimensions in addition to the temporal one. Some of these flatfish are physicists, and studying the laws of nature, they realize that to explain some of the processes that they observe daily, they must resort to the hypothesis of a third spatial dimension that is invisible and undetectable to them. Take rain, for instance—so strange as to make some drops, or rather some round and aqueous things, appear on the surface of the planet from out of nowhere—or the β decay of their radioactive elements. How extremely strange that seems: usually the neutrino or the electron disappear into nothingness after having been produced. One of the Dover sole physicists, a particularly visionary and acute scientist, manages to explain β decay by assuming that when the neutrinos or electrons seem to vanish, they are in fact passing into another dimension, the third, which is not visible to the Dover soles. Similarly, nothing prevents us from interpreting the complexity of our world, including the apparent disagreement between the various theories of the microcosm and macrocosm, by hypothetically postulating the existence of other, hidden dimensions to which the particles accede under certain conditions. Consider that one of the consequences of string theory is that the extreme weakness of the gravitational force may be because its quanta (the hypothetical gravitons) spend much of their time lounging about in the other seven spatial dimensions, turning up only once in a while in the three that we know. The extra dimensions can be big, like those of our three-dimensional world, or they can also be very small or rolled up. An example of small extradimensionality can be created by imagining a long tube seen from a distance. From short distances, the tube appears to be what it is: a three-dimensional cylinder. The same tube, however, seen from far away—or if you prefer, observed with a low-energy probe—is nothing more than a one-dimensional thread. This simple fact provides the justification and motivation behind one of the scientific objectives of the LHC (i.e., a machine that creates high-energy probes): to search for signs of extra, small dimensions, observing their hypothetical structure and the effects on the particles from very close up, which is equivalent to employing very high-energy collisions. Nevertheless, without experimental evidence, the strings are only a hypothesis—a plausible and fascinating one, but just a hypothesis. However, string theory seems to work on paper, in a formal sense (I mean from a mathematical standpoint), though there is certainly no shortage of criticism leveled at it. Alternative models have been proposed that do not require extra dimensions or microscopic vibrating threads. The hope is that any of those models can ultimately generate falsifiable predictions, as the philosopher of science Karl Popper used to say. Each theory must itself furnish the means with which its correctness can be verified, and the predictions must fall within the sphere of our technological and experimental capabilities if we are not to find ourselves back at square one again ...

Even theoretically, it looks impossible to go beyond Hyperunification—or at least we can say with certainty that it is very, very difficult to do so. We come up against the singularity, that corpuscle of the hyperhot universe that is on the point of exploding spectacularly, and we don't even know the origin of it. It has probably come from a giga-quantum-fluctuation generated from the vacuum and immediately magnified upon the Big Bang explosion, thus creating for itself the vertiginously expanding space and the time with which to order all physical events, one after another. In fact, what would be the point of talking about time if there were nothing around? There would be no reactions taking place, no oscillating atoms, no movement to measure: nothing. Time was most probably created, then, together with the universe, at the Big Bang. This implicitly provides an answer to one of the many million-dollar questions: what was there before the great explosion? We don't know, but to speak of "before" is certainly very complicated. The most sensible thing that occurs to me to represent this before is an analogy: it is like thinking about ourselves before we were born. We are clearly on a shifting ground riddled with pitfalls. And besides, it is hardly satisfactory to confront and then finish in a few words a complex cosmological issue, as if it was merely a corollary of particle physics. Perhaps the best way of concluding this retrospective journey in time is to revisit its key stages, the relevant milestones for the genesis of particles and forces. So let's return to the very beginning and then move swiftly to the present time.

At the time $t=0$, something unique occurs. From nothing, from the quantum vacuum, an enormous concentration of energy emerges. We have no idea how. This is followed by a huge explosion, the Big Bang. After 10^{-43} s, we find ourselves in the Planck era. The universe has a size of 10^{-33} cm, and a temperature of 10^{32} degrees centigrade—numbers that are completely beyond our comprehension and imagination. Unfortunately, even our efforts to generate theoretical hypotheses that explain the physics of that moment are put to the test. We even lack a mathematics capable of supporting the scientific theories to describe the newborn universe of that time; physics is almost totally unknown, and all speculations lead to completely incomprehensible situations for our tiny human mind. Gravity is born but we don't know how: probably combined with the other forces in the Theory of Everything. Perhaps, the space dimensions are manifestly 10 and not only 3 as we observe today. Quantum mechanics conflicts with gravity, time and space fluctuate in the so-called quantum foam: the concept of measurement has no meaning. The *former* is confused with the *after* and the *there* with the *here*. The same applies to the principle of cause and effect.

In the period between 10^{-43} and 10^{-36} s (quite a long lapse of time, right?), gravity separates from the other unified interactions due to the rapid decrease in energy density. The first particles and antiparticles are created from the breaking of the Hypersymmetry.

Between 10^{-36} and 10^{-32} s, a process begins that deserves an entire book in itself: that of cosmic inflation. This is a theory that explains our cosmological observations extremely well, and in particular the great uniformity in temperature (energy) of the cosmic microwave background radiation that we observe today. The universe begins a phase of extremely rapid and tumultuous expansion, as if propelled by a second Big Bang even more powerful than the first. The space of the nascent universe risks being torn as it expands at a speed much higher than that of light—there is no problem, as we have seen, with such a speed, since unlike matter that is subject to the speed limit determined by Einstein, space may expand at crazy, arbitrary velocities. The strong interaction separates from the electroweak one. The plasma of quarks and gluons is created. The size of the cosmos grows at extreme speed and multiplies by an unimaginable factor, ranging from 10^{30} to 10^{50}, causing the universe to pass from an infinitesimal length to about 10 cm in diameter: the most incandescent fireball that has ever been imagined or produced. Such an expansion is equivalent to that of a DNA molecule that instantly becomes as large as our entire galaxy!

Toward 10^{-32} s, the inflation expansion extinguishes itself, but the universe continues to grow like an enormous and dark firework. But around that moment, another, perhaps even more surprising thing happens. The symmetry between the number of elementary fermions and antifermions (quarks and leptons), probably created in equal numbers at the time of the Big Bang, breaks. Due to some reactions likely involving supermassive neutrinos, a slight asymmetry is generated, of some few parts out of a billion, in favor of matter. This "small" excess of matter survives the gigantic annihilation of the particle-antiparticle pairs, which generates the huge number of photons that we still find today as Cosmic Microwave Background (CMB) radiation. The surplus of matter will then constitute the entire universe, in which the antimatter appears almost completely absent today.

From 10^{-32} up to almost 10^{-11} s, there is the so-called Electroweak Era. W, Z, and H bosons are copious, the Higgs field begins to slow down all the other particles (or almost, anyway), supplying them with mass. We find ourselves in a period that we can nowadays explore with experiments: the maximum energy attained by the LHC—13 TeV—corresponds to an age of the universe of less than a picosecond (10^{-12} s).

In the interval between 10^{-12} and 10^{-6} s, the universe lives its Era of Quarks. These, along with electrons, neutrinos, and their antiparticles, are created in large quantities. The electroweak force separates into its constituents. The quarks prepare to build the hadrons. This happens during a relatively very long period of time, between 10^{-6} and 1 s, during the Era of Hadrons. Because the temperature is quite low ("only" 1,000 billion degrees), the quarks are eventually trapped to form all the types of hadrons, including protons and neutrons. Some of the electrons interact weakly with the protons, creating more neutrons and neutrinos. The latter, now free and interacting very weakly, begin their journey at the speed of light toward the four corners of spacetime that will lead them to our current time, enormously slowed by the expansion of the universe but still free to run about the cosmos.

Between a second and 3 minutes, the universe lives through the Era of Leptons. A large fraction of the hadrons (but, fortunately for us, not all of them ...) are annihilated, along with their antihadrons. Electrons and positrons dominate the universe, annihilating each other and being re-created by photon conversion. A great number of photons begin to fill space.

In the period between 3 and about 20 or so minutes after the Big Bang, we find a universe that is very active in creating the first light nuclei, especially of deuterium, helium, and lithium. These form in the thermonuclear fusion reactions triggered by the lower temperature of the universe—a few billion degrees at this stage. The fusion processes finally extinguish as the temperature continues to decrease, and the volume of the cosmos increases. The universe has already a size of several hundred light years.

From then on, for a long period of time of approximately 250,000 years, the universe swarms with photons, light nuclei, and electrons, with the latter still too energetic to be trapped and to form atoms. Another 100,000 years will be needed to create the first hydrogen and helium atoms (i.e., electrons bound to the preexisting nuclei). The temperature decreases still further, reaching a few thousand degrees. The initially opaque and dark cosmos now allows the photons to travel freely without being blocked by protons and electrons waiting in ambush. Space becomes transparent and large—1 million light years across. After about 380,000 years, the Fiat Lux occurs, with the faint red light of the first liberated photons. Before then, there was only pitch darkness, at least for a hypothetical observer sensitive to visible light ...

The evolution that followed is no longer just the domain of particle physics, but of astrophysics as well. Many fascinating and important things have happened in the nearly 13 billion years since the Fiat Lux. The light of the original photons is progressively extinguished due to the space expansion, and the universe plunges back into the so-called Dark Ages. After about 100 million years, gravity goes into action, acting on atomic matter made up principally of hydrogen gas. The quantum energy fluctuations of the first instants in the life of the universe, amplified by the expansion of spacetime, had created an unevenness in the spatial distribution of newly nascent matter, which was hence neither symmetrical nor homogenous. As a result, gravity has an easy task: attracting hydrogen atoms and obliging them to concentrate into centers of growth—it is the dawn of the creation of the first stars and galaxies. Such a process would not have been possible had the universe been completely uniform: every atom of matter would have been attracted in the same way as all the atoms in its vicinity, and for reasons of symmetry, there would have been no concentration. At many points in the universe, gravitational pull greatly increases the density of gas, and with it the temperature. Nuclear fusion reactions that had been extinguished some thirty minutes or so after the Big Bang are triggered again in the hyperhot centers of the nascent stars and ignite them. The splendor of light returns to the universe. Around the stars, which will then cluster into galaxies, planets form. Myriads of stars live their lives and then become extinguished, while new stars and whole new galaxies are continually being created. And so on, in an unceasing process of birth, growth, and death that is still active today, throughout the cosmos.

We thus pass from the eras in the life of the universe that lasted for minuscule fractions of a second to long periods of hundreds of millions, or even billions of years. For several billion years, the universe has been accelerating under the pressure of dark energy—a force stronger than that exerted by the gravitational one pulling it in the opposite direction, seeking to go back to the original situation of the Big Bang, perhaps by way of a Big Crunch. Around 4 billion years ago on Earth, perhaps in the proximity of an underwater volcano rich in energy and mineral elements, the first timid signs of biological life were created, going on under immense evolutionary pressure to eventually produce a creature of self-aware intelligence. For a few hundred thousand years—an instant in the life of the universe—*Homo sapiens* have been asking questions about the world around them and about themselves. And who knows how many other forms of life have existed in the billions upon billions of planets that populate the billions upon billions of galaxies! Due to the accelerated expansion generated by the dark energy and by the early cosmic inflation, today the universe has become immense: 93 billion light years in diameter, filled with approximately 10^{80} atomic particles—electrons, protons, and neutrons. These numbers are so huge as to be scarcely imaginable, and take even the scientist's breath away. Imagine traveling at the speed of our interplanetary probes (e.g., at 60,000 km/h). It would take no less than 75,000 years just to reach the second nearest star after the Sun, Proxima Centauri, which is scarcely 4.2 light years away from us!

The future of the universe is not wholly predictable. Much will depend on the specific cocktail of matter, dark matter, and dark energy, and on their characteristics. The most widely accredited theory is that of a progressive dilution of matter-energy, which will lead, albeit over a very, very long time—maybe as much as 10^{1000} years—to the so-called Great Cold: a slow death from starvation, after a most rich and turbulent life. Who knows. *Perhaps* everything will restart with another quantum fluctuation. *Perhaps* these fluctuations are happening somewhere at this very moment. *Perhaps* new universes are being continually created—separated from ours by abysses of absences of spacetime. Some are immediately extinguished, due to wrong constants of nature and having little resistance to natural selection. Others, *perhaps*, live (or will live) even more than our own, managing to develop their own self-awareness through the creation of intelligent beings who, sooner or later, will pose the same kind of questions as we have posed ourselves. *Perhaps*.

These are fascinating hypotheses and questions, devoid for the moment of valid scientific answers. For now, therefore, let's return to our particle physics laboratories and try to indicate the directions that we will be taking with future research. We have already said that making predictions in science is a risky business, and a difficult one. An unexpected discovery can change plans, programs, and paradigms: physics has shown us that this can and does happen—for instance, when we turned on the LHC and thought that if the Higgs had in fact existed, we would have to wait ten years or more to discover it, whereas here we are talking about its properties and already well invested in searching for Supersymmetry and extra dimensions. Nevertheless, in elementary particle physics, some kind of planning is also necessary. In speaking of the future, we should note that already today, fundamental physics research is unfolding in three main directions. On the one hand, we are heading toward the frontier of high energy represented by experiments with accelerators—at the LHC today, and tomorrow with who knows what as-yet-unbuilt machine of even higher energy. There is talk, for instance, about the Future Circular Collider (FCC), an accelerator for protons with a circumference of 100 km, and 100 TeV of energy, also to be situated in a tunnel on the outskirts of Geneva, or perhaps in China instead. We have seen that collisions made with increasingly high energy enable us to enter the unknown domain of extremely high-energy densities, corresponding in turn to times ever closer to the instant of the Big Bang. Today, we are at less than a picosecond: that amount of time seems insignificant, but we know that the Grand Unification is enormously distant—less than 10^{-35} seconds after the Big Bang itself—given the great number of powers of ten that separate the two temporal eras of the history of the universe. To earn a mere factor of 10 in approximating the Beginning of Everything requires exhausting technological, human, and financial efforts to such a degree as to make us think that we may realistically never reproduce in a laboratory that fateful moment and those infernal temperatures and energy densities. The hope is always of having at least indirect indications that may allow us to compare the various theories, and to discriminate among different hypotheses.

The other frontier concerns the study of very rare processes, with measurements of extreme precision capable of showing evidence for small weaknesses in the current theories. These results could allow us to discern the existence of more complex interpretative schemes—above all, for the physics that is technically inaccessible through the brute force of high-energy collisions. The physics of neutrinos is an important subject of this experimental effort, as is the so-called physics of flavor, which studies the very fine effects affecting conventional processes. Remember the loops run across by exotic particles in the Feynman diagrams (e.g., that relative to the effective Higgs mass)? Some particle decays that are not possible through lower-order processes because they violate conservation principles can occur instead in reactions described by higher-order diagrams involving loops of unknown particles. Measuring the contribution of these processes can be the indication of new physics, by studying known physics reactions with more acute eyesight and instruments than in the past.

The third challenge is tied to a relatively recent discipline, which combines knowledge and methods of particle physics, astrophysics, and cosmology: astroparticle physics. We are living through a moment in which an increased fund of knowledge has been able to develop through interdisciplinarity. In the case of astroparticle physics, the link between microcosm and macrocosm materializes in a common endeavor made up of experiments and theories that benefit from ample and diverse fields. On the to-do list of astroparticle physics, we find the search for dark matter; the study of cosmic rays and the particles emitted by distant cosmological sources, with experiments on the Earth's surface, underwater, and even beneath the ice of Antarctica; and physics in space, on satellites or space stations—and in the not too distant future, I hope, perhaps on the Moon and nearby planets like Mars. Here is a young discipline that is for the young—the product of that tendency toward the horizontal spanning of disciplines that were once separated.

Now I would very much like to conclude by discussing some themes related to future research in neutrino physics, a subject particularly close to my heart and one that can offer interesting opportunities for reflection on the wider future development of particle physics. Still on the subject of oscillations, there is an account that deserves to be cited, which in the last few years has had a marked resonance among the community of neutrino physicists. I am referring to the possible existence of the so-called sterile neutrinos. It all started in the 1990s, when the Liquid Scintillator Neutrino Detector (LSND) experiment at the Los Alamos National Laboratory found the signal of a possible oscillation channel that could not be attributed to the one between couples of standard neutrinos, v_e, v_μ, and v_τ. More precisely, what was being measured was a third value of Δm^2 that was different from the two already known, assigned to the oscillation of solar and atmospheric neutrinos, respectively. With three neutrinos, there can be only two values of Δm^2; the evidence of a third of Δm^2 necessarily implies the existence of at least one other neutrino! But if such a neutrino were to be conventional like its siblings, then it would have to interact with matter and consequently transform into its charged lepton partner in a weak charged current reaction. Nevertheless, we have no evidence at all for the existence of such a process, not least because the only currently known charged leptons are the electron, the muon, and the tau. So what follows from this? The hypothesis that has been elaborated is surely imaginative and fascinating. It assumes that there is at least one other "sterile" neutrino in nature, v_s, which does not couple with matter, hence its name, but which can mix with its siblings v_e, v_μ, and v_τ (which are the only existing conventional neutrinos, remember, as shown in figure 16.9 in chapter 16) through the oscillation mechanism. It gives me pleasure to point out that once again, it was Bruno Pontecorvo who coined the term *sterile neutrinos*, in 1969, to refer to right-handed neutrinos (RH) and left-handed antineutrinos (LH).

The result of the LSND experiment remained controversial for many years. It was refuted by certain other experiments, such as KARMEN in the UK, whereas another recent Fermilab project, MiniBooNE, appears to verify the exotic hypothesis of sterile neutrinos by observing events that are not easily attributable to a known experimental background. However, as always happens in physics, there is only one way to solve the mystery: experiment, experiment again, and carry on experimenting—to find out

FIGURE 17.11
A group of researchers standing in front of MicroBooNE, the multiwire and liquid argon detector, while it was being built at Fermilab (in 2013).

if the discovery is hiding around the corner or if there is a still unknown background. This is what my research group (figure 17.11) is carrying out in collaboration with other European and American institutes, in the context of a new project, still situated at Fermilab, called SBN (for "Short Baseline Neutrinos"). The experiment utilizes three separate detectors, MicroBooNE, SBND, and ICARUS, placed at an increasing distance from the source of the neutrinos and primed to detect the possible unexpected oscillation of muon neutrinos on a relatively short baseline—less than a kilometer, much shorter, for instance, than the 730 km of the CERN–Gran Sasso beam—as the result of a more complex oscillation scheme involving four neutrino species. The technique adopted by the detectors is that of the time projection chambers of liquid argon, a pioneering technology that was originally conceived at CERN in the 1980s by Carlo Rubbia. This makes it possible to obtain with electronic detectors of large size the same quality in the reconstruction of events as with the old bubble chambers, notably in terms of spatial resolution in the measurement of the particle tracks.

In a few years, with the joint effort of other similar projects that the scientific community is currently conducting with reactor antineutrinos, we will have the answer to the puzzle of sterile neutrinos. Will there be a further complication for neutrino physics? If so, what will be the role of this sterile neutrino in nature? Will it represent a window onto the discovery of new physics? Will we than have to conceive further, even more complex experiments to clarify the emerging scenarios? Or will we have discovered that it was down to an unknown background or a wrongly evaluated one? One thing is clear: the challenge that the neutrino continues to pose will not be ending any time soon ... In fact, there are many issues beyond sterile neutrinos that currently have their research focused upon our chameleon. We have already mentioned the great chapter of neutrino astronomy, capable of obtaining information on sources of extragalactic origin of extremely high energy—and one should not forget that there is still a good deal to be studied with regard to neutrino mixing and oscillations. It will be of great relevance to particle physics to understand why the almost identical mass eigenvalues are so small with respect to the masses of the other fermions: compared to that of the electrons, neutrino masses are at least a million times smaller! The problem is, above all, a theoretical one: what is the mechanism that generates a mass that is so small compared to the others, and why? It is no doubt another indication of a new and unknown physics. Will we obtain such indications from the forthcoming experiments?

Connected to the preceding chapter is an important question on the nature of the neutrino: Is it a Dirac particle, or a Majorana one? This matter is also very complicated, but it can be summarized by saying that in the first case, the neutrino behaves in an analogous way to its fermion siblings in being substantially different from its antiparticle. The electron is no doubt different from the positron due to the opposite electric charge, but the neutrino is electrically neutral, and a new possibility emerges for it. If Ettore Majorana were right, the neutrino would coincide in every respect with its antineutrino, and they would be two particles that were formally and substantially identical. Currently, a number of complex experiments of extreme sophistication are taking place to resolve this problem because it is of enormous importance for physics, for the understanding of the universe and of how it developed in its very first moments. Personally, I'm backing Majorana ... not due to chauvinism, but because the phenomenology that we could expect if the hypothesis were proved correct is very rich, and full of potentially revolutionary implications.

Another objective of great interest, also connected to the masses of neutrinos, relates to the measurement of the so-called mass hierarchy. On the basis of our present understanding of the PMNS matrix, we have derived with an excellent degree of precision relations that link the mass eigenstates (v_1, v_2, and v_3) with the flavor eigenstates (v_e, v_μ, and v_τ) (see chapter 14).

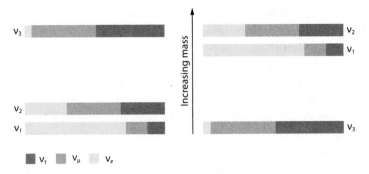

FIGURE 17.12
(Left) The so-called normal hierarchy of the masses of neutrinos; (right) the inverted one. Each mass eigenstate is expressed as a combination of the three flavor eigenstates, as derived from the measurements of the elements of the PMNS matrix.

Because the experiments on oscillations are not sensitive to the specific mass eigenvalues (m_1, m_2, and m_3), but only to the difference between their squares, ambiguities persist in the end. We do not know whether the order of the masses is $m_1 < m_2 < m_3$ or $m_3 < m_1 < m_2$. This is absolutely not a question of splitting hairs. On the contrary, the implications have wide-ranging significance and relate to the very nature of the neutrino itself. Let's look at the two hypotheses graphically illustrated in figure 17.12. The diagram on the left shows the so-called normal hierarchy with $m_1 < m_2 < m_3$. In this configuration, the three eigenstates v_1, v_2, and v_3 are each composed of a fraction of v_e, v_μ, and v_τ, in tribute to the values of the coefficients of the PMNS matrix. For instance, according to the experimental data, v_2 is a cocktail split almost democratically 1/3, 1/3, and 1/3 among the three flavors. In the diagram on the right, showing the so-called inverted hierarchy, the mass m_3, corresponding to the eigenstate v_3, is the lower one, with the specific combinations of the neutrino flavors staying the same. To distinguish between the normal and the inverted hierarchy is a goal that we should be able to achieve by comparing oscillation channels—for instance, by using neutrinos that propagate in matter (such as the rock of the Earth's crust) for sufficiently long distances (around 1,000 km). The formula for the oscillation in matter, more complex than in vacuum, is in effect sensitive to the ordering of the mass eigenvalues. Another possibility will be that of using antineutrinos originating from various nuclear reactors at various distances from the detectors. The propagation of neutrinos in dense matter influences the particles and antiparticles differently, and this difference depends on the specific mass hierarchy chosen by nature. If the hierarchy were normal, muon neutrinos would more probably oscillate into electron neutrinos, while muon antineutrinos would oscillate less into electron antineutrinos. The opposite would happen if the hierarchy were inverted.

The last of the main objectives of neutrino physics for the future will be verification of the potential existence of CP violation for neutrinos beyond that already established for quarks (remember what we said earlier and in chapter 9): it is possible that CP violation has rendered the original number of particles produced at the moment of the Big Bang larger than the number of antiparticles, by a mere factor of a few parts in 1 billion. The leptonic CP violation possibly related to neutrinos could have played a cosmological role even more important than that of the analogous violation for quarks in determining the matter-antimatter asymmetry. Experimentally, the identification of such an eventuality passes through the measurement of different oscillation probabilities for neutrinos and antineutrinos, or for an appropriate combination of measurements of different oscillation channels. Also consider that this particular violation of CP could have happened thanks to the interaction of superheavy neutrinos that existed only in the very first moments after the Big Bang. So much for the supposedly insignificant particles ...

To realize experiments in the future, we will have to work hard and hope that remarkable technological advances will materialize. The scientific plans on such themes are already well advanced. It was evident from the start that to reveal such subtle effects as those pertaining to CP violation, very intense neutrino beams and very efficient and massive detectors of many tens of thousands of tons at the very least would be necessary. This awareness evidently justified by the great complexity of the apparatuses and by the high costs behind their construction has caused scientists to alter the paradigm for proposals of future projects. Let me explain. Even the biggest among the experiments with neutrinos that we have spoken about have been built as a result of large international collaborations, consisting in some cases of hundreds of researchers—but also with a regional structure to the projects, and with costs that were sustainable by a few funding agencies and research institutes. For the next generation it is anticipated that costs will be at least ten to twenty times greater. This requires collaboration between a proportionally larger number of groups, coming from many countries and continents. Let's take the example of what happened at CERN with the LHC machine and its experiments: an overall cost of some billion dollars to build and put into operation experimental apparatuses at the limit of our scientific and technological knowledge, for a project that required a global effort from the community of particle physicists. Hundreds of international institutions were involved, and thousands of researchers efficiently coordinated within the scientific and political structure of CERN. It represents the success of a new way of understanding Big Science, ever more planned over the long term (even in terms of decades), with a complex *managerial* organization, made up of accurate forecasts, risk evaluations, and plan B's. All of this protects the creativity and ingenuity of the individual researcher, giving latitude for the unforeseen, the surprise of penetrating into the unknown without preconceptions—and with the unceasing curiosity of the physicists powering the whole thing forward.

When a few years ago the next generation of experiments on neutrino oscillations began to be discussed, it was decided that future detectors would also have to fulfill the role of observatories for neutrinos of astrophysical origins (astroparticle physics), and be capable of investigating proton decay by being massive. It seemed clear that neutrino physics was also approaching the gigantism of high-energy physics, of the LHC. Two projects began to emerge: the Long Baseline Neutrino Experiment (LBNE) in the United States, and (you guessed it) Hyper-Kamiokande in Japan, part of the ever-more-gigantic saga begun by Masatoshi Koshiba. Unfortunately, at the moment at which the two experiments were being proposed, the change of paradigm had not yet taken place. With a certain reluctance—though this is only my opinion—both the United States and Japan had the intention of proposing two complex national projects that would involve substantial international participation, but not to a degree that would undermine the lion's share enjoyed by the host nation. Reasons of international prestige, lobbying on the part of the respective groups of researchers, and the need to keep alive national interest in fundamental science are reasonable and understandable motivations, but they seemed to signal deafness to the message about the globalization of science that can be found in the example of CERN. Our ATLAS experiment at the LHC is no more German than it is Italian or Swiss, and I would add American, even though the United States is not one of the CERN member-states. To make a long story short, in the last few years, things have moved rapidly, and the paradigm shift to which I have alluded has actually happened and been implemented, in the United States, Japan, and in Europe. A comprehensive scientific document, the *P5 Report*, commissioned by the American funding agencies for fundamental physics and produced by an international group of physicists, arrived in 2014 at a series of conclusions that included succinct and clear recommendations for the politicians of science. Two of these recommendations relating to neutrino physics, numbers 12 and 13, read: "In collaboration with international partners, to develop at Fermilab a coherent program for the study of short- and long-band neutrino oscillations" and "To establish a new international collaboration for the planning and implementation of a new competitive infrastructure, the Long Baseline Neutrino Facility (LBNF), hosted by the United States." Behind these words, clear and innovative messages can be heard. The national experience should be shelved in favor of new projects with truly international origins: a genuine revolution for physics and science politics in the United States. One of the last remaining walls from the past was being toppled by pragmatism. The first recommendation was immediately taken up by converting the emergent SBN project into an international structure. The LBNF issue proved to be more complex. In this case, the sheer financial scale of the project, involving as it did some $2.5 billion, demanded caution and the construction of solid political-scientific institutional bases.

In May 2014, along with other colleagues in the international community of neutrino physicists, I was invited by the Fermilab director Nigel Lockyer to participate in a scientific committee that should have laid the groundwork for the creation of the new LBNF project. To ferry the various international partners toward the realization of an infrastructure at Fermilab (the neutrino *Super-beam*) and of an experimental site that was sufficiently far from it—about 1,300 km away—to turn on neutrino oscillations and detect subtle effects such as the violation of CP.

In a relatively short time, it has been possible to promote the creation of an international collaboration—of currently about 1,000 researchers from more than 175 institutes in more than thirty countries—dedicated to the creation of the Deep Underground Neutrino Experiment (DUNE) project. The experiment consists of a massive "far" detector (weighing 70,000 tons) positioned in the Sanford Underground Research Facility (SURF) in South Dakota, some 1,300 km from Fermilab (figure 17.13), and of a "near" apparatus installed at Fermilab itself, close to the neutrino beam source. The purpose of the latter is to measure the neutrino beam features where oscillations do not occur, to allow for a comparison with the response of the detector at SURF. The large, far detector will in this case also be based on the technology of the liquid argon time projection chambers. This gigantic and complex apparatus will be a Macro-MicroBooNE (so to speak), and will start delivering results around 2030. The depth of the DUNE underground site will guarantee the cosmic silence, effectively screening off the detector from everything except neutrinos of astrophysical or accelerator origin. The technology of the liquid argon time projection chambers will also ensure high sensitivity for the search for proton decay, and with a little luck, for

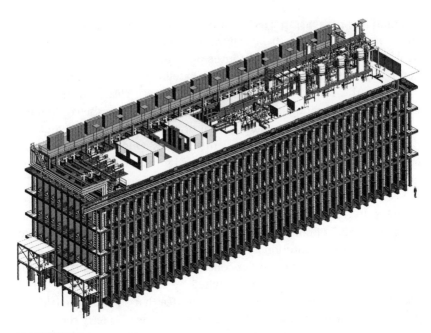

FIGURE 17.13
This is what one of the four cryostats in the "far" detector of the DUNE experiment will look like. It consists of a gigantic container about 20 meters high, 20 meters wide, and 65 meters in length, filled with more than 17,000 tons of liquid argon at a temperature of −184 degrees centigrade. The cryostats will be positioned underground, at a depth of about 1,600 meters. To give an idea of the scale, the figure of a typical researcher is shown next to the detector.

the second time in history after the Kamiokande result, it will allow the detection of neutrinos coming from a supernova—this time with a much higher detection performance. Needless to say, the collection of neutrinos from Fermilab will contribute to assessing the signal of the possible CP violation of, and to performing the measurement of the unknown hierarchy of the neutrino masses. In other words, it will provide enough work for generations of scientists. It moves me to think that many of those individuals who will make these great discoveries are still in high school today ... So here's wishing them the best of luck with that work!

18 BE CURIOUS

> Remember to look up at the stars and not down at your feet. Try to make sense of what you see and wonder about what makes the Universe exist. Be curious.
>
> —Stephen Hawking

Let me speak directly to the reader who has come with me this far.

At the end of our journey through the world of elementary particles, I would like to offer a final spur to reflection—one that issues directly from my own personal experience as a scientist. I have no conclusions to make here; in this book, I have sought only to suggest methods and provide the groundwork for understanding, offering you, I hope, prompts to pursue in greater depth the subjects that you find most interesting; to look for sources and bibliographical references, and to continue on the journey of discovery either by yourself or in the company of other guides.

In the preface, I told you that I was convinced that to be a researcher, to be a scientist, is one of the most wonderful professions in the world; something that I have felt even more keenly as a result of writing this book. I feel that I should dedicate it to all the curious, sincere, and honest young people in love with nature, ready to embark on a voyage to the Indies hoping to land in the Americas.

But I prefer to compare the work of a scientist to that of an artist or a poet, rather than to that of an explorer. We face the same practical problems daily—delusions, sacrifices, misunderstandings—but also experience many intangible and fragile sensations and satisfactions that are sometimes difficult to communicate to others.

Furthermore, is it not perhaps poetic to know that our small and insignificant planet wanders through an enormous universe that has existed for 13,800,000,000 years, perhaps born by chance and perhaps accompanied by an infinite number of siblings, each with its own laws of physics?

A universe that, in a fraction of an instant, determined everything that surrounds us today according to laws dictated by beauty.

A universe that for a few billion years has decided, surprisingly, to accelerate its expansion, pushed by a mysterious dark energy; and with billions upon billions of galaxies, the movement of which is conditioned by a matter that is equally dark and unknown.

A universe that, after an immensely long, unconscious life, thanks to human beings on planet Earth—and to who knows how many creatures scattered throughout its galaxies—has acquired self-awareness, and with the help of scientists over hundreds of years has questioned itself about its origin and evolution, and has had told the story of the fantastic things it did in its youth.

The universe, a huge container beyond which there is neither space nor time, filled to an outrageous degree by an invasive and all-pervasive field with an English name, and by an almost infinite number of less than microscopic objects that physicists have been studying tenaciously for more than 100 years, and which we have tried to understand the why and wherefore of together. Perhaps from studying them, one fine day in the future, we will manage to discover how everything began and how it will end (if there is an end) and to reveal the mysterious laws by which everything is governed.

Our minuscule particles and their way of relating to each other in the weave of spacetime have offered me the perfect excuse to speak to you of science and scientists, as well as of their work, perhaps from a less familiar perspective than usual, with the intention of inviting you to look at nature's marvels in a knowledgeable and curious way, and to become fascinated by those that still remain hidden. There are so many such wonders out there to be discovered.

ACKNOWLEDGMENTS

I am indebted to Sandro Bettini, Saverio Braccini, Leonardo Merola, and Michele Weber for their careful reading of the original Italian manuscript and their interesting suggestions, and to Paolo Zuccon and Callum Wilkinson for their review of the English version. A special thanks goes to Andrea Morstabilini, Paola Sala, and Irene Zocco of *Il Saggiatore*, and to Jermey Matthews of the MIT Press. I am also grateful to Helen Wheeler of Westchester Publishing Services for her project management.

Finally, a loving thanks to Paola Scampoli, for having been continuously on my side during the writing of this book and providing ideas, critiques, and corrections.

INDEX

Figures and tables are indicated by "f" and "t" respectively, following page numbers.

Acceleration, 98, 105, 179, 329–330. *See also* Particle accelerators
Aces, 189
AdA, 112, 113f, 114
ADONE, 112, 114, 208, 218–219, 218f, 221, 223
Alpha particles, 14–16, 17f
Alpha radiation, 12, 14–15
Alternate Gradient Synchrotron (AGS), 248
Alvarez, Luis, 85, 105, 246
Amaldi, Edoardo, 45, 73f
Amaldi, Ugo, 97, 340–342
Anderson, Carl, 78, 80, 82, 82f, 87–89, 88f
Angular momentum, 47, 50, 53–56, 54f
Antibaryons, 228, 230
Anticolors, 192–195, 196f, 198f
Antifermions, 168f, 346f
Antimatter, 79–80, 125–126, 160, 183, 369
Antimatter particles, 78, 80
Antimuons, 177, 187, 187f, 202
Antineutrinos, 73–74, 158f, 159, 159f, 233, 244f
  atmospheric neutrinos and, 267–268
  detection of, 246, 267–268
  electron antineutrinos, 233, 260, 284, 318
  electrons and, 158, 234
  protons and, 244
  sterile neutrinos and, 366–367
Antiparticles, 73–74, 79, 160, 167, 223, 225, 225f
  virtual particle-antiparticle pairs, 348, 350f
Antiprotons, 112, 119, 223, 314–316, 316f, 318f
Antiquarks, 119–120, 125–126, 192, 195, 197–198, 218–219, 221. *See also* Quark-antiquark pairs
Antisymmetricity, 190–191
Aristotle, 3, 21
Artificial neutrinos, 247–248, 267, 273, 275
Astroparticle physics, 364–365
ATLAS experiment, 204, 206f, 334–336, 336f, 339f, 370
Atmospheric neutrinos, 267–270, 272f, 273–274
Atomic bomb, 29, 76–77, 183, 242
Atomic nucleus, 43
  beta decay in, 73–74, 74f
  charged, 17
  collisions, 143–144, 143f, 350–351, 351f
  decay of, 73
  electric field generated by, 180, 180f

Atomic nucleus (cont.)
  electrons and, 15–17, 58–62, 166–167, 166f, 173, 173f
  electrons and protons in, 16–17, 43, 70–71, 75–76
  energy levels of particles in, 52–53, 58–59
  inside of, 226, 249
  mass of, 70–71, 226
  muons and, 83–85
  neutrinos and, 73–75
  neutrons and, 70–71, 75–76, 126–127
  in nuclear fission and fusion, 77
  protons and neutrons in, 76, 83, 111, 126–127, 133, 133f, 137, 141, 143–144, 192, 228–229
  Rutherford on, 12, 12f, 14–16, 16f, 64, 70–71, 126
  as spectator, 179, 182f
  spin of, 155–156
  strong interactions in, 192–193, 193f
  strong nuclear force in, 133
Atoms. *See also* Hydrogen atoms
  angular momentum of, 55–56
  Bohr model of, 41, 42f, 43, 70
  carbon, 76
  Dalton's theory of, 3–5, 4f
  Democritus on, 3, 5
  diameter of, 16
  Einstein on, 5–6
  electrically charged, 10, 10f
  electrons, force and, 132–133
  electrons in Bohr model of, 41, 42f, 43
  energy of, 13
  of gold, 15, 64
  helium, 12, 77
  internal structure of, 6–7
  ions, 8, 87
  Maxwell on, 5
  molecules and, 5–6
  momentum of, 55–56, 64
  in nuclear fusion, 77
  oxygen, 76
  positronium, 185
  as quantum entities, 55
  quarks and, 130
  Rutherford model, 15, 17f, 41, 70, 75
  silver, 54–56, 55f
  spin of, 56
  Thomson, J. J., model of, 10, 10f, 14–15, 17f
  visible light of, 63–64, 64f
Auger, Pierre, 45

Bahcall, John, 259–260, 260f, 269, 272
Baldin, Alexander, 242–243
Barish, Barry, 354
Baryons
  antibaryons and, 228, 230
  decay of, 203, 204f
  fermions and, 148f
  protons as, 228–229
  quantum numbers of, 228, 230
  quarks and, 118–119, 121f, 122–123, 123f, 125, 195, 196f
BEBC. *See* Big European Bubble Chamber
Becquerel, Henri, 11
Bellettini, Giorgio, 219
Bernoulli, Daniel, 5
Beta decay, 67–70, 155–156, 234, 234f, 236–237
  electron neutrino produced in, 254
  inverse, 244, 244f
  nuclear, 73–74, 74f
Beta electrons, 67–71, 69f, 155–156
Beta rays, 12–13, 74
Bettini, Sandro, 25
Big Bang, 32, 132, 142, 160–161, 164, 339
  Big Crunch and, 362
  electroweak unification theory on, 310
  energy density seconds after, 341
  gravity and, 359–360
  GUT and, 342, 364
  Higgs field and, 325–326

Index

high-energy interactions in particle accelerators and, 279
micro black holes and, 347–348
neutrinos from, 278–279, 284, 369
quarks, gluons and, 350
string theory and, 356
time and, 33–34, 359
Big Crunch, 362
Big European Bubble Chamber (BEBC), 89, 90f
Bjorken, James, 214
Black-body radiation, 37–40, 38f
B meson, 323
Bohr, Niels, 41, 42f, 43, 45–47, 69–70
Bohr magneton, 174–176
Born, Max, 46
Bose, Satyendra, 58
Bose-Einstein statistics, 58, 189–190
Bosons, 58–59, 166. *See also* Electroweak bosons; Higgs boson; *W* bosons; *Z* bosons
  fermions and, 58, 148f, 189–190, 343–344
  integer spin of, 58, 148f, 189–190, 343
  massless, 310
  as mediators, 167, 232–233, 233f, 238, 307–308, 310–311, 313–314, 314f
  Supersymmetric partners of, 346, 347f
  weak, 325
Bottom quark, 221, 222f, 223, 224f, 228, 336, 338
Bound particles, 71, 231, 233
Bremsstrahlung radiation, 178–179, 180f, 181–182
Brookhaven National Laboratory (BNL), 214, 215f, 217f, 248
Brookhaven synchrotron, 123
Brout, Robert, 323, 334
Brown, Robert, 6
Brownian motion, 6, 40–41
Bubble chambers, 94, 123, 286, 366
  BEBC, 89, 90f
  electron-positron pairs in, 91f

Gargamelle, 249–250, 251f, 312–313
  particle tracks in, 90–92, 91f
Butler, Clifford, 94

Cabibbo, Nicola, 237–238, 237f
Cabibbo-Kobayashi-Maskawa matrix (CKM), 238–240, 239f, 253, 255, 257, 272
Calorimeters, 291–292, 294–296, 295f, 298, 299f, 300
Carbon atom, 76
Carrelli, Antonio, 71
Cathode rays and tubes, 7–9, 9f
Cerenkov, Pavel, 262–263, 270f
CERN, 29, 33, 45, 65, 243–244, 279. *See also* Large Hadron Collider
  antimatter and, 183
  ATLAS experiment at, 204, 206f, 334–336, 336f, 339f, 370
  CHARM II experiment at, 311–313, 312f
  neutrino beam infrastructure, 276–277
  in OPERA experiment, 273–274
  Proton Synchrotron accelerator, 250
  Rubbia at, 366
  SPS of, 314–317, 316f
  Telegdi at, 252, 252f
  UA1 detector at, 314, 316f, 318–320, 318f, 319f
Chadwick, James, 13, 71
Charge, electrical, 7–9
Charged currents, 234, 236, 238–239, 239f, 313
Charged elementary particles, 10
Charged fermions, 168f, 169, 177, 346f
Charged leptons, 219, 223, 254, 279, 320, 338
Charged particles, 8, 89, 91–92
  acceleration of, 98, 105
  in detectors, 297–300
  photons and, 324–325
  positive, 98, 99f

Charged pion, 85, 86f, 92–93, 93f, 235, 235f
Chargino, 346
Charm-anticharm quantum system, 218
Charm antiquarks, 218–219
CHARM II experiment, 311–313, 312f
Charm quarks, 214, 218–221, 220f, 228, 238–239
Charpak, Georges, 288–289, 290f, 291
Chemistry, 5
Chen, Herb, 264–265
Christofilos, Nicholas, 107, 107f
CKM. *See* Cabibbo-Kobayashi-Maskawa matrix
Classical mechanics, 26–28, 163–164
Classical physics, 6, 9–10, 19, 37–39
  observer-object relationship in, 46
  quantum physics versus, 3, 39, 55
Cloud chamber, 88–89, 88f
CMB. *See* Cosmic Microwave Background
CMS experiment, 301, 301f, 334–335
Cobalt, 70, 73, 155–158
Cocconi, Giuseppe, 71
Cockcroft, John, 100, 101f
Coil, 54–55, 54f
Collisions, 87, 299–300. *See also* Large Hadron Collider
  atomic nuclei and, 143–144, 143f, 350–351, 351f
  decay and, 170–171, 204
  electrodynamic annihilation, 183–187, 184f, 185f, 186f, 187f
  electron-positron, 184–185, 208–210, 209f, 210f, 223, 320, 321f
  between hadrons, 125, 125f, 126, 141–142, 146–147, 316
  of high-energy electrons and positrons, 184–185, 208–210, 209f, 210f
  among high-energy particles, 32
  momentum in, 315–319
  of neutrinos, 312, 314f
  in particle accelerators, 32, 34, 65, 109, 111–112
  of particles, 29–30, 32, 34–35
  of particles, mass and, 109, 111–112, 147
  of pions, 126
  between protons, 143–144, 143f, 204, 206f, 335–336
  between protons and antiprotons, 223, 314–316, 316f, 318f
  between quarks, 315
  scattering angles, 129f
Color charge, 193f, 194–198, 197f, 198, 203
  conservation of, 309–310
  gluons, quarks and, 194–195, 194f, 196f
  of gluons, 310
  INFN experiments on, 208
  of quarks, 191–192, 194–195, 194f, 196f, 219
Color field, 194, 196f
Colors, 37
  anticolors and, 192–195, 196f, 198f
  gluons and, 192, 194–197, 194f, 198f
  of protons, 192–193, 193f
  in QCD, 191–195
  as quantum numbers, 190–191, 205, 208
  of quarks, 190–192, 194–195, 194f, 199f, 201–203
  strong interactions and, 192–194, 193f, 194f
Combinatorial mass value, 147
Conservation
  of color charge, 309–310
  of electric charge, 126, 309–310
  of energy, decay and, 69–70, 69f
  fermions and, 344
  of isospin, 228
  of matter and energy, 34–35
  of momentum, 136, 180, 318
  of parity, 153–154

Index

of quantum numbers, 228, 230
of strangeness, 124, 126
symmetry and, 163–164
violation of principles, 325
Controlled nuclear fusion, 77–78
Conversi, Marcello, 84–85, 85f
Copenhagen interpretation, 47
Cosmic inflation, 360
Cosmic Microwave Background (CMB), 278–280, 282, 360
Cosmic rays, 80–84, 92–94, 93f, 267–268, 280
Coulomb force, 15, 77, 132
Coulomb interaction, 173f, 166
Coupling constant
  in electromagnetic force, 191–192, 313
  electromagnetism and, 168–169
  electrons and, 177–178
  in Feynman diagrams, 168–169, 170f, 175f
  of fundamental interactions, 308, 309f, 340–341, 341f
  mediator particles and, 168
  QCD, energy and, 200f
  of quarks, 208
  strong interactions and, 168–169
  unification of electromagnetic and weak forces, 313
  weak, 170f, 238, 308, 311, 340, 341f
Courant, Ernie, 107, 107f
Cowan, Clyde, 157, 242, 244–246, 245f, 257
CP violation, 160, 240, 369, 371–372
Cronin, James, 248
Curie, Maria Skłodowska, 11–12, 11f, 76, 157
Curie, Pierre, 11–12, 11f, 76, 157
Cyclotron, 101–104, 102f, 103f, 108, 156–157

Dalton, John, 3–5, 4f, 13
Dark energy, 329, 362–363
Dark matter, 329, 346–348

Dautry, Raoul, 45
Davis, Ray, 246, 258–262, 260f
Davisson, Clinton, 44
de Boer, Wim, 340–341
de Broglie, Louis, 43–46, 48
de Broglie wavelength, 63, 63f, 109, 111
Decay, 87
  of baryons, 203, 204f
  beta, 67–70, 69f, 73–74, 74f, 155–156, 234, 234f, 236–237, 244, 244f, 254
  of charged pion, 92–93, 93f, 235, 235f
  of cobalt nucleus, 155–158
  collisions and, 170–171, 204
  energy conservation and, 69–70, 69f
  energy in, 68–69
  of hadrons, 146, 215–217, 217f, 238, 239f, 337–338
  of Higgs boson, 335–339, 336f, 338f, 339f
  of muons, 93, 144–145, 156–157, 233–234, 233f, 236–237, 247, 303–304
  of pions, 154, 156–157, 247, 254, 267, 276
  of protons, 229
  protons, neutrons and, 228–229, 231, 234–235, 234f
  quarks and, 234, 234f, 337–338
  radioactive, 50, 70, 76, 102, 155, 261, 283–284
  strangeness production and, 94–95, 95f
  strong interaction and, 146–148, 191, 203
  unstable, 30, 32, 144–146, 145f, 191
  weak, 144–145, 156, 231–232, 235, 238
  weak interaction and, 154, 156
  of Z bosons, 319, 319f, 321–322
Decay amplitude, 146, 191
Decuplet of baryons, 122–123, 123f
Deep Underground Neutrino Experiment (DUNE), 371–372, 372f
Democritus, 2–3, 5

Detectors. *See* Neutrino detectors; Particle detectors
Diffusion, elastic, 127
Diffusion, electron-nucleus, 173
Dilation, of time, 27–28
Dipole magnetic moment, 54–55
Dipole magnets, 107–108, 109f, 111f
Dirac, Paul Adrien Maurice, 58, 78–80, 78f, 162, 367
  Fermi-Dirac statistics, 58, 189–190
Doppler effect, 277–278
Down quarks
  isospin of, 228
  up quarks and, 119–121, 177, 202–203, 210–211, 221, 225, 225f, 228
Drell, Sidney, 211, 225, 252
Drell and Yan model, 211, 225, 252
DUNE. *See* Deep Underground Neutrino Experiment
Dyson, Freeman, 165

Eddington, Arthur, 353–354
Effective constants, 197f, 308
Eigenstates
  eigenvalues and, 50–52, 254–255
  flavor, 257, 258f, 367–368
  groups of, 237–238
  mass, 254–257, 258f, 367–368, 368f
  of neutrino flavors, 255
Eigenvalues, 50–52, 254–255, 256–257, 272, 367–368
Einstein, Albert, 23f, 47, 65, 331–333
  on atoms and molecules, 5–6
  Bose and, 58
  on general relativity, 25, 142, 162, 352–354
  "On the Electrodynamics of Moving Bodies" by, 21
  on photoelectric effect, 40–41
  Planck-Einstein formula, 40, 41, 43
  on relativity, 23–24
  on spacetime, 23–24, 28

  on special relativity, 13, 18, 21, 26, 29, 40
  on speed of light, 26, 360
  thought experiments of, 153
  on time, 22–23
Elastic diffusion, 127
Electric charge, 7–9
  of atom, 10, 10f
  of atomic nucleus, 17
  charged currents and, 234
  conservation of, 126, 309–310
  of cosmic rays, 81
  electric fields and, 135, 136f
  of electrons, 9–10, 10f, 15, 132, 135, 178–179
  electrons and protons, 75
  fractional, 189
  gravity and, 135–136
  of protons, 133, 133f, 135, 192–193, 193f
  of quarks, 119–120, 120f, 122–123
Electric fields, 7–9, 13, 19–20
  electric charges and, 135, 136f
  electrons, atomic nucleus and, 180, 180f
  of electrons, self-interaction with, 172, 172f, 173f, 175
  energy and momentum in, 138
  Higgs field and, 326
  magnetic fields and, 100–101, 308
  particle accelerators and, 98, 99f
Electric/gaseous detectors, 286
Electrodes, 98, 99f, 101
Electromagnetic annihilation, 183–187, 184f, 185f, 186f, 187f, 202, 211
Electromagnetic calorimeter, 298, 299f, 300
Electromagnetic fields, QED and, 166
Electromagnetic force, 131, 133–135, 133f, 134f, 139f, 140
  in atomic nucleus, 192–193, 193f
  in coupling constant, 191–192, 313

Index 383

in electroweak interactions, 250, 251f, 252
in hydrogen atoms, 232–233
photons and, 324–325
in QCD, 199–200
Electromagnetic interactions
charges of, 309
between electron and atomic nucleus, 166–167, 166f
of electrons, 138, 139f, 170f
fermions and, 167–169, 168f
Feynman diagrams of, 166–169, 166f, 168f, 170f
hadrons and, 208
between quark-antiquark pairs, 202, 202f
between two electrons, 138, 139f
Electromagnetic radiation, 7, 17, 105, 138, 282
Electromagnetic shower, 291–292, 292f, 293f, 294
Electromagnetic waves, 7–8, 12–13, 17, 19–20
in black body, 39–40
de Broglie wavelength of, 63f
frequency of, 39–41, 43
interference, 44–45, 45f
photons and, 41
speed of, 20–21
Electromagnetism, 17, 19, 308–309, 340
coupling constant, 168–169
fermions and, 169
Galilean invariance and, 20, 26–27
Electron antineutrinos, 233, 260, 284, 318
Electronic detectors, 127, 208, 220, 286, 366
Electron microscope, 64, 64f
Electron neutrinos, 219, 254, 224f, 231, 257–261
muon neutrinos and, 234–235, 250f, 268–269

oscillation and propagation of, 265, 266f, 267
Electron-positron collisions, 184–185, 208–210, 209f, 210f, 223
LEP accelerator, 320–323, 321f, 322f
Electron-positron pairs
in bubble chamber, 91f
decay and, 146
in electrodynamic annihilation, 183–187, 184f, 185f, 186f, 187f, 202
Feynman diagrams of, 181f, 182–186, 182f, 184f, 185f, 186f, 208, 209f
invariant mass of, 215–216, 217f
photons and, 180–182, 180f, 181f, 182f, 291, 292f
resonances and, 215
virtual, 172, 176, 177f, 196–197
virtual photon and, 173f, 185, 185f
Z boson decay into, 319, 319f
Electrons
angular momentum of, 53, 54f
antineutrinos and, 158, 234
atomic nucleus and, 15–17, 58–62, 166–167, 166f, 173, 173f
in atoms, force and, 132–133
beam of, in SLAC experiment, 127
beta, 67–71, 69f, 155–156
in Bohr model of atom, 41, 42f, 43
coupling constant and, 177–178
eigenvalues and eigenstates of, 50–51
electric charge of, 9–10, 10f, 15, 132, 135, 178–179
electric field and, 172, 172f, 173f, 175, 180, 180f
electromagnetic interaction between two, 138, 139f
electromagnetic interactions, 166–167, 166f, 170f
electromagnetism and, 17
energy of, 13–14, 17–18, 41, 50–51, 58–59, 75–76, 79
force, photons and, 62

Electrons (cont.)
  high-energy, 127, 177–179, 184–185, 185f, 318
  in hydrogen atom, 52, 53f, 61, 101, 122
  ionization, 89
  magnetic moment of, 174–176, 175f
  mass of, 34, 67, 105, 138–139, 175, 348
  mediator particles and, 138–140, 139f
  momentum of, 60–62, 61f, 79, 127, 138–141
  muon neutrinos and, 233–234, 251f
  muons, neutrinos and, 92–93, 93f
  muons and, 84, 170f
  neutrinos and, 78, 170f, 233, 233f
  nuclear, 71
  in particle accelerators, 105, 138
  as particles and waves, 44–45
  photoelectrons, 40
  photons and, 41, 60, 61f, 62, 64, 138–140, 139f, 166–167, 166f, 178–179
  position and trajectory of, 87
  position of, probability and, 49
  positrons, photons and, 179, 183–185, 184f, 185f
  positrons and, 35, 88–89, 91f, 135–136, 136f, 168f, 169, 172, 173f, 174, 208 (*see also* Electron-positron pairs)
  QED and, 174
  quantum trajectories of, 60–61
  quarks and, 129–130, 130f
  scattering, Feynman diagram of, 314f
  self-interaction with electric fields, 172, 172f, 173f, 175
  spin of, 57
  Thomson, J. J., on, 8f, 9, 9f, 68
  virtual, 179, 182
  virtual particles and, 174, 176, 177f
  virtual photons and, 166–167, 168f, 172, 172f, 173f, 182–183, 346f
  visible light and, 63–64

Electrons, protons and
  in atomic nucleus, 16–17, 43, 70–71, 75–76
  diameter of, 65
  electric charge of, 75
  hydrogen atom, 9–10, 76, 232–233
  ions and, 8
  mass of, 67, 105
  mediator particles and, 138
  neutrons, neutrinos and, 73–74
  positrons and, 89
  quarks and, 192–193, 193f
Electronvolt, 32, 79
Electroscopes, 80
Electroweak bosons, 311, 323–324, 326
  in early universe, 360–361
  Higgsinos, neutralinos and, 346
  mass of, 324, 335f, 340
  W bosons, 314, 315f, 318f
  Z bosons, 314, 315f, 318f, 320
Electroweak interactions, 315, 336–337
  electromagnetic force and, 250, 251f, 252
  strong interaction and, 339, 360
Electroweak unification, 310–311, 313–314, 321f, 339, 341f
Elementarity, 2–3, 5–6, 97
Elementary particles, 1–3, 6, 70, 231–232, 373–374
  angular momentum of, 54
  atoms versus, 16
  charged, 10
  creation and destruction of, 75
  definition of, 63–65
  detection of, 86–88
  fermions and bosons, 148f
  forces and, 131–143, 134f
  gravitational force and, 131–132, 142
  Heisenberg uncertainty principle and, 59–62
  indirect observation of, 86
  mass of, 32–34
  measurement of properties of, 86–87

Index 385

mediators, 136–142, 137f, 139f
pointlike, 63, 354–355, 356f
probability and, 48–50, 49f, 65
QED and, 165
quantum mechanics and, 46, 48–49
relativity and, 27, 30
repulsive and attractive forces
  between, 136–137, 137f
resonance and, 146
special relativity and, 25, 37, 78–79
spin of, 53, 157
Standard Model of, 320, 322f, 323,
  341f, 345–346, 345f
statistical system of, 57–58
strange, 94–95, 119–121
Supersymmetric, 344–346, 345f, 347f,
  348–349
symmetry and, 124–125, 164
table of, 225f
Elements, 2–3, 9–11, 53
Empedocles, 2
Emulsions, 90–94, 93f, 286
Energy
  of atoms, 13
  conservation, decay and, 69–70, 69f
  conservation, matter and, 34–35
  coupling constant, QCD and, 200f
  dark, 329, 362–363
  in decay, 68–69
  density, 33–34, 341
  eigenvalues and eigenstates of,
    50–52
  of electrons, 13–14, 17–18, 41, 50–51,
    58–59, 75–76, 79
  hadronic, 295–296
  of Higgs field, 326–327, 327f
  kinetic, 5–6, 28–29, 33f, 79, 279
  loss, ionization and, 89
  mass and, 29–30, 32–34, 77, 111,
    139–140
  matter, antimatter and, 125–126
  momentum and, 28–29, 61f, 88, 138,
    147

nuclear fusion and, 77–78
of particles, 29–30, 32–35, 33f, 52,
  65, 79
of particles, in atomic nucleus, 52–53,
  58–59
photons and, 13, 40, 43, 61f, 105,
  139–140
potential, 29, 79
special relativity, matter and, 126
time and, 33–34
Energy spectrum, 37–38, 38f, 40, 43,
  67–70, 69f
Englert, François, 323, 334, 335f
Entanglement, 48
Epicurus, 5
Ether, 2, 25–26
EUCLID experiment, 329
European Space Agency (ESA), 329
Exchange theorem, 67
Exclusion principle, 56–59, 62, 132,
  190, 343–344
Experimental background, of measurements, 65–66

FCC. See Future Circular Collider
FCNC. See Flavor-changing neutral
  currents
Fermi, Enrico, 57f, 58, 71–72, 122,
  280
  in Manhattan Project, 241–242
  on neutrinos, 73–75, 151, 242
  in Via Panisperna group, 73, 73f, 76,
    208
  on weak decay, 231–232
Fermi-Dirac statistics, 58, 189–190
Fermilab, 223, 250, 276–277, 365–366,
  366f, 370–372
Fermion-antifermion pairs, 346f
Fermion lines, 171
Fermions, 58–59, 80, 119, 166, 323
  annihilation of, 185
  antifermions and, 168f
  baryons and, 148f

Fermions (cont.)
  bosons and, 58, 148f, 189–190, 343–344
  charged, 168f, 169, 177, 346f
  conservation principles and, 344
  electromagnetic interactions, 167–169, 168f
  electromagnetism and, 169
  half-integer spin of, 58, 148f, 343
  mass of, 236, 254, 307–308, 324, 329, 335f
  mediator particles and, 167–168, 168f
  Pauli exclusion principle and, 190, 343–344
  table of, 223, 225–226, 225f, 234, 320
  vertices, 167–168
  virtual, 168f, 183–184
  weak interactions and, 168–169, 232–233
  weak nuclear force and, 133
Fermi's golden rule, 170–171, 191
Feynman, Richard, 151, 152f, 165, 165f, 252
Feynman diagram, 311, 364
  of *Bremsstrahlung* radiation, 178–179, 180f
  coupling constant in, 168–169, 170f, 175f
  of electromagnetic annihilation, 183–187, 184f, 185f, 186f, 187f
  electromagnetic interactions in, 166–169, 166f, 168f, 170f
  electron-positron pair, photons and, 181f, 182, 182f, 183–186, 184f, 185f, 186f
  electron-positron pairs, in high-energy collisions, 208, 209f
  of electron scattering, 314f
  fermions in, 167–169, 168f
  for Higgs boson, 337f, 338f
  lower-order and higher-order, 174, 177–178, 183, 184f, 186f, 197

  of magnetic moment of electron, 175, 175f
  muons in, 167, 168f, 170f
  positrons in, 167, 168f
  probability in, 168–171
  propagators in, 170–171
  QED and, 178
  of quark-antiquark pairs, 202f, 203f
  self-interaction in, 172, 172f, 173f
  of strong interaction between quarks, 169, 170f, 194–195, 194f
  strong interactions in, 169, 170f
  of Supersymmetry, 346, 347f
  virtual electron-positron pairs in, 172, 176, 177f
  virtual gluons and quarks in, 197, 197f, 198f
  virtual particles in, 167, 173–174, 176, 177f
  virtual photons in, 166–167, 168f, 172, 172f, 173f
  weak interactions in, 169, 170f, 233, 233f, 236
Flavor-changing neutral currents (FCNC), 213–214, 238–239
Flavor eigenstates, 257, 258f, 367–368
Flavor quantum numbers, 228
Flavors
  eigenstates, 257, 258f
  of neutrinos, 255, 258f, 265, 266f, 267, 271
  of quarks, 226, 237–239
Forces. *See also* Electromagnetic force
  Coulomb, 15, 77, 132
  elementary particles and, 131–143, 134f
  gravitational, 15, 131–132, 134–136, 134f, 142, 232–233, 353
  GUT of, 263, 340, 342–343, 342f
  Higgs, 326
  strong nuclear, 131, 133, 133f, 134f, 135, 155
  Superforce, 339–340, 342, 351

Index 387

unification of, 308, 339–340
weak nuclear, 74, 131, 133–135, 134f, 141–142
Frequency
of electromagnetic waves, 39–41, 43
of light, colors and, 37
of oscillation, resonance and, 146
radio, 105, 106f, 112
Friedman, Jerome, 127, 128f, 252
Fürstenau, Hermann, 340–341
Future Circular Collider (FCC), 363–364

Gaillard, Jean-Marc, 247–248
Galilean transformations, 20–21, 26–27
Galileo, 20
GALLEX experiment, 261–262
Gallium Neutrino Observatory (GNO), 261–262
Gamma rays, 12–13
Gargamelle bubble chamber, 249–250, 251f, 312–313
Gas-wire detectors, 286–288, 287f
Gauge theory, 178, 307. *See also* Quantum Chromo Dynamics; Quantum Electrodynamics
Gavrin, Vladimir, 261
Geiger, Hans, 14–15, 16f
Geiger counter, 286, 287f
Gell-Mann, Murray, 85, 94–95, 115, 120, 123, 191
Zweig and, 117–119, 118f, 127, 189
General relativity
Einstein on, 25, 142, 162, 352–354
gravity and, 352–354, 353f
Heisenberg uncertainty principle and, 354
quantum mechanics and, 25, 142, 352
Geoneutrinos, 283–284
Gerlach, Walther, 54–56, 55f
Germer, Lester, 44
GIM model, 213–214, 218, 220, 238–239

Glaser, Donald, 89
Glashow, Sheldon, 213–214, 250, 251f, 252, 309–310, 320
Glueballs, 195–196
Gluino, 344, 346
Gluons, 141–142
color charge and, 310
colors and, 192, 194–197, 194f, 198f
as mediator particles, 194, 194f
quark-antiquark pairs and, 200–203, 201f
quarks, plasma and, 350–351, 351f
quarks and, 194–195, 194f, 196f, 197, 197f, 199f, 201, 207f, 226–227, 235, 315–317
in strong interactions, 192, 235–236
virtual, 197–198, 197f, 202
GNO. *See* Gallium Neutrino Observatory
Gold, 14–15, 64
Goudsmit, Samuel, 56
Grand Unification Theory (GUT), 263, 340, 342–343, 342f, 349, 351, 364
Gran Sasso neutrino detector, 267–268, 274, 275f, 277, 284, 348, 349f
Gravitational force, 15, 131–132, 134–136, 134f, 142, 232–233
Gravitational waves, 25, 142, 354
Gravitino, 344
Graviton, 142
Gravity, 135–136, 352–354, 353f, 359–360
Greco, Mario, 219
Greenberg, Oscar, 190
Greinacher, Heinrich, 100
Gross, David, 199
Guralnik, Gerald, 323
GUT. *See* Grand Unification Theory

Hadronic calorimeters, 295–296, 299f, 300
Hadronic energy, 295–296
Hadronic shower, 294–295, 296f, 300

Hadrons
  charged currents, 238–239
  collisions between, 125, 125f, 126, 141–142, 146–147, 316
  decay of, 146, 215–217, 217f, 238, 239f, 337–338
  electromagnetic interactions and, 208
  electron-positron collisions and, 321f
  inside of, 227f, 248–249
  mass of, 226
  neutral currents, 238–239
  quark-antiquark pairs and, 203–204
  quarks and, 118–120, 121f, 122, 124–126, 125f, 189, 192, 196, 199–203, 219
  resonance and, 146, 147f, 215
  strong interaction of, 232–233
Hagen, Carl, 323
Half-integer spin, 56–58, 148f, 343–344
Harari, Haim, 221
Hardy, Godfrey, 161
Harmer, Don, 246
Heavy flavors, of quarks, 226
Heavy water, 76
Heisenberg, Werner, 46–47, 47f, 59, 75, 121
Heisenberg uncertainty principle, 61f, 71, 232
  decay and, 148
  in quantum mechanics, 59–62, 139–140, 174
  virtual particles and, 172, 345
Helicity, 157–158, 158f, 236
Helium atoms, 12, 77
Hertz, Heinrich, 7, 40
Hess, Victor, 80–81, 81f
Higgs, Peter, 323, 334, 335f
Higgs boson, 115, 323–324, 330–338, 332f, 360–361
  ATLAS experiment and, 334–336, 336f, 339f
  decay of, 335–339, 336f, 338f, 339f
  Feynman diagram for, 337f, 338f
  LHC and, 331, 333–334, 337f, 363
  mass of, 348–349, 350f
  quark-antiquark pairs and, 336–338
  self-interaction of, 348, 350f
  virtual particle-antiparticle pairs and, 348, 350f
  W and Z bosons and, 336–337
Higgs field, 326, 323–334, 327f, 332f, 340, 348, 350f, 360–361
Higgsinos, 344, 346
High-energy electron-positron collisions, 208, 209f
High-energy electrons, 127, 177–179, 184–185, 185f, 318
High-energy interactions, 198–199, 202, 279
High-energy neutrinos, 282
High-energy particles, 32, 65, 279–280
High-energy quarks, 198–199
Hofstadter, Robert, 126–127
Hooft, Gerardus 't, 310
Hubble telescope, 330, 331f
Hydrogen atoms, 15–16, 41, 362
  electromagnetic force in, 232–233
  electron and proton in, 9–10, 76, 232–233
  electron in, 52, 53f, 61, 101, 122
  protons and, 98, 101
Hyper-Kamiokande, 370
Hypersymmetry, 360
Hyperunification, 352, 352f, 354–355, 357, 359

IceCube neutrino detector, 282, 354
Iliopoulos, John, 213–214
Image, in mirror, 153, 158–159, 159f
INFN. *See* National Institute of Nuclear Physics
Integer spin, 58, 148f, 189–190, 343
Interaction probability, 114, 198–199, 242, 247, 282

Interactions, 148f, 149. *See also* Electromagnetic interactions; Strong interactions; Weak interactions
  Coulomb, 173f
  electromagnetic, between electron and atomic nucleus, 166, 166f
  electroweak, 250, 251f, 252, 315, 336–337, 339, 360
  fundamental, coupling constants of, 308, 309f, 340–341, 341f
  high-energy, 198–199, 202, 279
  neutrinos, 247, 263–264, 264f, 282, 311–313
  self-interactions, 172, 172f, 173f, 175, 348, 350f
  Standard Model of fundamental, 320, 322f, 323, 341f, 345f 345–346
Interference, waves and, 44–45, 45f
Interferometer, 26
Invariance principles, 163
Inverse beta decay, 244, 244f
Ionization, 87–90, 93, 101, 286–287, 288f, 289f
Ions, 8, 87
Iron, 135, 143–144
Isospin, 75, 120–122, 228
Isotopes, 76

Joint Institute for Nuclear Research (JINR), 242–243
Joliot-Curie, Frédéric, 80
Joliot-Curie, Irène, 80
Joyce, James, 117
J/Ψ particle, 216–221, 217f, 223, 319

Kajita, Takaaki, 270–271, 274f
Kamiokande neutrino detector, 262–263, 267–269, 281, 371–372
KamLAND experiment, 267, 271, 284
Kaon, 93–94, 125–126, 221
KARMEN experiment, 365
Kelvin (lord), 19
Kendall, Henry, 127, 128f

Kibble, Tom, 323
Kinetic energy, 5–6, 28–29, 33f, 79, 279
Kobayashi, Makoto, 221, 238
Koshiba, Masatoshi, 262–263, 270, 281–283, 283f, 370
Kowarski, Lew, 45

Lambda, 94–95, 126
Laplace, Pierre-Simon, 21–22
Large Electron-Positron collider (LEP), 320–323, 321f, 322f
Large Hadron Collider (LHC), 32, 34, 98, 114f, 115, 204, 206f, 268, 278
  calorimeters at, 294
  CMS experiment at, 301, 301f
  detectors, 285
  Higgs boson and, 331, 333–334, 337f, 363
  Supersymmetric particles and, 349
Laser Interferometer Gravitational-Wave Observatory (LIGO), 142
Latent image, 91–92
Lattes, César, 85
Lawrence, Ernest, 100–101
Lederman, Leon, 156–157, 221, 222f, 247–248, 249f, 323–324
Lee, Tsung-Dao, 154–157, 155f, 247–248
Left and right, 152–153, 155, 159–160
Left-handed helicity (LH), 157, 311, 317
Left-handed neutrino, 158–159, 158f, 159f
LEP, 320–323, 321f, 322f
Leptons, 83, 118, 216, 221
  charged, 219, 223, 254, 279, 320, 338
  quantum numbers, 229f, 230, 270
  quarks and, 223, 225, 225f
  in table of fermions, 223, 225–226, 225f
  tau, 274
  weak interactions, 236–237
Leucippus, 3
LH. *See* Left-handed helicity
LHC. *See* Large Hadron Collider

Light. *See also* Photons
  Cerenkov, 262–263, 270f
  as electromagnetic wave, 19–20
  frequencies of, colors and, 37
  speed of, 20, 25–28, 30–32, 33f, 44, 79, 97, 330, 360
  visible, 63–64, 64f
  wavelength of, 63
LIGO. *See* Laser Interferometer Gravitational-Wave Observatory
Linear particle accelerator, 100, 100f
Liquid Scintillator Neutrino Detector (LSND), 365
Livingston, Stan, 107, 107f
Lorentz, Hendrik, 24, 24f, 26–27
Lorentz invariance, 180–181
Lorentz transformation, 27–28, 157
Lothar Meyer, Julius, 2
LSND. *See* Liquid Scintillator Neutrino Detector
Luminosity, 114

Magellanic Cloud supernova, 280–282, 281f
Magnetic fields, 8–9, 9f, 13, 19–20, 55–56, 55f
  dipole magnets, particle accelerators and, 109f
  electric fields and, 100–101, 308
  spins and, 155–156, 158
  synchrotrons and, 103–104
Magnetic moment, of electrons, 174–176, 175f
Magnets, 54–55, 104, 107–108, 109f, 110f, 111f
Maiani, Luciano, 213–214
Majorana, Ettore, 71–72, 72f, 80, 367
Maki, Ziro, 257, 258
Manhattan Project, 76–77, 241–242
Markov, Moisey, 247
Marsden, Ernest, 14–15, 16f
Maskawa, Toshihide, 221, 238

Mass
  of atomic nucleus, 70–71, 226
  eigenstates, 254–257, 258f, 367–368, 368f
  eigenvalues, 256–257, 272, 367–368
  of electron-positron pairs, as invariant, 215–216, 217f
  of electrons, 34, 67, 105, 138–139, 175, 348
  of electroweak bosons, 324, 335f, 340
  of elementary particles, 32–34
  energy and, 29–30, 32–34, 77, 111, 139–140
  of fermions, 236, 254, 307–308, 324, 329, 335f
  general relativity and, 353, 353f
  of hadrons, 226
  of Higgs boson and, 348–349, 350f
  Higgs field and, 328–329, 332
  of J/Ψ, 218
  of mediator bosons, 307–308
  momentum and, 34
  of muon neutrino, 254–256
  of neutrinos, 78, 225f, 227, 253–257, 258f, 323, 368f
  particle collisions and, 109, 111–112, 147
  of particles, 29–34, 44, 52–54, 109, 111–112
  of protons, 34–35, 67, 105, 111, 141
  of protons and neutrons, 67, 234
  of quark-antiquark pairs, 201
  of quarks, 226
  of Supersymmetric particles, 344
  of W and Z bosons, 317, 326
Matrix element, 171
Matrix mechanics, 46
Matter
  antimatter and, 125–126, 160, 183, 369
  conservation of energy and, 34–35
  dark, 329, 346–348
  neutrinos through dense, 265, 368–369

Index   391

rigidity of, 131–132
"Russian doll" structure of, 129, 130f
stability of, 228–229
structure of, Dalton's atom and, 4–5
Matter waves, 43–45, 48
Matveev, Victor, 261
Maxwell, James Clerk, 5, 19–21, 25, 308, 340
McDonald, Art, 263–264, 271, 274f
McMillan, Edwin, 102–103, 106
Measurement, of particles
 accurate and precise, 13
 detection and observation, 86–87
 experimental background of, 65–66
 observer-object relationship in, 46–47
 position and momentum, 59–60
 position and trajectory, 87
 probabilities in, 65–66
 signal events in, 66
 statistical error in, 65–66
Mechanics, 20–22, 26–28, 46, 65, 163–164. *See also* Quantum mechanics
Mediator particles, 136–142, 137f, 168f
 bosons as, 167, 232–233, 233f, 238, 307–308, 310–311, 313–314, 314f
 coupling constant, 168
 electrons as, 138–140, 139f
 fermions and, 167–168, 168f
 gluons, 194, 194f
 photons as, 138–141, 139f
 of weak interactions, 141, 232–233, 310
Meitner, Lise, 13, 14f, 68, 76
Mendeleev, Dmitri, 2
Mesons, 118–120, 121f, 125, 154, 323
Mesotron, 83–84
Michelson, Albert, 19, 26
Micro black holes, 347–348
MicroBooNE, 366, 366f
Mikheyev-Smirnov-Wolfenstein effect (MSW), 265, 266f, 267
Milky Way galaxy, 280, 282–283, 330
Miller, David, 331
Millikan, Robert, 41

Minimal Supersymmetric Standard Model (MSSM), 345
Mirror image, 153, 158–159, 159f
Mixing, 238, 255, 257–258, 258f, 271–273
Momentum
 angular, 47, 50, 53–56, 54f
 of atoms, 55–56, 64
 in collisions, 315–319
 conservation of, 136, 180, 318
 of electrons, 60–62, 61f, 79, 127, 138–141
 energy and, 28–29, 61f, 88, 138, 147
 mass and, 34
 of photons, 180–181, 181f
 position of particles and, 59–60, 62
 of projectile particles, 65, 127
 spin and, 157–158, 158f, 189
Monte Carlo method simulations, 302–306, 305f
Morley, Edward, 19, 26
Motion, 5–6, 29, 40–41
MSSM. *See* Minimal Supersymmetric Standard Model
MSW. *See* Mikheyev-Smirnov-Wolfenstein effect
M-Theory, 356–357
Multiwire proportional chamber (MWPC), 288–289, 289f, 290f, 291, 298f
Muon-antimuon pairs, 177, 185, 187f, 202, 208–209, 209f, 215, 222f
Muon neutrinos, 235f, 247–248, 250
 beam of, 275–276
 electron neutrinos and, 234–235, 250f, 268–269
 electrons and, 233–234, 251f
 mass of, 254–256
 tau neutrinos and, 256f, 269, 274
Muons, 87, 137
 antimuons and, 177, 187, 187f, 202
 atomic nucleus and, 83–85
 cosmic, 268

Muons (cont.)
  decay of, 93, 144–145, 156–157, 233–234, 233f, 236–237, 247, 303–304
  decay of charged pion in, 235, 235f
  in electron-positron collisions, 208
  electrons, neutrinos and, 92–93, 93f
  electrons and, 84, 170f
  in Feynman diagrams, 167, 168f, 170f
MWPC. *See* Multiwire proportional chamber

Nagoya model, 257
Nakagawa, Masami, 257, 258
National Institute of Nuclear Physics (INFN), 112, 208, 218f, 219
NDE. *See* Nucleon Decay Experiment
Neddermeyer, Seth, 83
Negative electrodes, in particle accelerators, 98, 99f
Negative pions, 95f, 120–121, 211, 232
Neutral currents, 233f, 236
  flavor-changing, 213–214, 238–239
  in Gargamelle bubble chamber, 250, 251f
  hadronic, 238–239
  weak, 250, 251f, 252
Neutralinos, 346, 348
Neutral kaon, 94
Neutral pions, 93, 120, 232, 276
Neutretto, 247
Neutrino astronomy, 281–282, 367
Neutrino beam, 247–250, 275–277, 276f
Neutrino detectors, 242, 244–247, 249–250, 260–264, 297, 370–372
  for atmospheric neutrinos, 272f
  Gran Sasso, 267–268, 274, 275f, 277, 284, 348, 349f
  IceCube, 282, 354
  Kamiokande, 262–263, 267–269, 281, 371–372
  LSND, 365
  MicroBooNE, 366, 366f
  in OPERA experiment, 273–274, 275f
  photomultipliers in, 262–263, 264f, 270f, 271f, 273f, 281
  Super-Kamiokande, 229, 269–270, 270f, 271f, 273f, 274
Neutrino interactions, 247, 263–264, 264f, 282, 311–313
Neutrino oscillations, 243–244, 253
  of atmospheric neutrinos, 274
  detection of, 256–257, 370–371
  discovery of, 270–272, 274f
  of electron neutrinos, 265, 266f, 267
  flavor and, 271
  mechanism of, 256f
  mixing model of, 271–272
  MSW effect and, 265, 266f
  muon neutrinos, tau neutrinos and, 256f, 274
  resonances of, 265, 267
  Standard Model and, 323
Neutrino radiation, 279
Neutrinos, 13–14, 69f, 225f, 240, 364–365. *See also* Antineutrinos; Electron neutrinos; Muon neutrinos
  artificial, 247–248, 267, 273, 275
  atmospheric, 267–270, 272f, 273–274
  atomic nucleus and, 73–75
  from Big Bang, 278–279, 284, 369
  in CHARM II experiment, 312–313
  in colliders, 300
  collisions, 312, 314f
  dark matter and, 347
  through dense matter, 265, 368–369
  detection of, 244–245
  electrons, muons and, 92–93, 93f
  electrons and, 78, 170f, 233, 233f
  electroweak currents and, 252
  Fermi on, 73–75, 151, 242
  flavors of, 255, 258f, 265, 266f, 267, 271
  geoneutrinos, 283–284
  helicity, spin and, 157–158, 158f, 236
  high-energy, 282
  inverse beta decay and, 244, 244f

Index 393

leptons and, 254
Majorana, 80, 367
mass of, 78, 225f, 227, 253–257, 258f, 323, 368f
mixing, 255, 257–258, 258f
negative helicity of, 236
neutrons and, 70, 73–74
Pauli on, 67–68, 68f, 69f, 70–71, 73–75, 81–82, 84, 151, 231–232, 241–242, 246
photons and, 231, 278
Pontecorvo on, 242–244, 246–248
Reines and Cowan on, 242, 244–246, 245f
relic, 278–280
solar, 248, 258–265, 260f, 264f, 266f, 269
sterile, 365–367
from Sun, 81, 246, 258, 262, 283–284
from supernovae explosions, 280–282, 281f
tau, 223, 227, 255–256, 256f, 269, 274
in violation of parity, 157
weak interactions of, 134, 148f, 231–234, 241, 248–249
weak nuclear force and, 134
Z bosons and, 322f, 323
Neutrons, 67–68, 69f. *See also* Protons, neutrons and
in atomic nucleus, 70–71, 75–76, 126–127
neutrinos and, 70, 73–74
positrons and, 244–245
Newton, Isaac, 20–23, 22f, 27–28, 51–52, 132, 163–164
Nickel, 155, 158
Nishijima, Kazuhiko, 94–95
Niu, Kiyoshi, 220–221, 220f, 223
Niwa, Kimio, 223, 273
Noble gases, 286–287
Noether, Emmy, 163
November revolution, 213, 215f
Nuclear electrons, 71

Nuclear emulsions, 90–93
Nuclear exchange force, 75
Nuclear fission and fusion, 76–78
Nuclear isospin, 75
Nuclear physics, 15, 45–46, 48, 76, 103
Nuclear reactors, 29, 77, 242, 244, 246–247
Nuclear weapons, 45–46
atomic bomb, 29, 76–77, 183, 242
Nucleon Decay Experiment (NDE), 263
Nucleons, 75, 121, 127, 228
Nucleus. *See* Atomic nucleus

Occhialini, Giuseppe, 85
OPERA experiment, 273–275, 275f
Orbital angular momentum, 54, 189
Oscar D'Agostino, 73f
Oscillation, 105–107, 106f, 146, 256–257, 269. *See also* Neutrino oscillations
Oxygen atoms, 76

Pacini, Domenico, 81
Pais, Abraham, 94
Pancini, Ettore, 84–85, 85f, 213
Parity
conservation of, 153–154
violation of, 156–158, 156f, 159f, 164–165, 236, 246, 252
wave functions and, 153–154
weak interaction and, 154
Particle accelerators, 87, 97. *See also* Large Hadron Collider
AdA, 112, 113f, 114
ADONE, 112, 114, 208, 218–219, 218f, 221, 223
collisions in, 32, 34, 65, 109, 111–112
cyclotrons, 101–104, 102f, 103f, 108, 156–157
dipole magnets in, 107–108, 109f, 111f
electric fields and, 98, 99f
electrons in, 105, 138

Particle accelerators (cont.)
  electrostatic, 100
  linear, alternating voltage in, 98, 100, 100f
  luminosity of, 114
  neutrino beam in, 275–277, 276f
  oscillation in, 105–107, 106f
  particle trajectories in, 101, 102f, 103–108
  PETRA and, 204–205, 207f
  phase stability and, 105–106, 106f
  positively charged particles, electrodes and, 98, 99f
  proton beam in, 204, 276, 276f, 299
  protons in, 98, 99f, 101, 105, 107, 109, 111
  quadrupole magnets in, 107–108, 110f, 111f
  SLAC, 127, 129, 129f, 189, 214, 217f, 218–219
  SPEAR, 214, 216, 221
  SPS, 314–317, 316f
  strong focusing in, 107–108, 108f
  Synchrophasotron, 107, 107f
  synchrotrons, 102–108, 104f, 111f, 123, 314
Particle detectors, 94, 108–109, 285. *See also* Neutrino detectors
  calorimeters and, 291–292, 294–296, 295f, 298, 299f, 300
  charged particles in, 297–300
  in CHARM II, 312
  electric/gaseous, 286
  electronic, 127, 208, 220, 286, 366
  gas-wire, 286–288, 287f
  Geiger counters, 286, 287f
  hermetic, 318–319
  in high-energy collider, 299–300, 299f
  known physics, theories and, 301–302
  Monte Carlo simulations and, 302–306, 305f

MWPC, 288–289, 289f, 290f, 291, 298f
  onion-type, 299–301, 318
  positron, 87–92, 90f, 91f, 297, 298f
  in UA1 experiment, 314, 316f, 318
  XENON, 348, 349f
Particle physics. *See specific topics*
Particles. *See also* Elementary particles; *specific topics*
  alpha, 14–16, 17f
  antimatter, 78, 80
  antiparticles and, 73–74, 79, 160, 167, 223, 225, 225f
  bound, 71, 231, 233
  cathode rays and, 7–9
  energy and momentum of, 88
  energy in atomic nucleus, 52–53, 58–59
  energy of, 29–30, 32–35, 33f, 52, 65, 79
  fermions, 58–59
  high speeds in confinement, 61–62
  kinetic energy of, 33f
  mass of, 29–34, 44, 52–54, 109, 111–112
  Planck constant and, 40
  positions of, momentum and, 59–60, 62
  positions of, probability and, 48–50, 49f
  positions of, trajectories and, 62, 87
  projectile, 65, 127, 249
  speed, time intervals and, 35
  speed of light and, 30–31, 33f, 97
  speeds in confinement, 61–62
  spin of, 53, 54f, 56–58
  tunnel effect and, 49–50
  unstable, 30, 32, 62, 83, 92–94, 144–146, 145f
  wave functions of, 46–48, 49f
  waves and, 44–46
Particle tracks, 88–89, 88f, 90–92, 91f, 124f, 248
Partons, 129

Pauli, Wolfgang
  exclusion principle, 56–59, 62, 132, 190, 343–344
  on neutrinos, 67–68, 68f, 69f, 70–71, 73–75, 81–82, 84, 151, 231–232, 241–242, 246
Pearl, Martin, 221
Periodic table of elements, 2–3
Perlmutter, Saul, 329
Perrin, Francis, 78
Perrin, Jean Baptiste, 6, 78
PET. *See* Positron emission tomography
PETRA. *See* Positron-Electron Tandem Ring Accelerator
Phase stability, 105–106, 106f
Photino, 344
Photoelectric effect, 40–41
Photoelectrons, 40
Photomultipliers, 262–263, 264f, 270f, 271f, 273f, 281
Photons, 7–8, 353–354. *See also* Virtual photons
  in Big Bang, 278
  Bremsstrahlung, 179
  in bubble chamber, 91f
  Cerenkov light, 262–263
  charged particles and, 324–325
  as CMB radiation, 360
  in early universe, 360–362
  electromagnetic force and, 324–325
  electromagnetic waves and, 41
  electron-positron pair and, 180–182, 180f, 181f, 182f, 291, 292f
  electrons, positrons and, 179, 183–185, 184f, 185f
  electrons and, 41, 60, 61f, 62, 64, 138–140, 139f, 166–167, 166f, 178–179
  energy of, 13, 40, 43, 61f, 105, 139–140
  energy spectrum of, 43
  fermions and, 180
  as massless, 310, 326, 340
  as mediators, 138–141, 139f

  momentum of, 180–181, 181f
  neutrinos and, 231, 278
  Planck on, 44
  radiation of, 17, 179, 182
  speed of light, 44
  of visible light, 64
  wavelengths and, 44
  as waves, 43–44
  weak bosons and, 325
  in X-rays and gamma rays, 13
Physics. *See also specific topics*
  of Big Bang, 33–34
  classical, 3, 6, 9–10, 19, 37–39, 46, 55
  classical versus quantum, 3, 39, 55
  nuclear, 15, 45–46, 48, 76, 103
  particle, 1–2, 25, 87, 117, 132, 146, 213, 235, 246, 285, 289, 292, 364–365
  quantum, 3, 19, 39–40, 55, 75
  (*see also* Quantum mechanics)
Piccioni, Oreste, 84–85, 85f
Pion beam, 276
Pions, 87
  charged, 85, 86f, 92–93, 93f, 235, 235f
  collisions of, 126
  decay of, 154, 156–157, 247, 254, 267, 276
  negative, 95f, 120–121, 211, 232
  neutral, 93, 120, 232, 276
  Powell's, 137, 142
Piredda, Giancarlo, 313
Planck, Max, 19, 39–41, 39f, 44
Planck constant, 39–40, 51, 59, 175
Planck-Einstein formula, 41, 43
Planck length, 355–356
Plasma, 350–351, 351f
PMNS. *See* Pontecorvo-Maki-Nakagawa-Sakata matrix
Poincaré, Jules Henri, 24–25, 24f
Pointlike particles, 63, 354–355, 356f
Politzer, David, 199
Polonium, 11
Poltergeist experiment, 245

Pontecorvo, Bruno, 243f, 250f, 257–259, 258f, 271–272
  on Magellanic Cloud supernova, 281
  on neutrinos, 242–244, 246–248
  on sterile neutrinos, 365
Pontecorvo-Maki-Nakagawa-Sakata matrix (PMNS), 257–258, 265, 273–274, 367–368, 368f
Positions, of particles, 48–50, 49f, 59–60, 62, 87
Positron detectors, 87–92, 90f, 91f, 297, 298f
Positron-Electron Tandem Ring Accelerator (PETRA), 204–205, 207f
Positron emission tomography (PET), 102
Positronium, 185
Positrons, 78, 87. *See also* Electron-positron pairs
  in Anderson's cloud chamber, 88–89, 88f
  cosmic rays and, 82
  electron-positron collisions, 208, 209f, 210f, 223, 320, 321f
  electrons, photons and, 179, 183–185, 184f, 185f
  electrons and, 35, 88–89, 91f, 135–136, 136f, 168f, 169, 172, 173f, 174, 179, 208
  in Feynman diagrams, 167, 168f
  magnetic fields and, 88, 91f
  neutrons and, 244–245
Potential energy, 29, 79
Powell, Cecil, 85, 86f, 94, 154
Powell's pion, 137, 142
Powell's tau, 94, 221
Probability
  elementary particles and, 48–50, 49f, 65–66
  in Feynman diagrams, 168–171
  interaction, 114, 198–199, 242, 247, 282

  in measurements of particles, 65–66
  oscillation, 257, 269
Probability amplitudes, 52, 170–171, 175, 207f
Projectile particles, 65, 127, 249
Propagator, in Feynman diagrams, 170–171
Proton beam, 204, 276, 276f, 299
Protons. *See also* Electrons, protons and
  antineutrinos and, 244
  antiprotons and, 112, 119, 223, 314–316, 316f, 318f
  as baryons, 228–229
  in collisions, nucleus and, 143–144
  collisions of, 143–144, 143f, 204, 206f, 335–336
  collisions with antiprotons, 223, 314–316, 316f, 318f
  color of, 192–193, 193f
  decay of, 229
  electric charge of, 133, 133f, 135, 192–193, 193f
  forces when in contact, 134, 134f
  hydrogen atom and, 98, 101
  interior of, 226, 227f
  mass of, 34–35, 67, 105, 111, 141
  in particle accelerators, 98, 99f, 101, 105, 107, 109, 111
  quarks and, 119, 120f, 125–126, 129, 129f, 192–193, 193f, 234–235
  scattering, 143, 143f
  strong interactions of, 276
Protons, neutrons and, 73–76, 109, 120f, 129f, 130
  in atomic nucleus, 76, 83, 111, 126–127, 133, 133f, 137, 141, 143–144, 192, 228–229
  in collisions, 143–144
  colors of, 192
  concentration in stars and galaxies, 279
  decay, 228–229, 231, 234–235, 234f

Index 397

mass of, 67, 234
quarks and, 119, 120f, 121, 129f, 244
Proton Synchrotron accelerator, 250

QCD. *See* Quantum Chromo Dynamics
QED. *See* Quantum Electrodynamics
Quadrupole magnets, 107–108, 110f, 111f
Quanta. *See* Photons
Quantum biology, 50
Quantum Chromo Dynamics (QCD), 191–195, 196f, 210. *See also* Color charge; Colors
  color charge conservation in, 309–310
  coupling constant, energy and, 200f
  electromagnetic force in, 199–200
  QED and, 191, 194, 197–198, 309–310, 324
  quark-antiquark pairs in, 197, 197f, 200–201
  strong interactions and, 199–200, 204–205, 208
Quantum Electrodynamics (QED), 165–166, 165f, 172, 309. *See also* Feynman diagram
  charge of, 309–310
  electromagnetic annihilation in, 183–187, 184f, 185f, 186f, 187f
  electrons and, 174
  QCD and, 191, 194, 197–198, 309–310, 324
  renormalization in, 178, 307
Quantum mechanics, 1–2, 43
  angular momentum in, 50, 54
  elementary particles and, 46, 48–49
  elements, chemistry and, 53
  entanglement in, 48
  general relativity and, 25, 142, 352
  gravity and, 360
  Heisenberg uncertainty principle in, 59–62, 139–140, 174

  position of particles, probability and, 48–49
  radioactive decay and, 70
  Schrödinger's equation in, 46, 51–52, 53f, 60–61
  spin in, 53, 54f, 56–57
  tunnel effect in, 49–50
  wave functions in, 46–48, 49f, 50–52, 144
Quantum numbers
  of baryons, 228, 230
  colors as, 190–191, 205, 208
  conservation of, 228, 230
  flavor, 228
  of leptons, 229f, 230, 271
  of quarks, 223, 224f
  strangeness, 94–95, 120–122, 126, 228
Quantum particles, 49, 62, 141. *See also* Elementary particles
Quantum physics, 3, 19, 39–40, 55, 75
Quantum strings, 355
Quantum superposition principle, 50–51
Quantum trajectories, 60–62, 226–227
Quark-antiquark pairs, 198, 235
  electromagnetic interaction between, 202, 202f
  in electron-positron collisions, 208–210, 209f
  Feynman diagrams of, 202f, 203f
  gluons and, 200–203, 201f
  hadrons and, 203–204
  Higgs boson and, 336–338
  mass of, 201
  in QCD, 197, 197f, 200–201
  strong interaction in, 202, 202f, 224f
  virtual, 199f, 350f
  virtual photons and, 202
  Z boson decay into, 321–322
Quark doublets, 237–238, 253

Quarks
  aces and, 189
  antiquarks and, 119–120, 125–126, 192, 195, 197–198, 218–219, 221
  antisymmetricity and, 190
  atoms and, 130
  baryons and, 118–119, 121f, 122–123, 123f, 125, 195, 196f
  bottom, 221, 222f, 223, 224f, 228, 336, 338
  charm, 214, 218–221, 220f, 228, 238–239
  in CKM matrix, 238–240, 239f, 253, 255, 257, 272
  collisions between, 315
  color charges of, 191–192, 194–195, 194f, 196f, 219
  colors of, 190–192, 194–195, 194f, 199f, 201–203
  coupling constant of, 208
  decay and, 234, 234f, 337–338
  detection of free, 126
  discovery of, 123, 124f, 128f
  distance between, 200–201, 200f, 201f
  down, 119–121, 177, 202–203, 211, 221, 225, 225f, 228
  electric charge of, 119–120, 120f, 122–123
  electrons, protons and, 192–193, 193f
  electrons and, 129–130, 130f
  flavors of, 226, 237–239
  Gell-Mann and Zweig on, 117–119, 118f, 127, 189
  gluons, plasma and, 350–351, 351f
  gluons and, 194–195, 194f, 196f, 197, 197f, 199f, 201, 207f, 226–227, 235, 315–317
  hadrons and, 118–120, 121f, 122, 124–126, 125f, 189, 192, 196, 199–203, 219
  heavy flavors, 226
  high-energy, 198–199
  isospin of, 120–122
  leptons and, 223, 225, 225f
  mass of, 226
  as pointlike, 129, 130f
  protons, neutrons and, 119, 120f, 121, 129f, 244
  protons and, 119, 120f, 125–126, 129, 129f, 192–193, 193f, 234–235
  quantum numbers of, 223, 224f
  SLAC and, 127, 129, 129f
  spectator, 234–235
  strange, 119–121, 211, 221, 226
  strong interactions between, 194f, 195, 199–200, 200f
  top, 223, 224f, 228, 336–338
  up, 119–121, 125–126, 177, 202–203, 210–211, 221, 225, 225f, 228
  valence, 226, 227f
  virtual gluons and, 197–198, 197f

Rabi, Isidor, 84
Radiation
  black-body, 37–40, 38f
  Bremsstrahlung, 178–179, 180f, 181–182
  CMB, 278–280, 282, 360
  electromagnetic, 7, 17, 105, 138, 282
  photon, 17, 179, 182
  from supernovae explosions, 280
  synchrotron, 105, 178–179
  wavelength, 63
Radioactive decay, 50, 70, 76, 102, 155, 261, 283–284
Radioactivity, 11–15, 50, 81
Radio frequency, 105, 106f, 112
Radioisotope, 102, 103f
Radio waves, 277–278
Radium, 11
Rasetti, Franco, 73f
Reines, Frederick, 157, 242, 244–246, 245f, 257
Relative speeds, 20, 26–28, 30, 111–112
Relativistic time dilation, 27–28, 87, 146

Relativity, 22–24, 27–28, 30. *See also* General relativity; Special relativity
Relic neutrinos, 278–280
Renormalizability, 178, 307
Renormalization, 307, 348
Resonances, 146–148, 147f, 214–215, 217f, 265, 267
RH. *See* right-handed helicity
Richter, Burt, 214–217, 216f, 217f, 219–221, 222f
Riess, Adam, 329
Right and left, 152–153, 155
Right-handed antineutrino, 158–159, 158f, 159f
Right-handed helicity (RH), 157, 317
Rochester, George, 94
Röntgen, Wilhelm, 11
Rubbia, Carlo, 314, 315f, 366
Rutherford, Ernest, 13, 86, 94
  atomic model of, 15, 17f, 41, 70, 75
  on atomic nucleus, 12, 12f, 14–16, 16f, 64, 70–71, 126
  on bound particles, 71
  Cockcroft, Walton and, 100, 101f

SAGE experiment, 261
Sakata, Shoichi, 257, 258
Salam, Abdus, 250, 251f, 252, 309–310, 320
Sanford Underground Research Facility (SURF), 371
Savannah River reactor, 245–246
Scalars, 29
Scanning, 92
Scattering
  angles, 129f
  Coulomb, 166
  Feynman diagram of electron, 314f
  of high-energy electrons, 127
  protons, 143, 143f
Schmidt, Brian, 329
Schrödinger, Erwin, 46, 47f

Schrödinger's equation, 46, 51–52, 53f, 60–61, 78–79
Schwartz, Mel, 247–248, 249f
Schwinger, Julian, 165, 165f
Scintillators, 294, 295f
Segre, Emilio, 73f
Selectron, 344
Self-interaction, 172, 172f, 173f, 175, 348, 350f
Silver, 54–56, 55f, 91–92
Simulations, 302–306, 305f
SLAC. *See* Stanford Linear Accelerator Center
Smirnov, Alexei, 265
Smuon, 344
SNO experiment, 264, 264f, 271
Snyder, Hartland, 107, 107f
Solar neutrino puzzle, 258
Solar neutrinos, 248, 258–265, 260f, 264f, 266f, 269
Space, time and, 19–21, 168f, 359
Spacetime, 23–24, 28, 35, 256, 353, 353f
Spaghetti calorimeter, 295
SPEAR accelerator, 214, 216, 221
Special relativity
  Einstein on, 13, 18, 21, 26, 29, 40
  elementary particles and, 25, 37, 78–79
  energy, matter and, 126
  energy of particle and, 29
  Lorentz, Poincaré and, 24, 24f
  Newtonian mechanics and, 27, 28
  speed of particles and, 31
  undulatory mechanics, 65
Spectators, 179, 181, 182f, 234–235
Speed
  of electromagnetic waves, 20–21
  energy and, 34
  mass and, 28, 34
  momentum, electrons and, 60–62
  of particles, light and, 30
  relative, 20, 26–28, 30, 111–112
  time and, 31, 35
  of universe expansion, 329–330

Speed of light, 20, 25–28, 32, 44, 79
  Einstein on, 26, 360
  expansion of universe and, 330
  particles and, 30–31, 33f, 97
Spin. *See also* Isospin
  of atomic nucleus, 155–156
  of elementary particles, 53, 157
  half-integer, 56–58, 148f, 343–344
  helicity and, 157–158, 158f, 236
  integer, 58, 148f, 189–190, 343
  magnetic fields and, 155–156, 158
  momentum and, 157–158, 158f, 189
  of particles, 53, 54f, 56–58
  in quantum mechanics, 53, 54f, 56–58
Spontaneous symmetry breaking, 310, 323–324, 327f, 328
SPS. *See* Super Proton Synchrotron
Squarks, 346, 347f
Standard Model of elementary particles and of fundamental interactions, 320, 322f, 323, 341f, 345f 345–346
Standard Solar Model, 259
Stanford Linear Accelerator Center (SLAC), 127, 129, 129f, 189, 214, 217f, 218–219
Statistics
  Bose-Einstein and Fermi-Dirac, 58, 189–190
  exchange theorem of, 67
  system of elementary particles, 58
Steinberger, Jack, 247–248, 249f
Sterile neutrinos, 365–367
Stern, Otto, 54–56, 55f
Strangeness
  conservation of, 124, 126
  production, 94–95, 95f
  as quantum number, 94–95, 120–122, 126, 228
Strange particles, 94–95, 119–121
Strange quarks, 119–121, 211, 221, 226
String theory, 355–359, 356f, 357f
Strolin, Paolo, 273
Strong focusing, 107–108, 108f

Strong interactions
  in atomic nucleus, 192–193, 193f
  charges of, 309–310
  color charge and, 309–310
  colors and, 192–194, 193f, 194f
  coupling constant and, 168–169
  in decay, 146–148, 191, 203
  electroweak interaction and, 339, 360
  Feynman diagrams of, 169, 170f, 194–195, 194f
  gluons in, 192, 235–236
  of hadrons, 232–233
  neutral pions and, 232
  of protons, 276
  QCD and, 199–200, 204–205, 208
  in quark-antiquark pair, 202, 202f, 224f
  between quarks, 194f, 195, 199–200, 200f
Strong nuclear force, 131, 133, 133f, 134f, 135, 155
Stückelberg, Ernst, 165
Sun, 77
  neutrinos from, 81, 246, 258, 262, 283–284
  solar neutrinos, 248, 258–265, 260f, 264f, 266f, 269
Superforce, 339–340, 342, 351
Super-Kamiokande neutrino detector, 229, 269–270, 270f, 271f, 273f, 274
Supernovae, 280–282, 281f
Super Proton Synchrotron (SPS), 314–317, 316f
Supersymmetric particles, 344–346, 345f, 347f, 348–349
Supersymmetry, 342, 342f, 344–346, 346f, 347f, 348–350, 363
SURF. *See* Sanford Underground Research Facility
Suzuki, Atsuto, 267
Symmetry
  antisymmetricity, 190–191
  breaking, 310, 323–324, 327f, 328, 360

Index

conservation principles and, 163–164
elementary particles and, 124–125, 164
harmony and, 161–163
Hypersymmetry, 360
mathematical beauty and, 161–162, 164–165
parity violation and, 164–165
Supersymmetric particles, 344–346, 345f, 347f, 348–349
Supersymmetry, 342, 342f, 344–346, 346f, 347f, 348–350, 363
unification and, 344
Synchrophasotron, 107, 107f
Synchrotron radiation, 105, 178–179
Synchrotrons, 102–108, 104f, 111f, 123, 314
Szilárd, Leó, 100

Tau leptons, 274
Tau neutrinos, 223, 227, 255–256
  muon neutrinos and, 256f, 269, 274
Tau-theta puzzle, 154
Taylor, Richard, 127, 128f
Telegdi, Valentine, 252–253, 252f
Thermodynamics, 37, 38f, 58, 291
Thomson, George, 44–45
Thomson, Joseph John, 7–12, 8f, 9f, 44–45
  atomic model of, 10, 10f, 14–15, 17f
Thorne, Kip, 354
Time
  Big Bang and, 33–34, 359
  dilation of, 27–28
  Einstein on, 22–23
  energy and, 33–34
  intervals, 35
  intervals, elementary particles and, 35
  Newton on, 21–23
  relative speed and, 27–28
  relativistic dilation of, 27–28, 87, 146
  relativity of, 22–23, 27–28

space and, 19–21, 168f, 359
spacetime and, 23–24, 28, 35, 256, 353, 353f
speed and, 31
Ting, Samuel C. C., 214–216, 215f, 217f, 220–221, 222f
Tomonaga, Sin-Itiro, 165, 165f
Top quarks, 223, 224f, 228, 336–338
Totsuka, Yoji, 270, 272
Touschek, Bruno, 112, 208
Tunnel effect, 49–50, 231

UA1 and UA2 experiments, 314, 316f, 318–320, 318f, 319f
Uhlenbeck, George, 56
Unification, electroweak, 310–311, 313–314, 321f, 339, 341f
Universality of the weak interaction, 238
Universe. *See also* Big Bang
  dark matter of, 329, 346–348
  density of, 131
  evolution of, 360–363
  expansion of, 329–330, 360–363, 374
  future of, 363
  temperature of early, 34
Unstable particles, 30, 32, 62, 83, 92–94, 144–146, 145f
Up quarks, 125–126
  down quarks and, 119–121, 177, 202–203, 211, 221, 225, 225f, 228
  isospin of, 228

Valence quarks, 226, 227f
van der Meer, Simon, 314–315, 315f
Vectors, 28–29, 53, 157, 158f
Veksler, Vladimir, 102–103, 106
Veltman, Martinus, 310
Via Panisperna group, 73, 73f, 76, 208
Violation of CP, 160, 240, 369, 371–372
Violation of parity, 156–158, 156f, 159f, 164–165, 236, 246, 252

Virtual electron-positron pairs, 172, 176, 177f, 196–197
Virtual electrons, 179, 182
Virtual fermions, 168f, 183–184
Virtual gluons, 197–198, 197f, 202
Virtual particle-antiparticle pairs, 348, 350f
Virtual particles, 83, 140–141, 179
  electrons and, 174, 176, 177f
  in Feynman diagrams, 167, 173–174, 176, 177f
  Heisenberg uncertainty principle and, 345
  mediating bosons as, 313–314, 314f
Virtual photons, 140–141
  electron-positron pair and, 173f, 185, 185f
  electrons and, 166–167, 168f, 172, 172f, 173f, 182–183, 346f
  quark-antiquark pairs and, 202
  spectator atomic nucleus and, 182f
Virtual quark-antiquark pairs, 199f, 350f
Visible light, 63–64, 64f

Walton, Ernest, 100, 101f
Wave functions, 56
  fermion, 236
  parity and, 153–154
  Pauli exclusion principle and, 190
  in quantum mechanics, 46–48, 49f, 50–52, 144
  resonances and, 146
Wavelengths
  black-body radiation and, 40
  de Broglie, 63, 63f, 109, 111
  of electromagnetic radiation, 138
  of light, 63
  photons and, 44
  radiation, 63
Waves. *See also* Electromagnetic waves
  Doppler effect, 277–278
  gravitational, 25, 142, 354
  interference, 44–45, 45f

  matter, 43–45, 48
  particles and, 44–46
  photons as, 43–44
  radio, 277–278
W bosons, 233–236, 238, 244, 360–361
  discovery of, 314, 315f, 318f
  electroweak, 314, 315f, 318f
  Higgs bosons and, 336–337
  mass of, 317, 326
  production of, 316f, 317–318
  Z bosons and, 314, 315f, 316f, 317, 318f, 320, 326, 336–337
Weak bosons, 325
Weak charged currents, 234, 313
Weak coupling constant, 170f, 238, 308, 311, 340, 341f
Weak decay, 144–145, 156, 231–232, 235, 238
Weak interactions, 133–135, 231
  charged current, 239f
  charged pion decay, 235
  coupling constant and, 313
  decay and, 154, 156
  electroweak, 250, 251f, 252
  fermions and, 168–169, 232–233
  Feynman diagrams of, 169, 170f, 233, 233f, 236
  leptonic, 236–237
  mediators of, 141, 232–233, 310
  of neutrinos, 134, 148f, 231–234, 241, 248–249
  parity and, 154
  parity violation and, 236
  universality of, 238
Weak neutral currents, 250, 251f, 252
Weak nuclear force, 74, 131, 133–135, 134f, 141–142
Weinberg, Steven, 250, 251f, 252, 309–311, 320
Weinberg angle, 311, 314
Weiss, Rainer, 354
Weyl, Hermann, 161–163
Wheeler, John, 21

Widerøe, Rolf, 100–101, 112
Wigmans, Richard, 295
Wigner, Eugene, 163
Wilczek, Frank, 197, 199
Wino, 344, 346
Wire chamber, 297, 298f, 300
Witten, Edward, 356–357, 357f
Wolfenstein, Lincoln, 265
Wu, Chien-Shiung, 155–159, 156f, 159f, 252
Wulf, Theodor, 80

XENON detector, 348, 349f
X-rays, 11, 13, 92

Yan, Tung-Mow, 211, 225, 252
Yang, Chen Ning, 154–157, 155f, 247–248
Yukawa, Hideki, 83–85, 84f, 86f, 90, 137, 142

Z bosons, 232–233, 233f, 310, 360–361
  decay into electron-positron pair, 319, 319f
  decay into quark-antiquark pairs, 321–322
  discovery of, 314, 315f, 318f
  electroweak, 314, 315f, 318f, 320
  Higgs bosons and, 336–337
  LEP measurement of width, 322–323, 322f
  mass of, 317, 326
  neutrinos and, 322f, 323
  production of, 316f, 317
  W bosons and, 314, 315f, 316f, 317, 318f, 320, 326, 336–337
Zichichi, Antonino, 221, 223, 253
Zweig, George, 117–119, 118f, 127, 189, 219